2nd U.S. Precision Livestock Farming Conference

2nd U.S. Precision Livestock Farming Conference

Editors

Yang Zhao
Daniel Berckmans
Hao Gan
Brett Ramirez
Janice Siegford
Lingjuan Wang-Li
Robert T. Burns

MDPI

Basel • Beijing • Wuhan • Barcelona • Belgrade • Novi Sad • Cluj • Manchester

Editors

Yang Zhao
University of Tennessee
Knoxville, TN
USA

Daniel Berckmans
University of Tennessee
Knoxville, TN
USA

Hao Gan
University of Tennessee
Knoxville, TN
USA

Brett Ramirez
Iowa State University
Ames, IA
USA

Janice Siegford
Michigan State University
East Lansing, MI
USA

Lingjuan Wang-Li
North Carolina State University
Raleigh, NC
USA

Robert T. Burns
University of Tennessee
Knoxville, TN
USA

Editorial Office

MDPI
St. Alban-Anlage 66
4052 Basel, Switzerland

This is a reprint of articles from the Special Issue published online in the open access journal *Animals* (ISSN 2076-2615) (available at: https://www.mdpi.com/journal/animals/special_issues/4DMQ6B7A73).

For citation purposes, cite each article independently as indicated on the article page online and as indicated below:

Lastname, A.A.; Lastname, B.B. Article Title. *Journal Name* **Year**, *Volume Number*, Page Range.

ISBN 978-3-7258-1041-3 (Hbk)
ISBN 978-3-7258-1042-0 (PDF)
doi.org/10.3390/books978-3-7258-1042-0

Contents

About the Editors

Yang Zhao

Dr. Yang Zhao is an associate professor at the Animal Science department at the University of Tennessee. He served as the Proceedings Chair of the 2nd U.S. Precision Livestock Farming Conference, and the lead guest editor of this Special Issue. Dr. Zhao received his B.S. and M.S. degrees from China Agricultural University, and a PhD degree from Wageningen University in the Netherlands. Before joining The University of Tennessee, he was a faculty member of Agricultural and Biological Engineering at Mississippi State University for three and half years. Dr. Zhao serves as the chair and member in several academic and award committees of the American Society of Agricultural and Biological Engineers (ASABE) and is an associate editor of the *Transactions of the ASABE*. His research focuses on smart poultry farming, addressing challenges in poultry production regarding environment management, poultry behavior, and poultry welfare. His work has been published in over 150 scientific articles.

Daniel Berckmans

Dr. Daniel Berckmans has a master's and a PhD in Bioengineering. He worked as a Full Professor for 20 years at the research Division M3-BIORES (Measure, Model and Manage Bio-responses), Department of Biosystems at the Catholic University of Leuven, and he is an Adjunct Distinguished Professor at the University of Tennessee USA. Dr. Berckmans is also the founder, board member, and CTO of BioRICS NV that is focusing on continuous monitoring of human and animal mental and physical health. In education, he was teaching 8 different courses at the bachelors' and master's level. His research focus is on real-time monitoring of the health and well-being of individual humans and animals. He co-authored over 340 peer-reviewed publications and over 450 papers, participated in 64 PhD commissions in 9 countries, and promoted over 250 Master theses. Since 1982, 19 products for the world market have been co-developed with industrial partners. He is a co-inventor of 20 patents and coordinated several EU projects. Dr. Berckmans is worldwide considered as the spiritual father of the field of Precision Livestock Farming (PLF): real-time monitoring of animals with the technology of microphones, cameras, and sensors. He co-started the European Association for Precision Livestock Farming that organizes the 2-yearly European Conference for Precision Livestock Farming 'ECPLF' since 2003. Today the conference is organized in agreement with the "Asian Conference for PLF" (ACPLF) and the "US PLF Conference" (USPLF). Dr. Berckmans co-initiated the first European Master's in Human Health Engineering at the KU Leuven: technology for monitoring healthy people. He is co-founder of 3 spin-off companies, including BioRICS NV in 2006, SoundTalks in 2011, and BioRICS INC in San Francisco in 2016. He is a member of the recently started Scientific Council of the World Farmers Organization which represents a billion farmers worldwide.

Hao Gan

Dr. Hao Gan is an assistant professor in the Department of Biosystems Engineering and Soil Science at the University of Tennessee. He served on the Proceedings Committee of the 2nd U.S. Precision Livestock Farming Conference. Dr. Gan's research focuses on the development of robotic and sensing systems for precision agriculture and precision livestock farming. He serves as the chair and member of several technical committees of the American Society of Agricultural and Biological Engineers (ASABE), and is an active member of the International Society of Precision Agriculture (ISPA).

Brett Ramirez

Dr. Brett Ramirez is an associate professor in the Department of Agricultural and Biosystems Engineering at Iowa State University. He served on the Proceedings Committee of the 2nd U.S. Precision Livestock Farming Conference. Dr. Ramirez leads an applied research and extension program, which aims to provide science-based information to enable informed decision making that supports sustainable and efficient swine and poultry production systems. He is an active member in the American Society of Agricultural and Biological Engineers (ASABE) and the American Society of Heating, Refrigerating and Air-Conditioning Engineers (ASHRAE), as well as serving as an associate editor for the Journal of the ASABE.

Janice Siegford

Dr. Janice Siegford is a professor and associate chair in the Department of Animal Science at Michigan State University, where she leads the Animal Behavior and Welfare Group. She served on the Proceedings Committee of the 2nd U.S. Precision Livestock Farming Conference. Dr. Siegford's research examines the impact of management practices and housing environments on the behavior of welfare of production animals, with an emphasis on laying hens and swine. Dr. Siegford also works with colleagues in engineering and computer science to develop and validate non-invasive, automated methods for collecting behavior and welfare data from individual animals in their home environments. She is the president of the International Society for Applied Ethology and a member of the Poultry Science and World Poultry Science Associations. In 2020, she received the Humane Award from the American Veterinary Medical Association in recognition of her work to advance animal welfare through education and research.

Lingjuan Wang-Li

Dr. Lingjuan Wang-Li is a William Neal Reynolds Distinguished Professor in the Department of Biological and Agricultural Engineering at North Carolina State University. She served on the Proceedings Committee of the 2nd US Precision Livestock Farming Conference. Dr. Lingjuan Wang-Li specializes in Air Quality and Food Animal Production Systems Engineering. Much of her research work addresses the emissions, fate, and transport of air emissions from animal feeding operations, and food animal production systems environmental control and animal welfare. Dr. Lingjuan Wang-Li has authored and co-authored 69 peer-reviewed papers/book chapter, 128 conference papers, abstracts, and presentations. She was a recipient of 2010 National Science Foundation (NSF) Career Award and her research is currently funded by NSF and USDA NIFA programs. She was appointed by the secretary of USDA to the USDA Agricultural Air Quality Task Force (AAQTF) for three terms in 2013–2015, 2016–2017, and 2020–2023.

Robert T. Burns

Dr. Robert T. Burns is a Distinguished Professor in the Biosystems Engineering and Soil Science Department at the University of Tennessee, Knoxville. Robert served as the overall Conference Chair for the 2nd U.S. Precision Livestock Farming (USPLF2023) Conference. He has mentored over 70 graduate students, postdocs, and visiting professors/scholars, has authored 241 technical publications, and delivered over 650 presentations in over 20 countries. Dr. Burns is an American Society of Agricultural and Biological Engineers Fellow and a registered Professional Engineer.

animals

Editorial

Field Implementation of Precision Livestock Farming: Selected Proceedings from the 2nd U.S. Precision Livestock Farming Conference

Yang Zhao [1,*], Brett C. Ramirez [2], Janice M. Siegford [3], Hao Gan [4], Lingjuan Wang-Li [5], Daniel Berckmans [4] and Robert T. Burns [4]

1 Department of Animal Science, The University of Tennessee, Knoxville, TN 37996, USA
2 Department of Agricultural and Biosystems Engineering, Iowa State University, Ames, IA 50011, USA
3 Department of Animal Science, Michigan State University, East Lansing, MI 48824, USA; siegford@msu.edu
4 Department of Biosystems Engineering and Soil Science, The University of Tennessee, Knoxville, TN 37996, USA
5 Department of Biological and Agricultural Engineering, North Carolina State University, Raleigh, NC 27695, USA
* Correspondence: yzhao@utk.edu

Citation: Zhao, Y.; Ramirez, B.C.; Siegford, J.M.; Gan, H.; Wang-Li, L.; Berckmans, D.; Burns, R.T. Field Implementation of Precision Livestock Farming: Selected Proceedings from the 2nd U.S. Precision Livestock Farming Conference. *Animals* **2024**, *14*, 1128. https://doi.org/10.3390/ani14071128

Received: 26 March 2024
Accepted: 5 April 2024
Published: 7 April 2024

Precision Livestock Farming (PLF) involves the real-time monitoring of images, sounds, and other biological, physiological, and environmental parameters to assess and improve animal health and welfare within intensive and extensive production systems. Precision Livestock Farming systems have the potential to empower farmers to make better management decisions, based upon objective measures, during an animal production cycle. The potential benefits from PLF systems can only be realized if these systems are adopted in practice by livestock and poultry producers. The advancement of research in digital agriculture, and, more specifically, PLF technologies, over the past decade has been very impressive; however, the field implementation of PLF technology has not occurred at the same rate as research advancements.

The United States land grant university system is well equipped to interface between PLF research and on-farm adoption by disseminating research to farmers through the Extension Service component of the system. The on-farm implementation of PLF systems in the United States has the potential to provide increased animal production efficiency with improved animal health and welfare and human working conditions, while also increasing the economic, environmental, and social sustainability of animal production systems. Significant improvements in the production efficiency of all food animal sectors will be required in order to meet the anticipated growth in demand for animal proteins by a vaster worldwide population over the next 30 years. The commercial adoption of PLF systems is integral to achieving the increases in production efficiency needed from our livestock sectors and to providing needed data to animal production stakeholders.

For this reason, the theme of the 2nd U.S. Precision Livestock Farming conference (USPLF2023) was "Field Implementation of PLF". This emphasizes the significance of implementing technology in food animal production to meet the rising demand for animal products in the future. The USPLF2023 conference was held from 21 to 24 May 2023 at The University of Tennessee Conference Center in Knoxville, TN. The conference was organized by The University of Tennessee, University of Nebraska-Lincoln, and Tennessee State University.

The goal of the USPLF2023 conference was to help better equip researchers, Extension practitioners, technology providers, and livestock and poultry producers to adopt PLF systems on working farms across the United States and abroad. The USPLF2023 conference provided a timely and essential platform for industry professionals and academia to share knowledge, network, and collaborate on all aspects of animal management using innovative technology. The conference also made scholarships available for participation from farmers, minority-serving institutions educators, researchers, and Extension specialists, thus providing opportunities to a diverse group of attendees to benefit from the conference's valuable insights and discussions. In addition, the event hosted two other important meetings, namely the United States Department of Agriculture (USDA) National Institute of Food and Agriculture (NIFA) Inter-Disciplinary Engagement in Animal Systems Project Director meeting, and the meeting of the NC1211 Multi-state project (Precision Management of Animals for Improved Care, Health, and Welfare of Livestock and Poultry).

The USPLF2023 conference featured four primary research topics: Sensors and Sensing in PLF [1–5], Data Management and Algorithm Development [6–8], Measuring, Modeling, and Managing of Dynamic Responses [9–14], and Societal Impacts of PLF [15,16]. A total of 126 submissions were received for these topics from individuals representing universities, research institutions, and PLF companies from 13 different countries across Africa, Asia, Australia, Europe, North America, and South America. The submissions underwent two rounds of rigorous review by ad hoc reviewers to assess their relevance, originality, readability, and formatting. Out of these submissions, 114 were accepted for presentations at the conference, either orally or as poster presentations.

From the accepted submissions, the conference proceedings committee carefully selected the highest-ranked papers to be included in this Special Issue with the full endorsement of the Conference Proceeding Committee. Additionally, several voluntarily submitted proceedings were also considered and incorporated into this Special Issue. Each paper within this collection represents a significant contribution, showcasing the advancements made and the future potential of implementing PLF in the field.

This Special Issue is expected to be of immense value to individuals working in the interdisciplinary fields of animal science, agricultural engineering, computer science, electrical engineering, and veterinary science. It represents valuable resources for educators, researchers, Extension specialists, animal producers, and industry leaders who share a commitment to advancing the livestock and poultry industry in an innovative and sustainable manner.

Author Contributions: Conference organization, R.T.B., Y.Z. and D.B.; Conference Proceedings Committee, Y.Z., D.B., H.G., B.C.R., J.M.S. and L.W.-L.; writing—draft preparation, B.C.R. and Y.Z.; writing—review and editing, Y.Z., D.B., H.G., B.C.R., J.M.S., L.W.-L. and R.T.B. All authors have read and agreed to the published version of the manuscript.

Funding: USPLF2023 was sponsored in part by the United States Department of Agriculture (USDA) National Institute of Food and Agriculture (NIFA) (Grant # 2023-67015-39361), University of Tennessee Institute of Agriculture AgResearch, and industry partners.

Acknowledgments: We are grateful to all the authors who contributed their research progress, findings, and experiences to make this conference a reality. We express our heartfelt thanks to the reviewers for their valuable comments and contributions, which ensured the scientific quality and merit of the papers. The University of Tennessee AgResearch and Conference Committee provided essential support, and we express our deep appreciation. In addition, we extend our sincere thanks and gratitude to our sponsors and supporters, without whom, this conference would not have been a success.

Conflicts of Interest: The authors declare no conflicts of interest.

References

1. Menendez, H.M., III; Brennan, J.R.; Ehlert, K.A.; Parsons, I.L. Improving Dry Matter Intake Estimates Using Precision Body Weight on Cattle Grazed on Extensive Rangelands. *Animals* **2023**, *13*, 3844. [CrossRef] [PubMed]
2. Van Steenkiste, G.; Van Den Brulle, I.; Piepers, S.; De Vliegher, S. In-Line Detection of Clinical Mastitis by Identifying Clots in Milk Using Images and a Neural Network Approach. *Animals* **2023**, *13*, 3783. [CrossRef] [PubMed]
3. Islam, M.N.; Yoder, J.; Nasiri, A.; Burns, R.T.; Gan, H. Analysis of the Drinking Behavior of Beef Cattle Using Computer Vision. *Animals* **2023**, *13*, 2984. [CrossRef] [PubMed]
4. Jacobs, J.L.; Hersom, M.J.; Andrae, J.G.; Duckett, S.K. Training and Adaptation of Beef Calves to Precision Supplementation Technology for Individual Supplementation in Grazing Systems. *Animals* **2023**, *13*, 2872. [CrossRef] [PubMed]

5. Nyamuryekung'e, S.; Duff, G.; Utsumi, S.; Estell, R.; McIntosh, M.M.; Funk, M.; Cox, A.; Cao, H.; Spiegal, S.; Perea, A.; et al. Real-Time Monitoring of Grazing Cattle Using LORA-WAN Sensors to Improve Precision in Detecting Animal Welfare Implications via Daily Distance Walked Metrics. *Animals* **2023**, *13*, 2641. [CrossRef] [PubMed]
6. Nasiri, A.; Amirivojdan, A.; Zhao, Y.; Gan, H. Estimating the Feeding Time of Individual Broilers via Convolutional Neural Network and Image Processing. *Animals* **2023**, *13*, 2428. [CrossRef] [PubMed]
7. Magalhaes, E.S.; Zhang, D.; Wang, C.; Thomas, P.; Moura, C.A.A.; Holtkamp, D.J.; Trevisan, G.; Rademacher, C.; Silva, G.S.; Linhares, D.C.L. Field Implementation of Forecasting Models for Predicting Nursery Mortality in a Midwestern US Swine Production System. *Animals* **2023**, *13*, 2412. [CrossRef] [PubMed]
8. Paudel, S.; de Sousa, R.V.; Sharma, S.R.; Brown-Brandl, T. Deep Learning Models to Predict Finishing Pig Weight Using Point Clouds. *Animals* **2024**, *14*, 31. [CrossRef] [PubMed]
9. Nyamuryekung'e, S.; Cox, A.; Perea, A.; Estell, R.; Cibils, A.F.; Holland, J.P.; Waterhouse, T.; Duff, G.; Funk, M.; McIntosh, M.M.; et al. Behavioral Adaptations of Nursing Brangus Cows to Virtual Fencing: Insights from a Training Deployment Phase. *Animals* **2023**, *13*, 3558. [CrossRef] [PubMed]
10. van Erp-van der Kooij, E.; de Graaf, L.F.; de Kruijff, D.A.; Pellegrom, D.; de Rooij, R.; Welters, N.I.T.; van Poppel, J. Using Sound Location to Monitor Farrowing in Sows. *Animals* **2023**, *13*, 3538. [CrossRef] [PubMed]
11. Moon, J.; DuBien, J.; Ramachandran, R.; Liang, Y.; Dridi, S.; Tabler, T. Effects of a Sprinkler and Cool Cell Combined System on Cooling Water Usage, Litter Moisture, and Indoor Environment of Broiler Houses. *Animals* **2023**, *13*, 2939. [CrossRef] [PubMed]
12. Jaihuni, M.; Gan, H.; Tabler, T.; Prado, M.; Qi, H.; Zhao, Y. Broiler Mobility Assessment via a Semi-Supervised Deep Learning Model and Neo-Deep Sort Algorithm. *Animals* **2023**, *13*, 2719. [CrossRef] [PubMed]
13. Bery, S.; Brown-Brandl, T.M.; Jones, B.T.; Rohrer, G.A.; Sharma, S.R. Determining the Presence and Size of Shoulder Lesions in Sows Using Computer Vision. *Animals* **2024**, *14*, 131. [CrossRef] [PubMed]
14. Sefeedpari, P.; Pishgar-Komleh, S.H.; Aarnink, A.J.A. Model Adaptation and Validation for Estimating Methane and Ammonia Emissions from Fattening Pig Houses: Effect of Manure Management System. *Animals* **2024**, *14*, 964. [CrossRef]
15. Elliott, K.C.; Werkheiser, I. A Framework for Transparency in Precision Livestock Farming. *Animals* **2023**, *13*, 3358. [CrossRef]
16. Akinyemi, B.E.; Akaichi, F.; Siegford, J.M.; Turner, S.P. US Swine Industry Stakeholder Perceptions of Precision Livestock Farming Technology: A Q-Methodology Study. *Animals* **2023**, *13*, 2930. [CrossRef] [PubMed]

animals

MDPI

Article

Model Adaptation and Validation for Estimating Methane and Ammonia Emissions from Fattening Pig Houses: Effect of Manure Management System

Paria Sefeedpari *, Seyyed Hassan Pishgar-Komleh and Andre J. A. Aarnink

Wageningen Livestock Research, Wageningen University and Research, P.O. Box 135, 6700 AC Wageningen, The Netherlands; hassan.pishgarkomleh@wur.nl (S.H.P.-K.); andre.aarnink@wur.nl (A.J.A.A.)
* Correspondence: paria.sefeedpari@wur.nl

Simple Summary: Predicting emissions from livestock housing is crucial for helping farmers understand how various factors impact emissions at the pig house level. This knowledge empowers decision-makers and the farmers to make informed choices that encourage emission reduction and improve air quality in pig houses. To create innovative, low-emission pig housing systems, the development of a prediction model appears to be a straightforward solution. This paper introduces such a model for predicting methane and ammonia emissions from two commercial pig houses with two manure management systems: one with long storage in deep pits and the other with short storage, through daily flushing of a shallow pit with sloped walls (a reduced emitting surface) and partial manure dilution. This simulation model, which considers factors like animal weight, age, feed, and pen dimensions as input data, calculates manure quantity and quality, as well as methane and ammonia emissions over the growing period. In this model, the calculation of ammonia emission involves aggregating emissions from various sources identifiable within pig houses. These sources include manure channels, slatted floors, and solid floors, encompassing fouled areas, pigs, and pen partitions. This model successfully predicts the development of methane and ammonia emissions based on animal and farm characteristics.

Abstract: This paper describes a model for the prediction of methane and ammonia emissions from fattening pig houses. This model was validated with continuous and discrete measurements using a reference method from two manure management systems (MMS): long storage (LS) in deep pits and short storage (SS) by daily flushing of a shallow pit with sloped walls and partial manure dilution. The average calculated methane and ammonia emissions corresponded well with the measured values. Based on the calculated and measured results, the average calculated CH_4 emission (18.5 and 4.3 kg yr^{-1} per pig place) was in between the means from the continuous data from sensors (15.9 and 5.6 kg yr^{-1} per pig place) and the means from the discrete measurements using the reference method (22.0 and 3.1 kg yr^{-1} per pig place) for the LS and SS systems, respectively. The average calculated NH_3 emission (2.6 and 1.4 kg yr^{-1} per pig place) corresponded well with the continuous data (2.6 and 1.2 kg yr^{-1} per pig place) and the discrete measurements using the reference method (2.7 and 1.0 kg yr^{-1} per pig place) from LS and SS, respectively. This model was able to predict the reduction potential for methane and ammonia emissions by the application of mitigation options. Furthermore, this model can be utilized as a predictive tool, enabling timely actions to be taken based on the emission prediction. The upgraded model with robust calculation rules, extensive validations, and a simplified interface can be a useful tool to assess the current situation and the impact of mitigation measures at the farm level.

Keywords: methane emission; ammonia emission; modelling; manure; pig house

Citation: Sefeedpari, P.; Pishgar-Komleh, S.H.; Aarnink, A.J.A. Model Adaptation and Validation for Estimating Methane and Ammonia Emissions from Fattening Pig Houses: Effect of Manure Management System. *Animals* **2024**, *14*, 964. https://doi.org/10.3390/ani14060964

Academic Editor: Cesare Castellini

Received: 29 December 2023
Revised: 9 February 2024
Accepted: 8 March 2024
Published: 20 March 2024

1. Introduction

Emissions of methane (CH_4) and ammonia (NH_3) from agriculture have profound effects on climate change, air quality, and ecosystem health [1–3]. CH_4 is the second-most abundant anthropogenic greenhouse gas [4], arising preliminarily from animal digestion and manure storages [5]. In the Netherlands, livestock farming contributes to 75% of the CH_4 emission. In pig farming, most of the CH_4 (more than 80%) comes from the stored slurry (i.e., faeces and urine) in the pig houses. This is mainly caused by the long storage of slurry underneath the floor in the barns and from outside manure storage places [6]. An additional portion is emitted through the digestive system of pigs, particularly in the hindgut, contributing to enteric methane formation [7]. Therefore, reducing methane emission is crucial for livestock farming to meet the climate goals.

High NH_3 concentration is detrimental for a healthy indoor climate; its emission contributes to secondary particulate matter formation and leads to excessive nitrogen deposition in nature areas [2,8,9]. Worldwide, the primary source of NH_3 emission in agriculture originates from livestock housing, specifically the excretion of urine by livestock (Aarnink, 1993 [10–12]). Waste streams of the livestock sector contribute to 39% of the global NH_3 emission [13]. In Europe, around 80% of NH_3 production is attributed to livestock facilities [9], with pig production contributing to about 15% of the total. In the Netherlands, agriculture is responsible for 88% of NH_3 emission released into the air [6,14], with pig farming contributing to 20% of the NH_3 emission. The Dutch agricultural sector has marked a two-thirds reduction in NH_3 emission since 1990 due to various mitigating measures, including a ban on the application of manure to the land between September and February, the use of low-emission housing, covering outside manure stores, and diets with reduced protein [15]. This illustrates the ongoing importance of implementing additional measures to further minimize ammonia emission from pig production facilities.

The intensification of pig farming and the transition to indoor liquid manure storage have underscored the importance of precisely estimating CH_4 and NH_3 emissions influenced by the manure management system (MMS). This is essential for effectively addressing environmental challenges [16]. In recent years, efforts have been made to reduce CH_4 and NH_3 emissions from pig houses, leading to the development of low-emission housing systems [5]. One effective approach to reducing these emissions is frequent daily removal of manure by flushing the pit beneath the slatted floor [17–21]. This measure can be implemented in various forms and at different levels on farms.

In livestock production systems, the primary origin of NH_3 is the rapid hydrolyzation of urea in urine facilitated by the faecal enzyme urease and the breakdown of undigested proteins in manure [3,22]. The latter is a minor source of ammonia emission. This hydrolyzation process results in the formation of ammonium in an aqueous medium, in which the total ammoniacal N exists in equilibrium between the ionized NH_4+ and the unionized NH_3 forms [23]. This equilibrium is affected by temperature and pH, and higher levels cause an increased concentration of the unionized NH_3 form [24]. When the pH is below 7, nearly all ammoniacal N exists in its ionized form, while pH values above 7 lead to a significant rise in the unionized fraction [3].

The accurate quantification and understanding of methane and ammonia emissions from the livestock sector are crucial for assessing environmental impacts, developing sustainable farming practices, and addressing air and water quality concerns; however, evaluating the impact of numerous variants on emissions is a resource-intensive endeavour, both in terms of time and cost. Direct measurements at the housing level are time- and cost-consuming; these measured methods do not readily reflect emissions from various sources [19,25]. Models and algorithms with different levels of complexity have been developed for predicting CH_4 and NH_3 emissions [11,26–28]. In contrast, many other models tend to be static and empirical [29,30], often neglecting crucial aspects of animal metabolism, such as growth composition [31]. Moreover, empirical models have not been evaluated against experimental data [32]. To mitigate these challenges, a cost-effective alternative to direct measurements is to employ predictive models that allow the assessment of various interventions at the farm level while considering the factors affecting emissions release. A mathematical static model (MESPRO model) was developed to quantify the effect of practical measures (feed intake, diet composition, manure storage time, ambient and slurry temperature) on the characteristics and amount of slurry exertion from fattening pigs [31]. Subsequently, this model was developed into a dynamic model (ANIPRO model) to simulate the effect of ammonia volatilization in pig houses with partially slatted floors for fattening pigs [33]. Their model integrated aspects such as animal metabolism, diet composition, and the quantity and quality of urine and faeces, as well as the physical, chemical, and biological processes occurring in the urine pools on the floor and in the manure pit to predict ammonia emission. While the models were initially tested for fattening pigs with measured levels of ammonia emission, in 2018, they were extended to include other pig categories and were validated for the effect of various feeding measures on ammonia emission from fattening pigs, weaned piglets, and dry and pregnant sows [34]. However, these models were not previously tested for predicting CH_4 and NH_3 emissions affected by emission reduction measures related to manure management. Furthermore, it was observed that dynamically predicting methane emission influenced by frequent manure removal from indoor slurry pits in pig houses holds considerable potential for emission reduction. Therefore, the objectives of this study were to enhance the existing models by incorporating a dynamic methane prediction module and validate the model for the effect of pen design and manure management on methane and ammonia emissions from fattening pig houses.

2. Materials and Methods

2.1. Global Description of the Model

This existing model approach follows a mechanistic framework, striving to capture underlying processes as accurately as possible. The prediction model consists of three primary calculation modules: (1) the excretion model (MESPRO), (2) the ammonia emission model (ANIPRO), and (3) the methane emission model developed as part of this study. The model was developed and executed using MATLAB (R2018a). The model retrieves input data from Excel files. Within the main script, specific calculation rules are established to extract the respective data of each target scenario (for instance the pig housing types) separately, as delineated in the input data files. The simulations are conducted with daily time resolution over the course of a growing period (GP). Model calculations can be used for various pig categories, housing systems, and manure pit designs. The key calculation rules and the additions to the existing model are summarized here, while detailed descriptions can be found in the Supplementary Materials (SM) file.

The input variables of the model are length of production period, initial and final weight of pigs, total feed and water intake, feed composition, water use, weather data (temperature and relative humidity), climate set-up, building specifications, and storage time. The outputs, among others, are animal performance (animal weight, feed, and drinking water intake of each day), manure quantity and composition, and methane and ammonia emissions. The pH of slurry bulk was predicted based on measured data as a function of slurry total ammoniacal nitrogen, total inorganic carbon, and acetic acid concentrations (Aarnink et al., 2018 [34]). Furthermore, for a subset of parameters, calibration was executed using measurement data, i.e., the value of a parameter (e.g., a regression coefficient) was estimated based on the best fit on the measured data. This applies to three of the four parameters to estimate the Gompertz curves for animal weight and the cumulative feed and water intake (Section 2.2 of this study), the regression coefficient to calculate the evaporation coefficient for the floor and manure pit, calculation of NH_4-N from the total N-urinary excretion, surface temperature, air velocity of the top layer of manure and of the urine puddle on the floor, and the surface pH using regression lines based on lab measurements. The calculation rules have been extensively described in previous studies by [31,34]. A schematic representation of the model is shown in Figure 1.

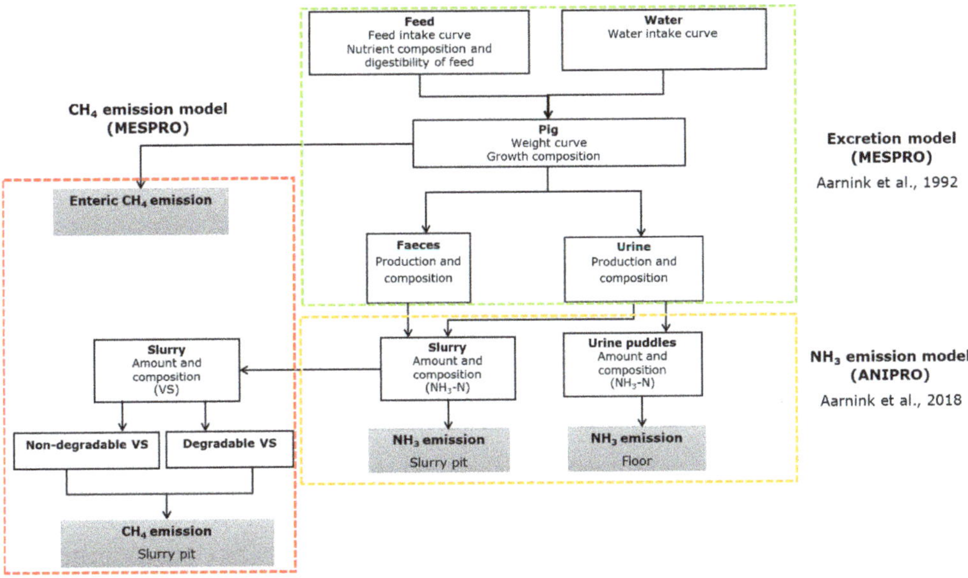

Figure 1. Schematic representation of the model concept for predicting ammonia and methane emissions (adopted from [31,34]).

2.2. MESPRO Module

As depicted in Figure 1, the model calculates both the quantity and composition of the slurry at the time of excretion and throughout the storage period within the pig house. To determine the emission on a daily basis and over the entire production period, calculation rules were established initially, focusing on estimating pig growth, feed consumption, and water intake. The patterns of growth, feed, and drinking water intake are described by the so-called Gompertz curve. Gompertz curve is a mathematical model used to describe the growth curves of animals including pigs [35,36]. This function is characterized by an initial exponential increase in weight, followed by a gradually decreasing rate of growth, eventually stabilizing at the final adult weight. The Gompertz curve is typically represented as an S-shaped curve (Equation (1)).

$$W_t = A + W_m . \exp\{-exp[-B(t - t^*)]\}/1000 \quad (1)$$

where W_t represents the weight of the animal at time t (kg), A is a constant that represents a specific weight (kg), W_m is the final adult weight of the animal (kg), B is a parameter that influences the shape of the growth curve and determines the speed of reaching the final weight (day^{-1}), t is the age of the animal (day), t^* represents the time when growth is maximal (day), 1000 is a conversion factor to convert from grams to kilograms.

The same Gompertz curve was employedto characterize the trajectory of both feed and drinking water intake over the growing period of the pigs. Parameters of the model (A, B and t^*) were determined using a regression analysis on the measured data. The fourth parameter (W_m in Equation (1), and its substitutes FI_m (total feed intake) and DWI_m (total drinking water intake) in feed and water intake curves, respectively) were calculated based on the measured values of the start and end weight and the total feed and water intakes during the growth period. The estimated final weight of the adult pigs was found to be a rough approximation; therefore, this parameter was assigned as variable in our calculation tool, while the other values were fixed at the values shown in Table 1. By making one parameter variable, it was possible to fit the curve to the achieved final weight, which is an input in our model. Incorporating these curves into the calculation tool and using the final weight, total feed intake, and water intake as input data enabled the estimation of parameters associated with the asymptotic values of these variables in the calculation model.

Table 1. Estimated parameters of the Gompertz curve (see Equation (1)), by regression on measured data, for calculating weight, feed, and drinking water intake of fattening pigs. The standard deviations are shown in brackets.

Parameter [1]	Weight	Feed Intake	Drinking Water Intake
A	0 [2]	−41.9 (13.1)	−149 (108)
$W_m/FI_m/DWI_m$ [3,4]	164.2 (3.9)	608 (48.3)	1432 (115.1)
B	0.0146 (0.0010)	0.0111 (0.0017)	0.0103 (0.0042)
t^* [5]	110.4 (4.3)	154.7 (10.8)	147.4 (23.9)

[1] For an explanation of the variables, see Equation (1). [2] The constant was set to zero, because including it in the regression formula did not provide a better fit. [3] FI_m and DWI_m are the parameters for feed and drinking water intake, comparable to W_m for the growth curve. [4] These values are fitted in the model to the input data (the initial and final weight of the pigs corresponding to W_m, and the total feed and water intakes corresponding to FI_m and DWI_m, respectively). [5] t^* is the time when growth is maximal (day).

The daily feed and water intake over the entire production period was used to determine the quantity and composition of excreted manure. Furthermore, considering the production stage of the pigs, the uptake of metabolizable energy, and the (calculated) growth curve, the nutrient retention in the animals was determined. Utilizing the digestion coefficients of the feed (particularly protein) and the retention of nutrients, the quantity and quality of manure was calculated. Additionally, employing a water balance, the concentration of the nutrients was calculated [31,34].

To determine the volume of slurry discharged into the manure pit, the calculation process was initiated by estimating the combined excretion of slurry and water. The total slurry production was determined based on the excretion of dry matter (DM_{exc}), after accounting for the gas produced from the hydrolysis of organic matter (biogas) and the CO_2 formed by the hydrolysis of urea ($Urea_{CO2}$). The total dry matter was estimated by combining the undigested organic matter (ash) with the organic matter (OM_{exc}) excreted through urine and the difference between inorganic matter (ash) intake and ash excretion.

2.3. ANIPRO Module

Within this model, ammonia emission is calculated by summing up the ammonia emission from different sources that can be distinguished within a pig pen, including manure channels, slatted floors, solid floors (which account for fouled pigs and pen partitions), and other relevant sources. The model incorporates physical attributes and indoor climate conditions of pig houses (Equation (2)). The ammonia emission from the different sources (floor and manure pit) in pig houses is calculated using the following formula [33]:

$$E_{NH3} = \frac{k_{NH3} \cdot A \cdot f \cdot [NH_4N]}{H} \quad (2)$$

where E_{NH3} is the ammonia emission (mol/s); k is the mass transfer coefficient, which is related to the air velocity and temperature of the emitting surface (m/s); f is the fraction of unionized ammonia in the solution, which is influenced by pH and temperature (T) and is a dimensionless value; $[NH_4N]$ is the total ammoniacal nitrogen concentration (mol/m^3); H is Henry's constant, which is also temperature-dependent and a dimensionless constant. By multiplying the calculated source strengths (emission per square meter) by the emissive surface (A), which can represent either the floor (urine puddles) or the manure pit (m^2), the ammonia emission from each source and the total ammonia emission of the entire pig house are determined. The emitting pit surface is calculated based on the shape and dimension of the manure pit or the floor. The emitting surface of the manure pit with straight walls is equal to the dimensions of the manure pit (length × width). In a manure pit with sloping walls, the emitting surface is influenced by the height of the manure. This height can be calculated from the manure production of the pigs and the shape of the manure pit. In this model, a distinction is made between soiling of the solid floor, soiling of the concrete slatted floor (at the front of the pen), and soiling of the metal, triangular slatted floor (at the back of the pen). By multiplying the calculated source strengths (emission per square meter) by the respective surface areas, both the total ammonia emission and the ammonia emission from each emitting surface can be determined individually [34]. Additional information for the estimation of the parameters of Equation (2) are listed in the SM file (Equations (S1)–(S9)).

2.4. Extended MESPRO Module

The MESPRO module constitutes calculation rules for the prediction of biogas production as a result of anaerobic conditions in manure pits. Methane formation is influenced by various factors, including manure temperature, storage duration, organic matter content of the manure, and the residual aged manure in the storage acting as an inoculum [28]. In this module, CH_4 emission from liquid manure is calculated using an algorithm proposed by [29]. This model is based on the Arrhenius equation and uses the residual volatile solids (VS) and a temperature response function for methanogenesis (Equation (3)).

$$F_t = (VS_d + 0.01VS_{nd})\, e^{(lnA - \frac{E_a}{RT})} \quad (3)$$

where F_t is specific methane production rate (g CH_4 kg VS^{-1} h^{-1}), VS_d and VS_{nd} represent the fast and slowly degradable fractions of volatile solids (VS) (kg kg^{-1} VS), A is the Arrhenius parameter (g CH_4 kg VS^{-1} h^{-1}), E_a is the activation energy (kJ mol^{-1}), R is the gas constant (kJ mol^{-1} K^{-1}), and T is temperature (K). The algorithms used to calculate the production of VS in manure and the corrected methane emission rate for the in situ temperature are presented in the SM file (Equations (S10)–(S14)). The model parameters (E_a, lnA, VS_d) utilized in this study are presented in Table 2.

Table 2. Key parameters of the methane estimation model, based on previous studies (E_a and lnA) and this study (VS_d).

Parameter	Petersen et al. (2016) [32]	This Study
VS_d (kg kg^{-1} VS)	0.51	0.83
E_a (kJ mol^{-1})	81.0 *	81.0 *
lnA (g CH_4 kg^{-1} vs. h^{-1})	31.3	31.3 **

* adopted from [37]; ** adopted from [32]; VS_d: degradable volatile solids; E_a: Apparent activation energy (kJ mol^{-1}); A: the Arrhenius parameter.

In order to predict the CH_4 emission per day of the growing period, the model uses the amount of residual manure in proportion to the amount of discharge from the inside pit to an outside destination. On a daily time step, the loading of actual manure level inside the pig house is simulated according to the initial height of manure at the beginning of the growing period, removal rate, and the last day of emptying the pit (day of the growing period). In other words, the model first estimates the composition of the remaining manure in the pit and subsequently simulates the methane production rates per day of storage in the manure pit underneath the slatted floor. Thus, the model assumes a linear relationship between the volume of manure and CH_4 emissions.

2.5. Model Application

To show the capacity of the model to predict NH_3 and CH_4 emissions from commercial pig houses, we compared model simulations to measured data. Experiments were carried out in two commercial fattening pig facilities over one year, from October 2020 to October 2021, including four growing periods, each lasting around 90 days. These experiments were conducted in two separate rooms, each representing a distinct manure management system (MMS). The first experimental room represented a conventional MMS with long-term storage (LS system) of manure in deep pit underneath the (partly) slatted floor.

The second experimental room was equipped with an adapted slurry pit for short-term storage (SS system) of manure by daily flushing from the pit underneath the (partly) slatted floor, dilution of the front channel, and reduced manure-emitting surface (Figure 2). These two systems were examined for the mitigation potential of methane and ammonia emissions affected by the manure removal frequency, pen design, and dilution of manure with water.

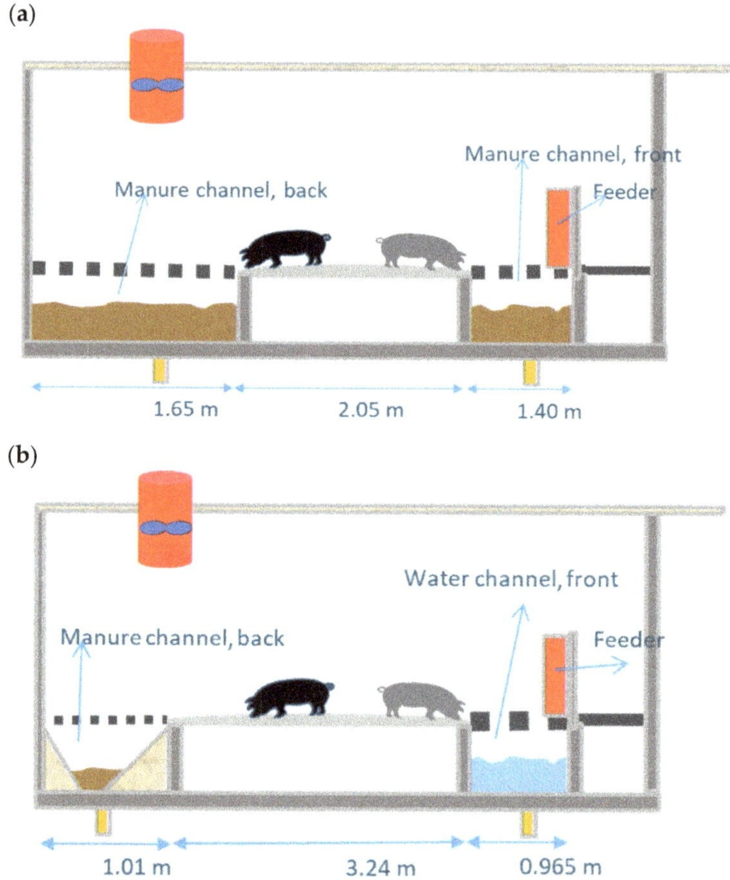

Figure 2. Schematic representation of the fattening pig houses (**a**) long-term storage (LS) and (**b**) short-term storage (SS) of manure inside the pig house. In the LS system, both front and back channels were filled with manure, while in the SS system, the back and front channels were filled with manure and manure diluted with water (cleaning water and spoiled drinking water).

The measured data comprised continuous readings of NH_3 and CH_4 concentrations (using sensors), ventilation rates, and temperature. Discrete measurements included manure production, manure composition, pen fouling (with urine), the percentage of the solid floor and the sloping walls that is moistened with urine, manure temperature, and height in the pit. Measurements for determining the emission of methane and ammonia gases were conducted using a reference method [38,39] at specific times (two randomly selected days per growing period) over one year (October 2020–2021) [40]. The characteristics of the farms and a schematic representation of the housing systems are presented in Table 3 and Figure 2.

Table 3. Overview of the experimental pig rooms with two different manure management systems.

Characteristics	LS [1]	SS [2]
No. of animal places	54	78
Growth range (kg)	23.6–115.6	22.6–114.0
Room length (m) × width (m)	11.28 × 5.90	15.55 × 6.00
No. of pens	6	6
Pen length (m) × width (m)	5.10 × 1.88	5.22 × 2.59
Depth of manure pit	1.20	0.50
Area per animal (m^2 pig^{-1})	1.00	1.00
Material slatted floor (back-front slatted floor)	Metal triangular—Concrete	Metal triangular—Concrete
Material solid floor	Concrete	Concrete
Slatted floor/Solid floor (%)	60/40	38/62
Slope of manure pit wall (°)	90	45
Manure removal interval (d)	45 [3]	1
Feeding/Drinking system	Dry feeder/Nipple	Dry feeder/Nipple

[1] LS: long term storage of manure; [2] SS: Short term storage; [3] Mean of emptying interval.

The ammonia concentration of the outgoing air was measured continuously using a Dräger Polytron C300 sensor (Dräger, Lübeck, Germany). The data could be read remotely. The methane and carbon dioxide concentrations in the ventilation air of the departments were continuously measured by means of an ABB monitor (ABB-Uras26, ABB, Frankfurt, Germany). The outgoing concentrations were measured hourly for both LS and SS systems. The data was then written to a data-logging system (CR1000X; Campbell Scientific Inc., Logan, UT, USA), which could be read remotely.

The pig houses were mechanically ventilated by means of an automatically controlled climate computer. The actual ventilation flow was then measured with a measuring fan in the ventilation shaft with the same diameter as the fan. Each revolution of the measuring fan gave a number of pulses, and this number was continuously logged by the climate computer of the pig farm. Temperature (°C) and relative humidity (%) were measured in each department near the ventilation shafts by means of a sensor (Vaisala HMP60; Vaisala GmbH, Hamburg, Germany). These data were stored on a data-logging system (CR1000X; Campbell Scientific Inc., Logan, UT, USA).

2.6. Data Analysis

The CH_4 and NH_3 emissions were calculated from the measured concentration and ventilation rate according to a reference method [38,39]. For both pig rooms (j = 1, 2), the emission, E_{ij} (kg yr^{-1} per pig place), was calculated per measurement day using the concentration of CH_4 and NH_3 in the outgoing air (C_{outij}) and the incoming air C_{inij}, (both in mg m^{-3}), the average ventilation flow V_{ij} (m^3 h^{-1} per pig place), and the density of the gas (ρ) to convert the concentration (ppm) to mg m^{-3} (0.667 and 0.71 kg m^{-3} for CH_4 and NH_3). This was then multiplied by 24 and 365 and divided by 10^6 to calculate the kg of methane and ammonia emissions per pig place per year (Equation (4)). For the reference measurement method, a vacancy factor of 3% was used to calculate the annual emission associated with the production cycles) [41].

$$E_{ij} = \left(C_{outt_{ij}} - C_{in_{ij}}\right) \cdot \rho \cdot V_{ij} \cdot 24 \cdot 365/10^6 \quad (4)$$

2.7. Model Validation

To assess the agreement between predicted and measured values and the accuracy of the model, the Mean Absolute Error (MAE) and Root Mean Square Error (RMSE) were calculated. Additionally, Y = X graphs were constructed, and the coefficient of determination (R^2) was provided. Lower values of MAE and RMSE imply higher agreement between the predicted and measured values. However, higher values of R^2 show a stronger correlation between predicted and measured values. The MAE, RMSE; and R^2 were calculated using the following equations (Equations (5)–(7)).

$$\mathrm{MAE} = \frac{\sum_1^n |V - \hat{V}|}{n} \tag{5}$$

$$\mathrm{RMSE} = \frac{\sqrt{\sum_1^n (V - \hat{V})^2}}{n} \tag{6}$$

$$\mathrm{R}^2 = 1 - \frac{\sum_1^n (V - \hat{V})^2}{\sum_1^n (V - \overline{V})^2} \tag{7}$$

where MAE is the mean absolute error, RMSE is the root mean square error (both in unit of the parameter), V is the measured value, $\hat{V}$ is the predicted value, n is the number of values, R^2 is the coefficient of determination, and $\overline{V}$ is the mean of the measured values.

3. Results and Discussion

3.1. Indoor Climate Parameters

A summary of the calculated and measured room temperature, manure temperature, and relative humidity for the LS and SS systems are presented in Table 4. The trend of the calculated and continuous-measured values of these parameters over one year are visualised in Figure S1 in the SM file. Based on the obtained results, the average predicted temperature was comparable with the calculated values for the two examined systems (MAE of 1.5 and 1.1 °C, RMSE of 4.0 and 3.8 °C, and R^2 of 0.8 and 0.92 for the LS and SS systems, respectively). The mean predicted room temperatures were 22.3 °C for LS and 21.4 °C for SS, corresponding with measured values of 23.0 °C and 21.0 °C, respectively. The average temperature of the manure was measured as 24.5 °C and 22.3 °C, respectively. The lower temperature ranges for the SS system can be explained by the short storage courses and daily removal of manure in this system. The manure temperature was underestimated by the model (MAE of 6.7 and 6.8 °C, RMSE of 6.5 and 7.5 °C, and an extremely low R^2 (<1) for the LS and SS systems, respectively). The manure temperature was calculated by using empirical relationships with the measured outgoing air temperature. This significant difference in manure temperatures suggests the need to consider temperature variation at inside storage facilities in countries such as the Netherlands, where long-term manure storage inside the pig houses is a common practice. Furthermore, the high variation between the measured and predicted temperature of manure can be due to issues such as limited data availability and measurement errors. In this study, manure temperature was measured manually in a limited number of observations (six times over one year). The calculated relative humidity was relatively underestimated by the model compared to the continuous measured data (MAE and RMSE > 10% and R^2 of 0.57 and 0.63 in both systems).

Table 4. Average calculated and measured room temperature, manure temperature, and relative humidity for two manure management systems (MMS) in fattening pig rooms. Standard deviations are given between parentheses.

Variable	MMS	Calculated (Model)	Measured (Continuous)	Measured (Discrete) [3]	MAE [4]	RMSE [5]	R^2
Room temperature	LS [1]	22.3 (1.6)	23.0 (1.8)	-	1.5	4.0	0.80
(°C)	SS [2]	21.4 (2.7)	21.0 (2.6)	-	1.1	3.8	0.92
Manure temperature	LS	19.5 (1.3)	-	24.5 (1.7)	6.7	6.5	<1
(°C)	SS	18.7 (2.2)	-	22.3 (2.4)	6.8	7.5	<1
Relative humidity	LS	59.7 (4.1)	67.6 (4.4)	-	9.4	13.1	0.57
(%)	SS	56.1 (5.3)	67.9 (5.0)	-	13.1	15.5	0.63

[1] LS: long storage system; [2] SS: Short storage system; [3] Reference method [40]; [4] Mean absolute error; [5] Root mean square error between calculated and measured values based on daily differences.

3.2. Methane Emission

A summary of the calculated and measured volatile solids (VS) and methane (CH_4) emission from the LS and SS systems are presented in Table 5. The calculated vs. content of the manure was lower than the measured value (ca. 10% with MAE of around 2.6 g kg^{-1} RMSE in the range of 11.2–12 g kg^{-1} and average R^2 of 0.20 for both systems). The average calculated CH_4 emission (18.5 and 4.3 kg yr^{-1} per pig place) was in between the means from the continuous data (15.9 and 5.6 kg yr^{-1} per pig place) and the means from the discrete measurements using the reference method (22.0 and 3.1 kg yr^{-1} per pig place) for LS and SS, respectively. The largest RMSE was observed between the predicted and continuous measurements, with a value of 4.6 kg yr^{-1} per pig place.

Table 5. Average calculated and measured volatile solids and methane emission for two manure management systems (MMS) in fattening pig rooms. Standard deviations are given between parentheses.

Variable	MMS	Calculated (Model)	Measured (Continuous)	Measured (Discrete) [3]	MAE [4]	RMSE [5]	R^2
Volatile solids-manure	LS [1]	68.5 (5.9)	-	76.2 (11.5)	2.6	12.5	0.16
(g kg^{-1})	SS [2]	68.0 (7.3)	-	77.4 (9.0)	2.8	11.2	0.21
CH_4 emission(kg yr^{-1} per pig place)	LS	18.5 (5.3)	15.9 (7.9)	22.0 (5.0)	3.1	4.6/2.8	0.64
	SS	4.3 (2.4)	5.6 (3.7)	3.1 (1.3)	1.9	3.3/3.0	0.13

[1] LS: long storage system; [2] SS: Short storage system; [3] Reference method [40]; [4] Mean absolute error; [5] Root mean square error between calculated and measured (continuous/discrete) values based on daily differences.

The development of calculated and measured methane emission and the height of manure in the pit representing the manure volume are presented in Figure 3a,b. Based on this graph, the predicted volume of the stored manure and the corresponding methane emission fitted well with the measurements. The breaks seen in this graph are due to the partial emptying of the pit on certain days during each growing period. No continuous measurement data were available for the first growing period (GP1) due to unreliable sensor calibrations. Based on the results of the in vivo measurements (the reference method) for SS, a reduction potential for methane emission of 86.0% (±5.6%) compared with LS can be expected.

Comparing the measured CH_4 emission (continuous and point measurements) with the calculated values implies that the model predictions fitted quite well with the measurements. The mean of the measured (reference method) and calculated methane emission of the LS system were 22.0 ± 5.0 and 18.5 ± 5.3 kg yr^{-1} per pig place, respectively. Those for the SS system were 3.1 ± 1.3 and 4.3 ± 2.4 kg yr^{-1} per pig place, respectively. The variation observed between continuous CH_4 emission measurements obtained from the sensors and discrete measurements using the reference method can be attributed to several factors. The primary factor to note is that the sensors utilized in our experiment were new and had been calibrated by the manufacturer, without specific calibration for our study. Hence, it is reasonable to attribute any discrepancies to potential errors in calibration. The secondary factor: the main objective of sensor measurements was to monitor emission patterns with a lesser focus on measuring absolute emissions. Furthermore, ensuring the proper functioning of the sensors is imperative and requires ongoing monitoring.

During the growing period, methane emission was mainly influenced by the level of manure stored in the pit, featuring lower emission rates after each (partial or full) emptying of the manure pit (Figure 3b). For SS, the manure pit was emptied every day, while the water channel was emptied once per growing period. This partially explains the increasing pattern of methane emission in the SS system (blue lines), besides the expected increase in methane production from enteric fermentation. In the third growing period (GP3), it is likely that the water channel of the SS system was emptied more frequently, in line with farm practices implemented on the commercial farm, though these data have not been registered and therefore have not been accounted for in the model inputs. Consequently, this change may have contributed to the observed reduction in CH_4 emission, as detected by both the sensor and the reference method, in contrast to the predictions generated by the model.

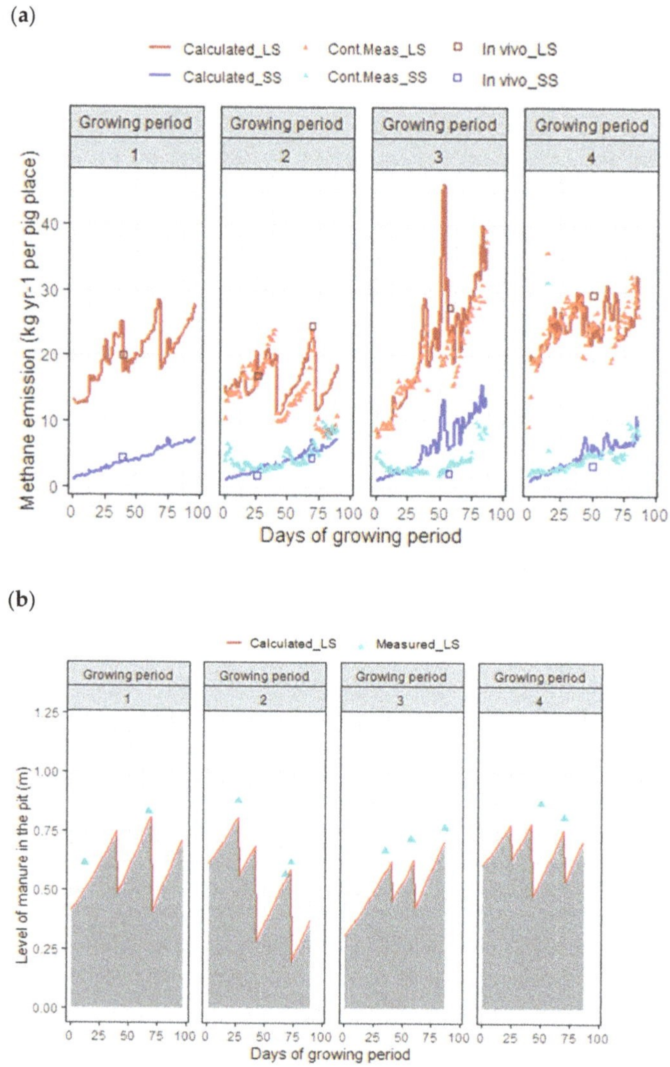

Figure 3. Calculated (line) and measured (point) methane emission (kg yr^{-1} per pig place) (**a**) and height of the manure accumulated in the pit (**b**) per growing period (GP) for two manure management systems (MMS) in fattening pig rooms. LS: long storage in pit with straight walls (the red symbols and line) and SS: short storage in pit with sloped walls (the blue symbols and line). All data represents the mean daily values converted to annual levels. The 'Cont.Meas' category indicates the continuous measurements using sensors. The 'in vivo' data indicates the discrete reference measurements. In growing period 1, no sensor measurements were recorded. The start and end dates of the growing periods were: GP1: 8 October 2020–12 January 2021; GP2: 21 January 2021–20 April 2021; GP3: 27 April 2021–21 July 2021; GP4: 27 July 2021–21 October 2021.

For a better prediction level, detailed farm records of the removal volume, its frequency, and the residual manure in the pit are necessary inputs to the model. In this study, the height of manure in the deep pit and the level of the water channel were not regularly measured and observed. The model, therefore, assumes a linear correlation between the estimated volume of the manure, estimated by the height and area of the manure pit, and the CH_4 emission. In SS, it is expected that, with a more frequent emptying system in the water channel and thorough cleaning of the channels after flushing, CH_4 emission can be reduced to even a larger extent. Due to the daily and (almost) complete removal of the manure, the anaerobic conversion of the organic matter in the manure to methane and carbon dioxide barely occurs [42]. By limiting the growth and activity of methanogenic communities and by reducing the amount of organic matter by the more frequent removal of manure to outside storage, lower methane emission rates can be expected [43]. Another study has recently shown that methane emission is highly dependent on the frequency of manure removal and less dependent on temperature of the manure in pig house storage with increased manure removal frequencies [20]. Therefore, the present findings indicate that daily manure removal can significantly diminish methane emission, to the extent that the only source of CH_4 is the enteric methane from the pigs. The enteric methane from pigs was reported as approx. 1.5 kg pig^{-1} year^{-1} produced by pigs themselves [44]. Thus, it is to be expected that, with SS, most methane emission from manure could be prevented.

Another general observation is that, in addition to manure management practices, CH_4 emission in the LS system was influenced by seasonal effects (Figure 3a). Specifically, the average methane emission during the summer months (GP3 and GP4, encompassing June and July) exceeded that of the other two growing periods occurring in the autumn and winter seasons. The current model predicts the manure temperature based on the room temperature, and the room temperature is calculated according to the outside temperature. We stress, therefore, the importance of accurately predicting the manure temperature in relation to the manure volume in the storage for accurate prediction of CH_4 emission.

The linear relationships between the continuous-measured and calculated methane emission are shown in Figure 4. The results show that the model has a better prediction for LS than SS (R^2 of ca. 64% and 13%). This difference in prediction can be explained by the higher number of assumptions about the LS system (for example, the cleaning level of the manure pit, as well as the emptying intervals of the water channel, which may not be fully aligned with the practical situation of the examined farm). By removing the predicted values of the third growing period (GP3), the R^2 increases (by a factor of 2). Overall, improving the inputs of the model with accurate activity data from farm practices can increase the accuracy of the prediction.

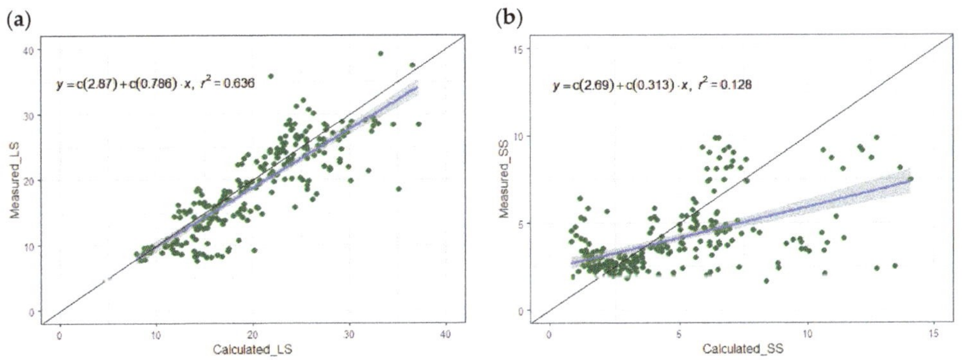

Figure 4. The continuous-measured (Y) and calculated (X) methane emission (kg yr^{-1} per pig place) in (**a**) LS: long storage in pit with straight walls; (**b**) SS: short storage in pit with sloped walls. The solid black line represents the 1:1 line. The points represent the measured data points and the blue line is the regression line between (X) and (Y) showing the best-fit line throughout the data points. All data represents the mean daily values converted to annual emission.

3.3. Ammonia Emission

A summary of the calculated and measured ammonia emission per source (manure pit and soiled floor surface) for one year are presented in Table 6. The average values of most of the measured parameters were comparable with the calculated values. The mean calculated ammonia emission for LS and SS were 2.6 (±1.0) and 1.4 (±0.8) kg yr^{-1} per pig place, respectively. The calculated and measured parameters were in good agreement, confirming the accuracy of the model predictions. The MAE of NH_3 emission was 0.8 and 0.5 kg yr^{-1} per pig place. Additionally, the RMSE was 1 and 0.5 kg yr^{-1} per pig place with an R^2 of 0.45 and 0.67 for the LS and SS systems, respectively. From these results, it is also evident that the NH_3 emission emitted from the manure pit is higher than from the floor (on average, 80% for LS and 70% for SS were released from the manure pit). This highlights the impact of mitigation measures on reducing NH_3 emission from the inside storage pits compared to the floor. Measures controlling the indoor climate of the pig houses and management factors affect ammonia formation on the floor surface. For example, in pig houses, the risk of soiling becomes greater as the areaof solid floor increases. This risk can be limited by a good pen design and by maintaining a good indoor climate (e.g., by air or floor cooling in the summer) [34]. Many studies have shown that lower ammonia emission can be achieved with partly slatted floors, provided that the solid part of the floor remains clean [45–47].

Table 6. Average calculated and measured ammonia emission sources for two manure management systems (MMS) in fattening pig rooms. Standard deviations are given between parentheses.

Variable	MMS	Calculated-Model	Measured - Continuous	Measured - Discrete [3]	MAE [4]	RMSE [5]	R^2
NH_3 emission—Floor	LS [1]	0.5 (0.3)	-	-	-	-	-
(kg yr^{-1} per pig place)	SS [2]	0.5 (0.3)	-	-	-	-	-
NH_3 emission—Manure pit	LS	2.2 (0.9)	-	-	-	-	-
(kg yr^{-1} per pig place)	SS	0.9 (0.6)	-	-	-	-	-
Total NH_3 emission	LS	2.6 (1.0)	2.6 (0.9)	2.71 (0.4)	0.8	1.2/1.1	0.45
(kg yr^{-1} per pig place)	SS	1.4 (0.7)	1.2 (0.9)	1.01 (0.2)	0.5	0.8/0.3	0.67

[1] LS: long storage system; [2] SS: Short storage system; [3] Reference method [40]; [4] Mean absolute error; [5] Root mean square error between calculated and measured (continuous/discrete) values based on daily differences.

The total ammonia emission obtained from the model predictions (lines) in comparison with the sensor measurements (triangular points) and the reference method (in square and circle) are represented in Figure 5. This graph also demonstrates the reduction potential in ammonia emission from the manure pit affected by the manure management system and pen design (the red line and points for LS and the blue ones for SS). The average ammonia emission reduction measured by the reference method corresponds to 62.7% (±7.4%), with an average emission rate of 2.63 and 1.0 kg yr^{-1} per pig place (Booijen et al., 2023) [41]. The discrepancy observed between continuous NH_3 emission measurements obtained from the sensors and discrete measurements using the reference method can be attributed to several factors, as mentioned above. The new sensors, utilized in this study at the commercial farm, were not calibrated within this study. Additionally, potential inaccuracies in the NH_3 sensors, particularly in measuring very low (<2 ppm) and very high concentrations, may contribute to the variation. During the last two growing periods (GP3 and GP4), the predicted NH_3 emission surpassed both the continuously measured data and reference measurements. This disparity could be linked to the predicted ambient and manure surface temperatures during the warm season.

From the results, it can be concluded that the prediction of ammonia emission was well-aligned and reasonably accurate for both systems when compared with the reference points. Overall, the reduction measures in the SS system, including reduction of the slurry pit surface with sloped pit walls, frequent manure removal, and dilution with water, demonstrated relatively the good reduction potential of NH_3 emission compared to the reference system.

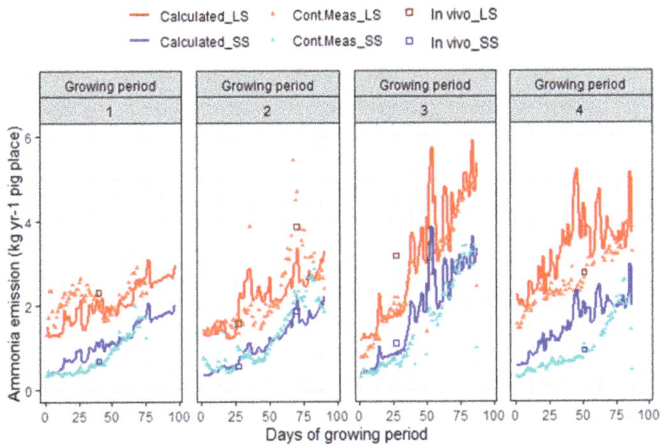

Figure 5. Calculated (line) and measured (point) ammonia emission (kg yr^{-1} per pig place) for two manure management systems (MMS); LS: long storage in pit with straight walls (the red symbols and line); SS: short storage in pit with sloped walls (the blue symbols and line). Category 'Cont.Meas' indicates the continuous measurements using sensors. Category 'in vivo' indicates the discrete reference measurements. Breaks are due to problems with the sensor or outlier detection. The start and end dates of the growing periods were: GP1: 8 October 2020–12 January 2021; GP2: 21 January 2021–20 April 2021; GP3: 27 April 2021–21 July 2021; GP4: 27 July 2021–21 October 2021.

The linear relationships between the daily measured and calculated ammonia emission per MMS are shown in Figure 6. The slopes of the linear regression line and R^2 were around 0.63 and 0.45 for LS and 0.83 and 0.67 for SS. This relationship also suggests that as emission rates increase, the variability in emission becomes higher, leading to lower prediction accuracy.

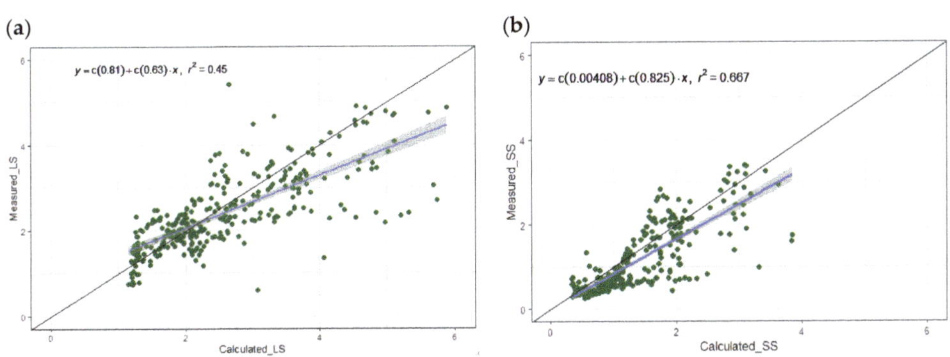

Figure 6. The measured (Y) and calculated (X) ammonia emission (kg yr^{-1} per pig place) of two MMS (**a**) LS: long storage in pit with straight walls; (**b**) SS: short storage in pit with sloped walls. The solid black line represents the 1:1 line. The points represent the measured data points and the blue line is the regression line between (X) and (Y) showing the best-fit line throughout the data points. Data represents the average daily values converted to annual level.

3.4. Model Parameters and Implications for the Predicted Emission

A model approach in estimating the influence of mitigation measures has been introduced in this study. The most critical component in these predictions is water excretion, due to the error caused by the incorrect estimation of water evaporation from manure and fouled floors with urine, which is affected by the air velocity and temperature at the surface. Within the model, the air velocity above the evaporation surface is assumed to be constant.

Model predictions can be improved by better prediction of the pH of the manure. The current model lacks an adequate prediction of manure pH, which is a crucial factor influencing NH_3 emission. This is partly due to the fact that a large number of factors are affecting the pH; in particular, the carbonate content of urine and manure is hard to predict. As an intermediate solution, the pH of the urine and manure was measured and used as an input for the model. The pH of the top layer of the manure was determined from the pH of the bulk of the manure based on a lab-scale analysis at the University of Southern Denmark in Odense [34]. Further development of the model should focus on the accurate estimation of the pH by using a measurement set-up to measure the surface pH of the top 0.1 mm of manure, comparable to a practical situation, or using the measured pH of the bulk manure as an input for the emission model.

Another important point for improvement of the model is the assumption about the emitting surface. In this current version of the model, the ammonia emission per m^2 of contaminated concrete slatted floor was assumed to be the same as that of the concrete solid floor. This current model could be further improved by developing a dynamic urination model for determining the variation in ammonia emission over time (on an hourly basis), as suggested by [33].

Temperature and air velocity above the emitting surface were estimated from the measured temperature and ventilation quantity at the exhaust, which, although currently easily measured, can be nevertheless improved by air flow models to better estimate the temperature and air velocity above different emitting surfaces. Extra effort is required to incorporate and apply these features into this model in a simple way. Further development of the methane emission model should be focused on temperature variation by the depth at inside storages and degradability of the organic matter over the storage period. Furthermore, the accuracy of the model is heavily reliant on precise and adequate input data. In this work, the regular recording of the manure level height in the storage was lacking. It is anticipated that with a larger dataset, the predicted methane emission will be improved.

It should be pointed out that the model currently uses the temperature-related equation for the estimation of methane emission. This relationship is dependent on the lnA value [32]. The lnA value used in this study was obtained from the study by [32,37]. Therefore, for a better estimation of methane emission using the approach used in this study, it is recommended to parameterize the model by the determination of lnA for country-specific and system-specific (linked to various manure management systems) pig farms in the Netherlands.

4. Conclusions

This paper discussed the prediction of CH_4 and NH_3 emissions through a model-based approach. The study compared the results obtained by integrating two comprehensive models (ANIPRO + MESPRO) with values derived from continuous data measurements and a discrete reference method. The model approach was applied to two fattening pig houses implementing emission reduction measures and compared with a conventional housing system for pigs in the Netherlands. The main conclusions drawn from this study are as follows:

1. The average calculated CH_4 and NH_3 emissions on an annual basis correspond well with measured values for the examined measures.
2. The measurements confirmed the reduction potential of the studied measures for CH_4 and NH_3 emissions from pig houses. The model could predict these effects with an acceptable degree of accuracy.
3. The obtained results suggest that improving the calculation rules of the model for better estimation of variables affecting ammonia emission, such as the pH, temperature, and air velocity, will lead to a better prediction of emissions.

4. The model attributes provide valuable means for assessing the impact of mitigation measures on CH_4 and NH_3 emissions. This provides a robust basis for assessing the impact of management and housing strategies on CH_4 and NH_3 emissions from pig houses, which, in turn, helps support more sustainable practices in pig farming.

Supplementary Materials: The following supporting information can be downloaded at https://www.mdpi.com/article/10.3390/ani14060964/s1. The supporting material is published online alongside this paper, including ädditional equations and a figure of the result section (Figure S1 Calculated (line) and continuous-measured (point) values of indoor temperature (°C), relative humidity (%) and ventilation rate (m^3 h^{-1} per pig place). The red lines and points represent the reference department and those in blue represent the trial department. Each segment indicate one growing period (GP); 8 October 2020–12 January 2021; GP2: 21 January 2021–20 April 2021; GP3: 27 April 2021–21 July 2021; GP4: 27 July 2021–21 October 2021.) [32–34,48–52].

Author Contributions: Data preparation, P.S. and A.J.A.A.; Calculations and data analysis, P.S., S.H.P.-K. and A.J.A.A.; Writing and revision of the paper, P.S., A.J.A.A. and S.H.P.-K. All authors have read and agreed to the published version of the manuscript.

Funding: This study was funded under the 2018 Joint Call of the ERA-NETs FACCE ERA-GAS, SusAn and ICT-AGRI on "Novel technologies, solutions and systems to reduce greenhouse gas emissions in animal production systems" and by the Dutch Research Council (NWO) (budget No. 8910) and the Ministry of Agriculture, Nature and Food Quality in the Netherlands.

Institutional Review Board Statement: Ethical review and approval were waived for this study, due to the fact that the pigs received normal, commercially available diets and housing and management were according to Dutch law for commercial pig production. There were no interventions that required a license according to the Animal Experiments Act.

Informed Consent Statement: Informed consent has been obtained from the owner of the animals involved in this study.

Data Availability Statement: The data presented in this study are available on request from the corresponding author. The data are not publicly available due to the data still being processed to produce other papers.

Acknowledgments: We extend our gratitude to C.M. Groenestein and M. Booijen for their valuable inputs, as well as to the laboratory researchers at Wageningen Livestock Research for their support during the measurement campaigns of this study. Additionally, we express our appreciation to the entrepreneurs of the pig farms where these measurements were conducted.

Conflicts of Interest: The authors declare no conflicts of interest.

References

1. Zong, C.; Li, H.; Zhang, G. Ammonia and greenhouse gas emissions from fattening pig house with two types of partial pit ventilation systems. *Agric. Ecosyst. Environ.* **2015**, *208*, 94–105. [CrossRef]
2. de Vries, W. Impacts of nitrogen emissions on ecosystems and human health: A mini review. *Curr. Opin. Environ. Sci. Health* **2021**, *21*, 100249. [CrossRef]
3. Le Dinh, P.; van der Peet-Schwering, C.M.C.; Ogink, N.W.M.; Aarnink, A.J.A. Effect of Diet Composition on Excreta Composition and Ammonia Emissions from Growing-Finishing Pigs. *Animals* **2022**, *12*, 229. [CrossRef] [PubMed]
4. Myhre, G.; Shindell, D.; Pongratz, J. Anthropogenic and natural radiative forcing. In *Climate Change 2013: The Physical Science Basis. Contribution of Working Group I to the Fifth Assessment Report of the Intergovernmental Panel on Climate Change*; Stocker, T.F., Qin, D., Plattner, G.-K., Tignor, M., Allen, S.K., Boschung, J., Nauels, A., Xia, Y., Bex, V., Midgley, P.M., Eds.; Cambridge University Press: Cambridge, UK, 2014.
5. O'Mara, F.P. The significance of livestock as a contributor to global greenhouse gas emissions today and in the near future. *Anim. Feed Sci. Technol.* **2011**, *166*, 7–15. [CrossRef]
6. van der Zee, T.; Bannink, A.; van Bruggen, C.; Groenestein, K.; Huijsmans, J.; van der Kolk, J.; Lagerwerf, L.; Luesink, H.; Velthof, G.; Vonk, J. *Methodology for Estimating Emissions from Agriculture in the Netherlands. Calculations for CH_4, NH_3, N_2O, NOx, NMVOC, PM10, PM2.5 and CO_2 Using the National Emission Model for Agriculture (NEMA)–Update 2021*; National Institute for Public Health and the Environment: Bilthoven, The Netherlands, 2021.
7. Jørgensen, H.; Knudsen, K.E.B.; Theil, P.K. *Enteric Methane Emission from Pigs*; INTECH Open Access Publisher: London, UK, 2011.

8. Van Damme, M.; Clarisse, L.; Whitburn, S.; Hadji-Lazaro, J.; Hurtmans, D.; Clerbaux, C.; Coheur, P.-F. Industrial and agricultural ammonia point sources exposed. *Nature* **2018**, *564*, 99–103. [CrossRef] [PubMed]
9. Wyer, K.E.; Kelleghan, D.B.; Blanes-Vidal, V.; Schauberger, G.; Curran, T.P. Ammonia emissions from agriculture and their contribution to fine particulate matter: A review of implications for human health. *J. Environ. Manag.* **2022**, *323*, 116285. [CrossRef] [PubMed]
10. Aarnink, A.J.A. Factors affecting ammonia concentration in slurry from fattening pigs. In Proceedings of the First International Symposium on Nitrogen Flow in Pig Production and Environmental Consequences, Wageningen, The Netherlands, 8–11 June 1993; pp. 413–420.
11. Webb, J.; Misselbrook, T.H. A mass-flow model of ammonia emissions from UK livestock production. *Atmos. Environ.* **2004**, *38*, 2163–2176. [CrossRef]
12. Groenestein, C.M. *Environmental Aspects of Improving Sow Welfare with Group Housing and Straw Bedding*; Wageningen University and Research: Wageningen, The Netherlands, 2006.
13. Galloway, J.N.; Dentener, F.J.; Capone, D.G.; Boyer, E.W.; Howarth, R.W.; Seitzinger, S.P.; Asner, G.P.; Cleveland, C.C.; Green, P.; Holland, E.A. Nitrogen cycles: Past, present, and future. *Biogeochemistry* **2004**, *70*, 153–226. [CrossRef]
14. Anonymous. Emissies Van de Luchtverontreinigende Stoffen Per Sector [Emission of Air Pollutants Per Sector]. Emissieregistratie, Rijksoverheid [National Government, Emission Registration]. Available online: https://www.emissieregistratie.nl/data/overzichtstabellen-lucht/luchtverontreinigende-emissies (accessed on 20 November 2023).
15. van Bruggen, C.; Bannink, A.; Bleeker, A.; Bussink, D.; van Dooren, H.; Groenestein, C.; Huijsmans, J.; Kros, J.; Lagerwerf, L.; Oltmer, K. *Emissies Naar Lucht Uit de Landbouw Berekend met NEMA voor 1990–2021*; Wettelijke Onderzoekstaken Natuur & Milieu: Wageningen, The Netherlands, 2023.
16. De Vries, J.W.; Aarnink, A.J.A.; Groot Koerkamp, P.W.G.; De Boer, I.J.M. Life cycle assessment of segregating fattening pig urine and feces compared to conventional liquid manure management. *Environ. Sci. Technol.* **2013**, *47*, 1589–1597. [CrossRef]
17. Amon, B.; Kryvoruchko, V.; Fröhlich, M.; Amon, T.; Pöllinger, A.; Mösenbacher, I.; Hausleitner, A. Ammonia and greenhouse gas emissions from a straw flow system for fattening pigs: Housing and manure storage. *Livest. Sci.* **2007**, *112*, 199–207. [CrossRef]
18. Groenestein, C.M.; Smits, M.C.J.; Huijsmans, J.F.M.; Oenema, O. *Measures to Reduce Ammonia Emissions from Livestock Manures: Now, Soon and Later*; Wageningen UR Livestock Research: Wageningen, The Netherlands, 2011.
19. Philippe, F.-X.; Nicks, B. Review on greenhouse gas emissions from pig houses: Production of carbon dioxide, methane and nitrous oxide by animals and manure. *Agric. Ecosyst. Environ.* **2015**, *199*, 10–25. [CrossRef]
20. Feng, L.; Guldberg, L.B.; Hansen, M.J.; Ma, C.; Ohrt, R.V.; Møller, H.B. Impact of slurry removal frequency on CH_4 emission and subsequent biogas production; a one-year case study. *Waste Manag.* **2022**, *149*, 199–206. [CrossRef] [PubMed]
21. Vechi, N.T.; Jensen, N.S.; Scheutz, C. Methane emissions from five Danish pig farms: Mitigation strategies and inventory estimated emissions. *J. Environ. Manag.* **2022**, *317*, 115319. [CrossRef] [PubMed]
22. Canh, T.T. *Ammonia Emission from Excreta of Growing-Finishing Pigs as Affected by Dietary Composition*; Grafisch Service Centrum: Wageningen, The Netherlands, 1998.
23. Cortus, E.L.; Lemay, S.P.; Barber, E.M.; Hill, G.A.; Godbout, S. A dynamic model of ammonia emission from urine puddles. *Biosyst. Eng.* **2008**, *99*, 390–402. [CrossRef]
24. Cai, Y.; Gallegos, D.; Zheng, Z.; Stinner, W.; Wang, X.; Pröter, J.; Schäfer, F. Exploring the combined effect of total ammonia nitrogen, pH and temperature on anaerobic digestion of chicken manure using response surface methodology and two kinetic models. *Bioresour. Technol.* **2021**, *337*, 125328. [CrossRef] [PubMed]
25. Aarnink, A.J.A.; Demeyer, P.; Rong, L.; Aarnink, A.J.A.; Demeyer, P.; Rong, L. A Simple Model as Design Tool for Low-Ammonia Emission Pig Housing. In *Technology for Environmentally Friendly Livestock Production*; Springer: Berlin/Heidelberg, Germany, 2023; pp. 11–21.
26. Bjerg, B.; Cascone, G.; Lee, I.-B.; Bartzanas, T.; Norton, T.; Hong, S.-W.; Seo, I.-H.; Banhazi, T.; Liberati, P.; Marucci, A. Modelling of ammonia emissions from naturally ventilated livestock buildings. Part 3: CFD modelling. *Biosyst. Eng.* **2013**, *116*, 259–275. [CrossRef]
27. Bjerg, B.; Norton, T.; Banhazi, T.; Zhang, G.; Bartzanas, T.; Liberati, P.; Cascone, G.; Lee, I.B.; Marucci, A. Modelling of ammonia emissions from naturally ventilated livestock buildings. Part 1: Ammonia release modelling. *Biosyst. Eng.* **2013**, *116*, 232–245. [CrossRef]
28. Dalby, F.R.; Hafner, S.D.; Petersen, S.O.; Vanderzaag, A.; Habtewold, J.; Dunfield, K.; Chantigny, M.H.; Sommer, S.G. A mechanistic model of methane emission from animal slurry with a focus on microbial groups. *PLoS ONE* **2021**, *16*, e0252881. [CrossRef]
29. Sommer, S.G.; Petersen, S.O.; Møller, H.B. Algorithms for calculating methane and nitrous oxide emissions from manure management. *Nutr. Cycl. Agroecosyst.* **2004**, *69*, 143–154. [CrossRef]
30. Li, C.; Salas, W.; Zhang, R.; Krauter, C.; Rotz Al Mitloehner, F. Manure-DNDC: A biogeochemical process model for quantifying greenhouse gas and ammonia emissions from livestock manure systems. *Nutr. Cycl. Agroecosyst.* **2012**, *93*, 163–200. [CrossRef]
31. Aarnink, A.J.A.; Van Ouwerkerk, E.N.J.; Verstegen, M.W.A. A mathematical model for estimating the amount and composition of slurry from fattening pigs. *Livest. Prod. Sci.* **1992**, *31*, 133–147. [CrossRef]
32. Petersen, S.O.; Olsen, A.B.; Elsgaard, L.; Triolo, J.M.; Sommer, S.G. Estimation of methane emissions from slurry pits below pig and cattle confinements. *PLoS ONE* **2016**, *11*, e0160968. [CrossRef] [PubMed]

33. Aarnink, A.J.A.; Elzing, A. Dynamic model for ammonia volatilization in housing with partially slatted floors, for fattening pigs. *Livest. Prod. Sci.* **1998**, *53*, 153–169. [CrossRef]
34. Aarnink, A.J.A.; Van de Pas, P.A.; Van der Peet-Schwering, C.M.C.; Hol, A.; Binnendijk, G.P.; Le Dinh, P.; Hafner, S.D.; Ogink, N.W.M. *Rekentool Voor Het Bepalen van de Effecten van Voer- En Management-Maatregelen op de Ammoniakemissie bij Varkens: Ontwikkeling en Validatie [Calculation Tool to Estimate the Effects of Feeding and Management Measures on Ammonia Emission in Pigs]*; Report 1086; Wageningen Livestock Research: Wageningen, The Netherlands, 2018; p. 118.
35. Wellock, I.J.; Emmans, G.C.; Kyriazakis, I. Describing and predicting potential growth in the pig. *Anim. Sci.* **2004**, *78*, 379–388. [CrossRef]
36. Aarnink, A.J.A.; Huynh, T.T.T.; Bikker, P. Modelling heat production and heat loss in growing-finishing pigs. In Proceedings of the CIGR-AgEng Conference, Aarhus, Denmark, 26–29 June 2016; pp. 26–29.
37. Elsgaard, L.; Olsen, A.B.; Petersen, S.O. Temperature response of methane production in liquid manures and co-digestates. *Sci. Total Environ.* **2016**, *539*, 78–84. [CrossRef] [PubMed]
38. Groenestein, C.M.; Mosquera Losada, J.; Ogink, N.W.M. *Protocol Voor Meting van Methaanemissie uit Huisvestingssystemen in de Veehouderij 2010 = Measurement Protocol for Methane Emission from Housing Systems in Livestock Production 2010*; Report Wageningen UR Livestock Research; Wageningen UR Livestock Research: Wageningen, The Netherlands, 2011; p. 32.
39. Ogink, N.; Mosquera Losada, J.; Hol, A. *Protocol Voor Meting van Ammoniakemissie uit Huisvestingssystemen in de Veehouderij 2013a= Measurement Protocol for Ammonia Emission from Housing Systems in Livestock Production 2013a*; Wageningen UR Livestock Research: Wageningen, The Netherlands, 2017.
40. Booijen, M.; Wagensveld, J.; van Riel, J.; de Mol, R.; Aarnink, A. *Emissiereductie Methaan, Ammoniak en Geur in Varkensstallen Met Dagontmesting*; Wageningen Livestock Research: Wageningen, The Netherlands, 2023.
41. Mosquera Losada, J.; Hol, J.M.G.; Winkel, A.; Lovink, E.; Ogink, N.W.M.; Aarnink, A.J.A. *Fijnstofemissie Uit Stallen: Vleesvarkens = Dust Emission from Animal Houses: Growing and Finishing Pigs*; Wageningen UR Livestock Research: Wageningen, The Netherlands, 2010.
42. Mosquera Losada, J.; Hol, J.M.G.; Groenestein, C.M. *Emissies Uit de Biologische Veehouderij: Processen en Factoren*; Wageningen UR Livestock Research: Wageningen, The Netherlands, 2012.
43. Habtewold, J.; Gordon, R.; Sokolov, V.; VanderZaag, A.; Wagner-Riddle, C.; Dunfield, K. Targeting bacteria and methanogens to understand the role of residual slurry as an inoculant in stored liquid dairy manure. *Appl. Environ. Microbiol.* **2018**, *84*, e02830-17. [CrossRef] [PubMed]
44. Monteny, G.J.; Groenestein, C.M.; Hilhorst, M.A. Interactions and coupling between emissions of methane and nitrous oxide from animal husbandry. *Nutr. Cycl. Agroecosystems* **2001**, *60*, 123–132. [CrossRef]
45. Koerkamp, P.W.G.G.; Metz, J.H.M.; Uenk, G.H.; Phillips, V.R.; Holden, M.R.; Sneath, R.W.; Short, J.L.; White, R.P.P.; Hartung, J.; Seedorf, J. Concentrations and emissions of ammonia in livestock buildings in Northern Europe. *J. Agric. Eng. Res.* **1998**, *70*, 79–95. [CrossRef]
46. Sun, G.; Guo, H.; Peterson, J.; Predicala, B.; Laguë, C. Diurnal odor, ammonia, hydrogen sulfide, and carbon dioxide emission profiles of confined swine grower/finisher rooms. *J. Air Waste Manag. Assoc.* **2008**, *58*, 1434–1448. [CrossRef]
47. Philippe, F.-X.; Cabaraux, J.-F.; Nicks, B. Ammonia emissions from pig houses: Influencing factors and mitigation techniques. *Agric. Ecosyst. Environ.* **2011**, *141*, 245–260. [CrossRef]
48. Albright, L.D. Environment Control for Animals and Plants. American Society of Agricultural Engineers: St. Joseph, MN, USA, 1990.
49. Beeking, F.F.E.; Ingelaat, F.B.J.M.; van Beek, G. *De Specifieke Vochtafgifte Van Leghennenmest [Specific Moisture Production from Layer Litter]*; Report 607; COVP-DLO: Beekbergen, The Netherlands, 1994; 29p.
50. Soest, P.V. Use of detergents in the analysis of fibrous feeds. II. A rapid method for the determination of fiber and lignin. *J. Assoc. Off. Agric. Chem.* **1963**, *46*, 829–835. [CrossRef]
51. Triolo, J.M.; Sommer, S.G.; Møller, H.B.; Weisbjerg, M.R.; Jiang, X.Y. A new algorithm to characterize biodegradability of biomass during anaerobic digestion: Influence of lignin concentration on methane production potential. *Bioresour. Technol.* **2011**, *102*, 9395–9402. [CrossRef]
52. Rigolot, C.; Espagnol, S.; Robin, P.; Hassouna, M.; Béline, F.; Paillat, J.-M.; Dourmad, J.-Y. Modelling of manure production by pigs and NH_3, N_2O and CH_4 emissions. Part II: Effect of animal housing, manure storage and treatment practices. *Animal* **2010**, *4*, 1413–1424. [CrossRef]

animals

Article

Determining the Presence and Size of Shoulder Lesions in Sows Using Computer Vision

Shubham Bery [1], Tami M. Brown-Brandl [1,*], Bradley T. Jones [2], Gary A. Rohrer [2] and Sudhendu Raj Sharma [1]

[1] Department of Biological Systems Engineering, University of Nebraska-Lincoln, Lincoln, NE 68583, USA; sbery2@huskers.unl.edu (S.B.); raj.sharma@unl.edu (S.R.S.)
[2] Genetics and Breeding Research Unit, USDA-ARS U.S. Meat Animal Research Center, Clay Center, NE 68933, USA; brad.jones@usda.gov (B.T.J.); gary.rohrer@usda.gov (G.A.R.)
* Correspondence: tami.brownbrandl@unl.edu

Simple Summary: Shoulder lesions present a significant welfare concern, particularly among breeding sows. They are frequently linked to decreased mobility and loss of body condition during lactation. This research explores using RGB cameras and the potential of different computer vision methods for detecting and estimating their size. Findings indicate that these techniques hold promise in effectively identifying and quantifying lesion size. This could empower producers to proactively monitor sow welfare, facilitating timely detection and intervention for these lesions.

Abstract: Shoulder sores predominantly arise in breeding sows and often result in untimely culling. Reported prevalence rates vary significantly, spanning between 5% and 50% depending upon the type of crate flooring inside a farm, the animal's body condition, or an existing injury that causes lameness. These lesions represent not only a welfare concern but also have an economic impact due to the labor needed for treatment and medication. The objective of this study was to evaluate the use of computer vision techniques in detecting and determining the size of shoulder lesions. A Microsoft Kinect V2 camera captured the top-down depth and RGB images of sows in farrowing crates. The RGB images were collected at a resolution of 1920 × 1080. To ensure the best view of the lesions, images were selected with sows lying on their right and left sides with all legs extended. A total of 824 RGB images from 70 sows with lesions at various stages of development were identified and annotated. Three deep learning-based object detection models, YOLOv5, YOLOv8, and Faster-RCNN, pre-trained with the COCO and ImageNet datasets, were implemented to localize the lesion area. YOLOv5 was the best predictor as it was able to detect lesions with an mAP@0.5 of 0.92. To estimate the lesion area, lesion pixel segmentation was carried out on the localized region using traditional image processing techniques like Otsu's binarization and adaptive thresholding alongside DL-based segmentation models based on U-Net architecture. In conclusion, this study demonstrates the potential of computer vision techniques in effectively detecting and assessing the size of shoulder lesions in breeding sows, providing a promising avenue for improving sow welfare and reducing economic losses.

Keywords: shoulder lesions; ulcers; sows; deep learning; YOLO; U-Net

Citation: Bery, S.; Brown-Brandl, T.M.; Jones, B.T.; Rohrer, G.A.; Sharma, S.R. Determining the Presence and Size of Shoulder Lesions in Sows Using Computer Vision. *Animals* **2024**, *14*, 131. https://doi.org/10.3390/ani14010131

Academic Editor: Jin-Ho Cho

Received: 24 October 2023
Revised: 23 December 2023
Accepted: 26 December 2023
Published: 29 December 2023

1. Introduction

Shoulder lesions, amongst lactating sows, are commonly seen in the swine industry. Lesions commonly develop during the first two weeks of farrowing [1]. These are formed due to the deficiency of oxygen to the underlying shoulder tissue caused by pressure incited from the flooring. The tissues lose blood supply and die similar to human pressure ulcers [2]. The anatomy of a sow's scapula bone has a large ridge known as the scapular spine. When the differences in the structure of the scapula in sows were analyzed, sows with a prominent scapula spine were found to be at a higher risk of lesion formation [3]. When a sow lies on its side, extra pressure is exerted on the tissue surrounding the spine making it vulnerable to ulcer formation [4]. Though the lesions are described as shoulder sores, they develop near the dorsal aspects of the spine of the scapula, not at the scapulohumeral joint [5]. The dorsal aspect of the spine is a common location of ulcer development, but lesions may develop over any bony prominence such as the tarsus or cubital joint. The severity of lesions can vary from mild lesions to bone-deep ulcers if left untreated. They are associated with poor animal welfare because of the pain and increased infection risk that can lead to euthanizing the animal. In addition to the welfare-associated problem, shoulder lesions increase production costs due to the treatment and increased culling rate [6]. Bone-deep ulcers often negatively affect the sow's carcass value as they lead to failure of the final quality check. Therefore, treating these ulcers early to reduce welfare and economic concerns becomes highly important. Early lesion detection will result in timely veterinary treatments, such as the application of zinc oxide to the affected area. Most research regarding shoulder lesions has primarily concentrated on potential causes, associated effects, and strategies to prevent or manage ulcerations. Implementing automatic monitoring technology for shoulder lesions has the potential to alleviate the workload of farm workers and facilitate prompt treatment. In recent times, innovative animal husbandry solutions have emerged, which enable continuous and automated tracking of individual animals' well-being through the utilization of diverse sensors and cameras [7]. However, limited research has been conducted on the monitoring of these lesions. One previous study [8] investigated the use of thermal cameras to measure the related temperature increase caused by an inflammatory response. This approach allowed for the early detection of lesions up to 7 days before visual signs appeared. These advancements not only enhance farm productivity but also enable the early detection of emerging health concerns.

While thermal imaging may be able to detect the beginning of a lesion, RGB cameras have a cost advantage. Potentially, RGB images, in combination with machine learning techniques, may be able to detect lesions shortly after they become visible and could potentially track their progression. Machine vision techniques like deep learning-based models have become increasingly effective in solving various problems. For example, convolutional neural networks (CNNs) like Resnet101, Xception, and MobileNet have already been successfully implemented to classify different postures in swine [9,10]. YOLO (You Only Look Once) has been used for the early detection of estrus behavior in cattle by modifying the spatial pyramid pooling (SPP) module in the architecture [11]. Although there has not been much work conducted in lesion detection in animals using RGB cameras, studies have been performed using deep learning models to detect and localize ulcers from the images of diabetic human feet [12]. Recently, image segmentation based on deep learning has become one of the main image segmentation methods. Different variations of U-Net architecture have been commonly used for detecting various kinds of lesions like skin [13,14], lungs [15], brain [16], etc. Another study presented a two-stage deep learning method for accurately segmenting skin lesions from dermoscopic images based on YOLO–DeepLab networks [17]. However, the images these studies used for model training had a significantly larger lesion-to-background area ratio alongside uniform illumination without much noise.

With shoulder lesions in sows, it is likely the lesion portion of the image could only be a few pixels as cameras are mounted at a height to avoid interference with day-to-day activities. In addition, for training any deep learning model from scratch, one needs to have large amounts of supervised data; for example, the pictures of sows manually annotated with bounded boxes around the lesion region to make the model understand the difference between lesion and all other objects in an image. To make these algorithms better at generalizing, focus should not only be on the dataset size but also on the data quality so that the model can extract different patterns and features from the data during the training phase.

Many applications with livestock species have the limitation of a small, annotated dataset. To tackle the limitations of a small-sized dataset, transfer learning has been used to minimize this gap. It is a machine learning technique which reuses a model trained for one task by applying it to another, related task. It uses the weights or features of a model that has been trained on much bigger datasets like MS COCO or ImageNet and feeds them into the target network. Then, modifications are performed in the initial and final layers to accommodate predictions on the custom dataset. The middle layers of any CNN learn general features of objects like specific edges, color patches, etc. and are applicable across many datasets and tasks. The target model freezes its middle layers, uses the mid-layer features from the base model, and retrains its initial and last layers to learn specific features related to the custom dataset. Jensen and Pedersen used transfer learning to localize and count pigs in slaughterhouses [18].

Therefore, the objectives of this study were to

- Compare the performance of deep learning models, including two versions of YOLO (5s, 5m and 8s, 8m) and two versions of FRCNN (R50 Backbone and X-101 Backbone) in the localization of shoulder lesions in various stages of development.
- Compare two traditional imaging segmentation methods (Adapting thresholding and Otsu's method thresholding) and two deep learning-based U-Net architectures (Vanilla and Attention U-net) to segment lesion pixels and estimate size.

The paper is structured as follows: Section 1, "Introduction", outlines previous work; Section 2, "Materials and Methods", includes details on the data collection (Section 2.1), lesion localization (Section 2.2), and segmentation processes (Section 2.3), along with size referencing techniques (Section 2.4); Section 3, "Results", presents the findings on lesion localization (Section 3.1) and segmentation (Section 3.2) and addresses the challenges encountered (Section 3.3); Section 4 compares the findings with other studies under "Discussion" and the paper concludes in Section 5, "Conclusions", which summarizes the key findings and insights.

2. Materials and Methods

2.1. Data Collection

The experiment was conducted at the U.S. Meat Animal Research Centre (USMARC) located outside Clay Center, NE, USA. All animal husbandry protocols were performed in compliance with federal and institutional regulations regarding proper animal care practices and were approved by the USMARC Institutional Animal Care and Use Committee (2015–2021). The facility at USMARC is a farrow to finish swine production unit. This study utilized 360 sows of a Yorkshire–Landrace cross breed. This dataset was from a study designed to test behavioural and production characteristics of sows housed in three different crate sizes [19]. The images were captured in one of the two farrowing facilities. The farrowing facility housed three farrowing rooms, with each room containing twenty farrowing crates, totaling sixty crates per farrowing cycle. The facility was well lit, with the lights being on for 12 h a day, from 5:30 a.m. until 5:30 p.m. An aluminium theatre triangle truss, 21.6 m in length, was placed above each row of crates. The bottom end of the truss was at an approximate height of 2.6 m above the ground. A time-of-flight depth sensor with an integrated digital camera (Kinect V2™, Microsoft, Redmond, WA, USA) was centred above each crate and mounted on the truss's bottom at a height of 2.55 m from the floor. Sensors were enclosed within a waterproof housing to protect the cameras during pressure washing and disinfecting. Digital and depth images were collected every 5 s. Only the RGB images (resolution 1920 × 1080) were used for the experiment. The setup can be seen in Figure 1. One mini-PC with Windows 10 Home Edition (Windows 10 Home, Microsoft, Redmond, WA, USA) was connected to a single camera. A total of 20 mini-PCs were connected to an external disk station (DS1517+, Synology Inc., Bellevue, WA, USA), each having five 10 TB hard disk drives (ST10000VN0004, Seagate Technology LLC, Cupertino, CA, USA). Only images collected during the lights-on period of the day were used. The room was illuminated with 20 T-8 fluorescent bulbs.

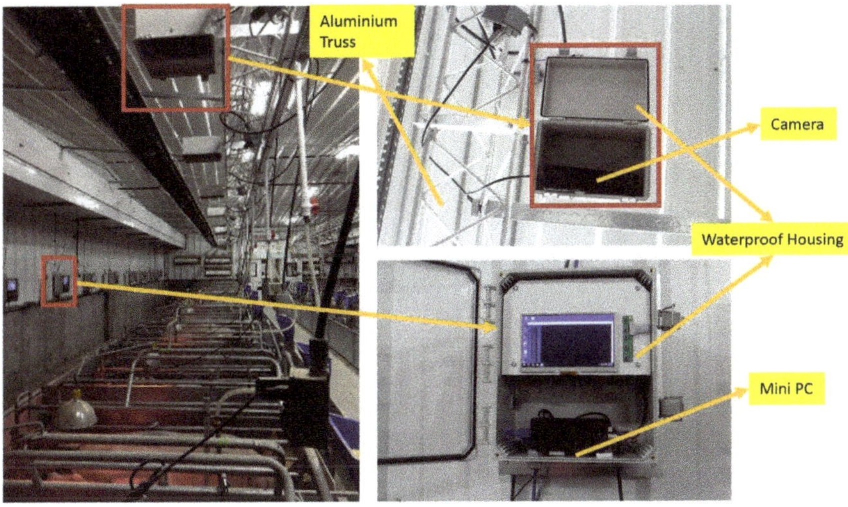

Figure 1. Image collection system utlizing a Kinect V2™ mounted on an aluminum truss.

An image capturing program was developed in MATLAB (R2017a, The Math-Works, Inc., Natick, MA, USA). This program was designed to capture and store one digital image approximately every 5 s, providing a view of lesion development across the four-week period of sow confinement in crates. Out of all the sows, 70 animals developed lesions. Within this timeframe, a total of 824 images were filtered, each depicting the progression of shoulder lesions in 70 sows. The dataset was split in the 80:20 train to test ratio.

Table 1 offers insight into the distribution of images across different weeks of lactation. Notably, data from the first week, where lesions are typically not visually evident, was also included as part of the dataset to enhance the model's capability to handle false positives. This increase in data size aimed to enhance the model's capability to handle false positives.

Table 1. Distribution of images by week in the dataset.

Week	Number of Images
1	93
2	293
3	269
4	169

Throughout this study, the progression of lesions became apparent, with most lesions manifesting visual signs during the 1–2-week phase. These initial stages are characterized by minor wounds limited to the outer layer (epidermis) marked by reddening of the affected area. As time progressed to week 3, more severe lesions emerged, causing abrasions and reaching the lower skin layer (dermis) while growing in diameter and forming granulated tissue. By the fourth week, the most severe cases exhibited signs of deep bone lesions. Figure 2 shows the lesion progression of a severe lesion. It is important to highlight that not all the sows followed the same timeline for lesion development. Benign cases might undergo healing during later stages, often after forming scabs within the initial 2 weeks.

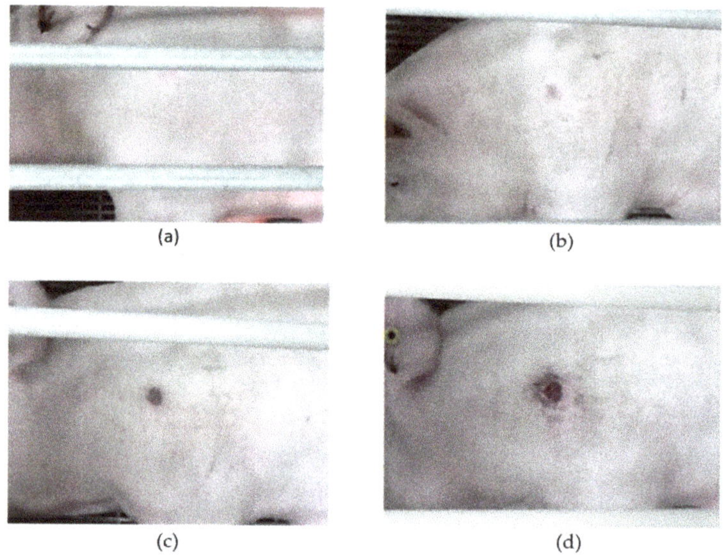

Figure 2. Lesion progression over the farrowing cycle. (**a**) Lesion in the first week—minimal visible signs. (**b**) Lesion in the second week—top layer affected. (**c**) Lesion in the third week—scab formed. (**d**) Lesion in the fourth week—increased lesion diameter with possible dermis damage.

2.2. Lesion Localization

Figure 3 shows a sample image. To minimize computation for training and inference, cropping was applied because the cameras' positions were fixed, and the animals were confined to the center of the crate. Precise identification of lesion boundaries was crucial for size determination. This was achieved using deep learning-based object detection models, which could detect lesions from sow images for localization and output bounding box coordinates of the lesion within the frame.

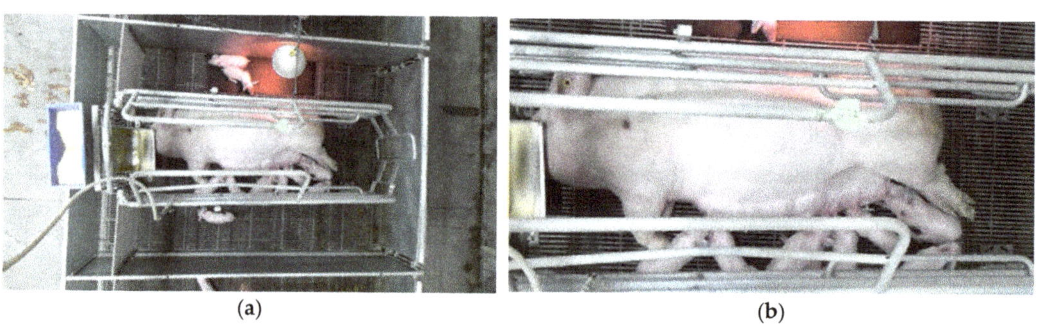

Figure 3. (**a**) Sample RGB images collected. (**b**) Cropped RGB image.

Two object detection architectures were tested, named YOLO (You Only Look Once) and Faster R-CNN (Region-based Convolutional Neural Network).

- YOLO (You Only Look Once) approaches object detection as a regression problem. These models are renowned for their exceptional balance between speed and accuracy, making them ideal for real-time applications. This is crucial in a production setting where rapid and reliable detection of shoulder lesions is essential. The availability of various model sizes (small and medium) allowed for the tailoring of the model to the computational resources available and the optimization of either speed or accuracy. It processes entire images in one step, providing direct predictions for bounding boxes and class probabilities. This study explored two different versions of YOLO developed by Ultralytics [20] as seen in Figure 4, namely v5 and v8, with varying architecture sizes (small "s" and medium "m").

- Faster-RCNN is a two-stage detector, which uses a Region Proposal Network (RPN) instead of using a selective search algorithm to output object proposals. Region-of-Interest (ROI) pooling is applied to make all proposals the same size. Then, processed proposals are passed to a fully connected layer that classifies the objects in the bounding boxes. Two different backbone models, ResNet-50 and ResNet-101, pre-trained on ImageNet classification tasks were implemented to leverage their deep residual learning framework. This is particularly advantageous for capturing complex features of shoulder lesions, which might be missed by shallower networks like YOLO.

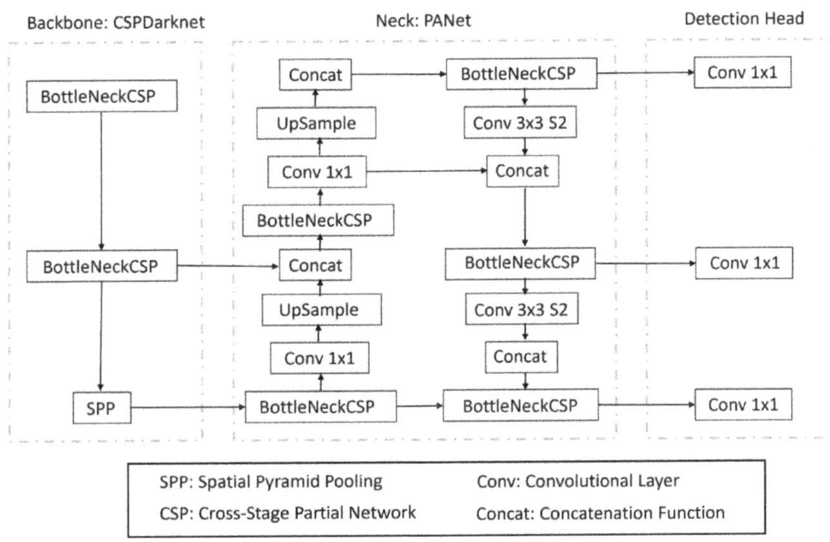

Figure 4. Ultralytics YOLO base architecture used for lesion localization.

Due to limitations in the dataset size, transfer learning was employed. This method fine-tuned a pre-trained detector with COCO or ImageNet weights to adapt it to a custom dataset. Both the YOLO and FRCNN models utilized the PyTorch framework (Version 1.9). Data preparation was conducted using OpenCV (Version 4.6.0), NumPy (Version 1.23), and Pandas (Version 1.5.0). Matplotlib (Version 3.6.0) was employed for data visualization. All models were trained for 300 epochs, with early stopping set to 50 consecutive iterations without performance improvement to prevent overfitting. The hyperparameters for FRCNN included a batch size of 8, a learning rate of 0.00025, and ReLU activation. For FRCNN, the default image size range of (800, 1333) pixels was used. For YOLOv5 and YOLOv8, default hyperparameters from [21] were used and the input images were resized to 1280 by 720. Object detection performance was measured by using mean average precision (mAP), which takes classification and localization into account while evaluating. The predicted bounding boxes are compared with the ground truth coordinate boxes and if the overlap between them is more than the threshold value of Intersection-over-Union (IoU), then it is considered a True Positive (TP) otherwise it gets classified as a False Positive (FP). If the model fails to detect anything when the object is there, it is a False Negative (FN). After this, precision and recall of a model are calculated. Then, the mean of all the average precision values ranging across different IoU thresholds from 0.50 to 0.95 in increments of 0.05 was calculated as shown in Equation (1); for this research, $n = 1$, as the model had to detect a single class.

$$\mathrm{mAP} = \frac{1}{n}\sum_{i=0}^{n} \mathrm{APi} \quad (1)$$

where: n = number of classes.

2.3. Lesion Pixel Segmentation

Following the localization of shoulder lesions within the frames using object detection, the next crucial step involved segmenting the lesion pixels within the cropped bounding boxes. Accurate pixel-level segmentation was essential for precisely quantifying the size and extent of shoulder lesions. To obtain the segmented lesion pixels, both traditional and DL-based techniques were tested on the detected regions.

Image-processing-based binarization techniques were first used to separate out the lesion pixels from the sow's body, as lesions were darker in color when compared to the rest of the body. Python's OpenCV module was used to implement two automatic image thresholding techniques, namely Otsu's method and Gaussian adaptive thresholding. Otsu's method returns a single intensity threshold that divides pixels into two classes: foreground and background. The base value is calculated by maximizing or minimizing the intensity variance between both classes. In adaptive thresholding, different thresholds are calculated for different parts of the image for segregating all the pixels. This can handle variations in lighting due to shadows from the crate bars and multiple lesion clusters, ensuring robust pixel segmentation across the entire dataset.

Traditional image binarization might not work consistently when there is a lot of shadow noise from crate bars in the image [20] or when the variation in the lesion and the sow's skin pixels is insufficient for it to differentiate between them, especially in earlier stages of development. Deep learning-based techniques successfully addressed these challenges where traditional image processing falls short. U-Net [22] is one such architecture that has been extensively implemented in the biomedical field for pixel-level segmentation. It is based on an encoder–decoder architecture. The vanilla U-Net alongside a variation proposed in [23] were implemented, where attention gates were introduced within the CNN architecture to make the network focus on the target object, suppressing irrelevant information within the region. Both models were initialized using ImageNet weights with frozen VGG16 backbone. ReLU was used as the activation function, with a batch size of 8, binary cross entropy as the loss function, a learning rate of 0.001, and an Adam optimizer. Both models were trained for 100 epochs.

The Dice coefficient was used for evaluating the performance of all the pixel segmentation approaches. It assesses how well a predicted binary mask aligns with a ground truth binary mask. The Dice coefficient produces a value between 0 and 1, where 0 indicates no overlap and 1 indicates a perfect match. ImageJ was used to manually annotate the lesion pixels and obtain the corresponding binary masks.

$$Dice\ Coefficient = \frac{2\cdot|TP|}{2\cdot|TP| + |FP| + |FN|} \quad (2)$$

where

- TP represents the number of true positive pixels, i.e., pixels that are correctly classified as lesions in both the ground truth and predicted masks.
- FP represents the number of false positive pixels, i.e., pixels that are classified as lesions in the predicted mask but not in the ground truth.
- FN represents the number of false negative pixels, i.e., pixels that are lesions in the ground truth but not in the predicted mask.

2.4. Size Referencing

To estimate the area covered by lesions in measurable units, a calibration was conducted using the crate's anti-crush bar, which was positioned at the same level as the lying sow and at a similar shoulder height, serving as a reference object, as shown in Figure 5. The bar had a diameter of 25.4 mm and was 15 pixels wide. These values were utilized to calculate the area of one pixel in millimeters, which amounted to 2.87 mm^2. By applying Equation (3), the lesion's area was converted from pixels to mm^2.

$$Lesion\ Area\ (\mathrm{mm}^2) = Number\ of\ lesion\ pixels \times 2.87\ \frac{\mathrm{mm}^2}{\mathrm{pixel}} \quad (3)$$

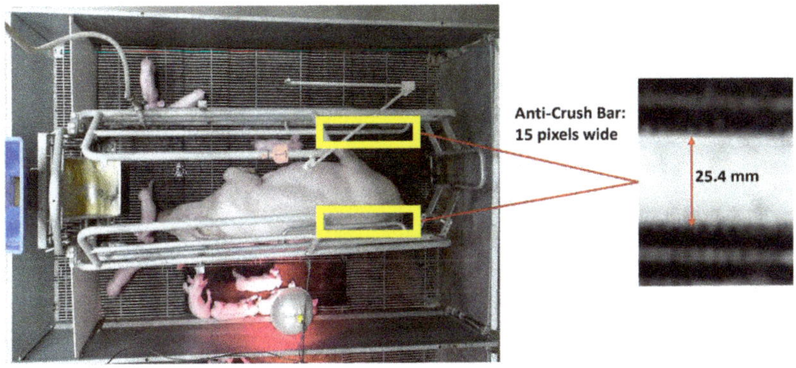

Figure 5. Example image to determine lesion size, the anti-crush bar was used for size referencing. The anti-crush bar was 25.4 mm in diameter equating to 15 pixels in these images.

The entire pipeline for lesion size estimation is shown in Figure 6, starting with image cropping to lesion localization, passing the cropped bounding box to binarization models, and then performing size referencing to estimated size.

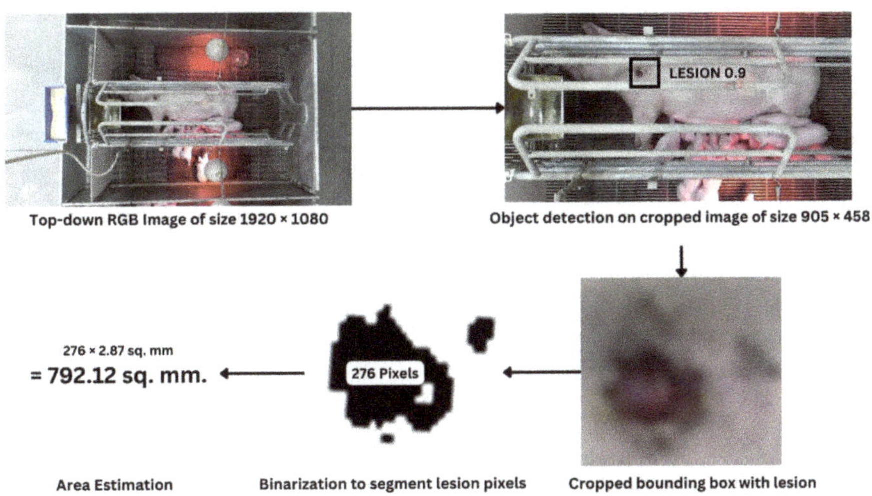

Figure 6. The entire pipeline showing different steps of sow shoulder lesions from image capturing to area estimation.

3. Results

3.1. Lesion Localization

The models were trained using a 32 GB NVIDIA Tesla V100 GPU. The main goal of this research was to identify shoulder lesions and determine their size. To achieve this, various CNN-based models for lesion localization and pixel segmentation employing image processing techniques were used. For lesion localization, Faster R-CNN (FRCNN) was trained with two different ImageNet backbones and two YOLO versions: YOLOv5 and YOLOv8, each with varying architectures. YOLO models outperformed FRCNN, as indicated in Table 2.

Table 2. Comparison of performance metrics for YOLOv5, YOLOv8, and Faster R-CNN (FRCNN) object detection models. The table presents mean average precision (mAP) metrics at two different scales: mAP at a threshold of 0.5, and mAP at a threshold of 0.5:0.95, representing the average mAP calculated at IoU thresholds from 0.5 to 0.95 in steps of 0.05.

Architecture	mAP@0.5	mAP@0.5:0.95::0.05
YOLOv5s	0.91	0.46
YOLOv5m	0.92	0.48
YOLOv8s	0.84	0.31
YOLOv8m	0.81	0.35
FRCNN–R50 Backbone	0.26	0.12
FRCNN–X-101 Backbone	0.56	0.14

YOLOv5 models excelled in lesion localization, exhibiting fewer false positives, and detecting lesions in early stages of development compared to FRCNN and YOLOv8. As a result, YOLOv5m, with an mAP@0.5 of 0.92 and mAP@0.5:0.95::0.05 of 0.48, was selected for the localization task. This performance difference may be attributed, in part, to dataset size. FRCNN's two-stage nature might require a larger dataset for optimal learning, whereas YOLO's single-stage architecture yielded promising results even with smaller datasets, in line with [24]. After using a Python script to crop the detected lesion regions, segmentation techniques were applied to calculate the lesion area in terms of pixels.

3.2. Lesion Segmentation

Binarization techniques were employed to isolate lesion pixels within the cropped bounding boxes. Regarding traditional image binarization techniques, Otsu's method exhibited better performance than Adaptive thresholding. The subpar performance of adaptive thresholding can be attributed to its sensitivity to local lighting conditions, resulting in multiple clustered segmented regions within a single cropped frame, impacted by non-uniform lighting in the crates.

During the second week of lesion progression, deep learning-based methods outperformed traditional image processing techniques. The vanilla U-Net achieved a Dice coefficient of 0.71, followed closely by its attention gates-based counterpart at 0.68, both surpassing Otsu's method at 0.65 as shown in Table 3.

Table 3. Dice coefficients for segmentation techniques across various weeks of lesion progression.

Segmentation Method	Dice Coefficient		
	2nd Week	3rd Week	4th Week
Adaptive Thresholding	0.38	0.20	0.16
Otsu's Method	0.66	0.83	0.82
Vanilla U-Net	0.72	0.63	0.61
Attention U-Net	0.68	0.63	0.62

However, as lesions became more pronounced, and the distinction between foreground and background pixels became clearer, Otsu's method outperformed all other techniques. It achieved Dice coefficients of 0.83 and 0.81 for lesions in the third and fourth weeks of development, respectively, significantly surpassing the performance of DL-based segmentation models.

The U-Net-based models showed superior performance in the early stages of the analysis. However, they encountered difficulties as the analysis progressed, as seen in Figure 7. The issue arose because these models were proficient at identifying pixels with lighter intensities as part of the lesions. Yet, as the lesions advanced, the pixels within them became darker. Despite this change, the U-Net models continued to identify the lighter skin pixels as lesions, which ultimately limited their effectiveness compared to Otsu's method.

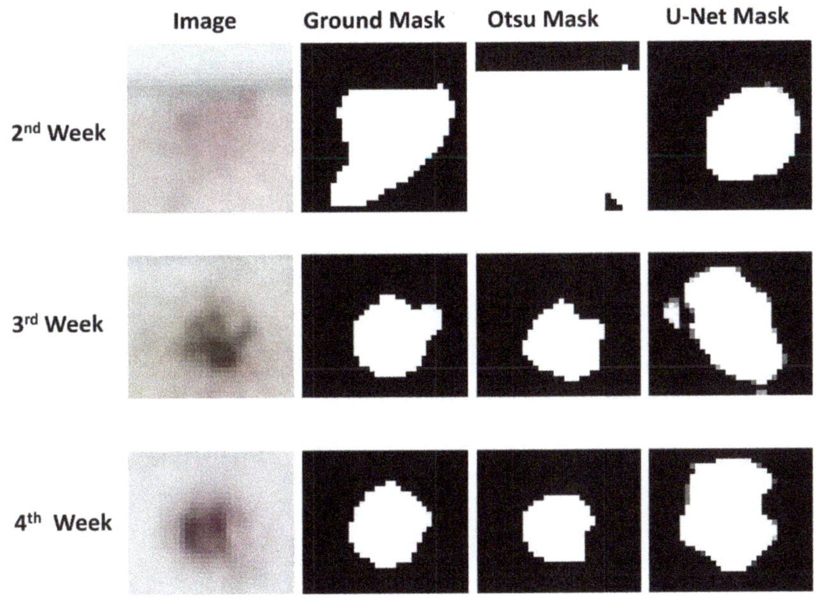

Figure 7. Comparing Otsu's Method and the U-Net segmentation method with ground truth data over three weeks of lactation.

3.3. Challenges

Non-uniform crate lighting and sow movement were two big challenges faced in size estimation. Lesions were only apparent when the sow was lying on its sides. The sows' movement impacted how the lesion area was viewed by the camera, as seen in Figure 8. The lesion pixel's color intensity was impacted by the movement which can highly influence the segmentation process leading to variation in estimated lesion pixel count.

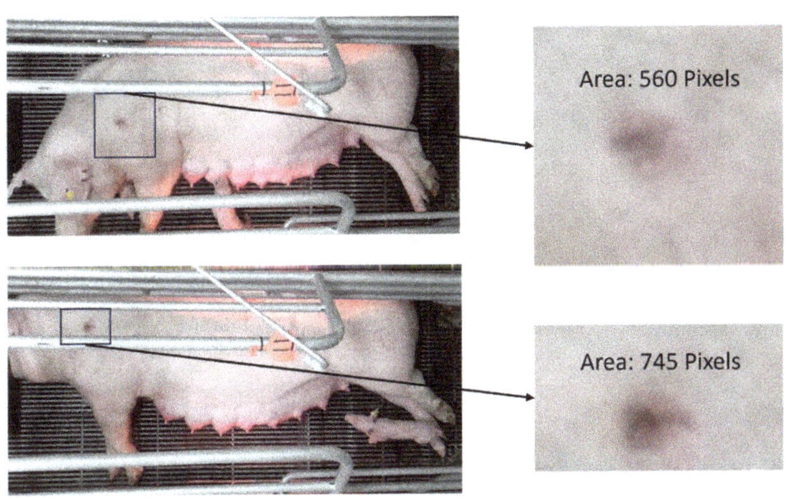

Figure 8. Frames captured five seconds apart where the sow changed its posture impacting the lesion appearance. Both images are sized 905 by 458 pixels.

In addition to the sow's movement, the bars of the crate caused occlusion, which at times made it challenging to accurately localize the lesion as it could become hidden behind these bars. Furthermore, shadows cast by these bars affected the binarization process as shown in Figure 9a. Deep learning-based methods proved more effective at segmenting pixels that included shadows, while traditional binarization techniques like Otsu's struggled because they sometimes misclassified lesion pixels as shadows, mainly due to the limited difference in pixel intensity between the two (Figure 9b).

(a)

(b)

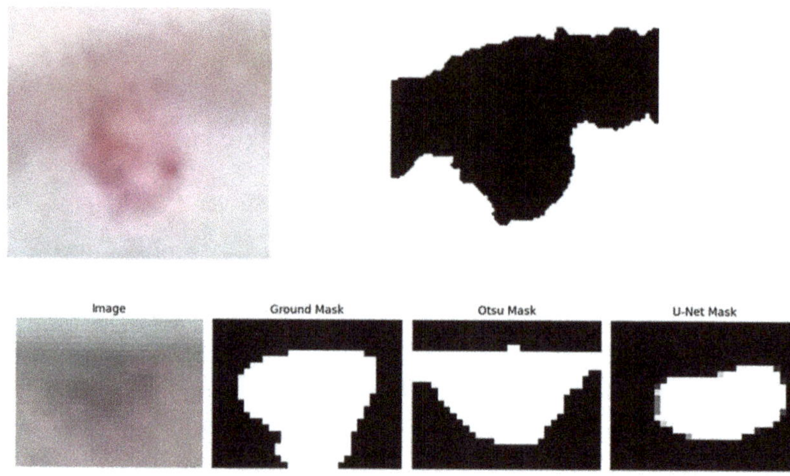

Figure 9. Shoulder lesion detection as impacted by (**a**) the shadow cast by the crate bar causing issues in binarization. (**b**) Comparing Otsu's method and U-Net when the crate bar is present in the cropped region.

The binarization techniques performed poorly when the captured image had a sow being treated for lesions using a topical solution like zinc oxide. The zinc oxide ointment was the same color intensity (deep yellow) as a lesion. Algorithms mistake the applied solution for lesion pixels outputting a larger lesion area. Lesions in this study are notably relatively minor and continued to resolve not progress throughout the imaging period. This is in part due to the installation of new crates in the facility. These crates provided a large sow space for the animals. Also, all animals in this study were fourth parity or less, leading to smaller a body structure of the animals. Both factors contributed to provide appropriate comfort and improved the ability of the sows to rise, likely minimizing lesions. The floor slats were galvanized metal. The galvanization process leaves a slightly abrasive surface. This abrasiveness disappears with subsequent use but was present at the time of this study and was a contributing factor in the ethology and development of these lesions.

4. Discussion

This work proposed a method that advances the detection and segmentation of shoulder lesions in sows. While thermal imaging surpasses RGB in early detection, its high cost limits practicality in farm settings [8,25]. The combination of deep learning and image processing techniques offers a cost-effective alternative. Specifically, YOLOv5 models excelled in localizing lesions within frames, with the YOLOv5m model performing the best with an mAP@0.5 of 0.92 and mAP@0.5:0.95::0.05 of 0.48. It strikes a balance between speed and accuracy, making it suitable for real-time applications where both these factors are crucial, as seen in [26–28].

Deep learning-based segmentation models exhibited superior noise handling capabilities, effectively addressing issues like inconsistent illumination and shadows. But, U-Net models were unable to segment pixels well in the later stages, which is in contrast with [13,29,30]. It is worth noting that these studies implementing U-Net for skin lesion segmentation used microscopic images collected under uniform lighting and consistent physical conditions, unlike the conditions in a sow barn. Extending the dataset to include more images from the latter stages of lesion progression may improve the binary mask accuracy, provided there are no hardware constraints. Otsu-based binarization demonstrated strong performance in the later stages of lesion detection with Dice coefficients of 0.83 and 0.82 in the third and fourth week, respectively, where DL-based U-Net models initially performed better.

Otsu's method had speed advantages and did not require expensive hardware implementation compared to deep learning architectures as also mentioned in [31]. To enhance its performance in the early stages, pre-binarization steps, such as increasing image contrast through histogram equalization, could further improve segmentation, which was also seen in [32]. While this may result in some information loss of lesion pixels, it can define clearer boundaries for the lesion area, ultimately improving binarization performance.

5. Conclusions

In summary, this study used RGB images collected on 360 lactating sows housed in typical farrowing crates. This study aimed to determine if images could be used to monitor shoulder lesion formation. This study first compared the performance of two different deep-learning models in the localization of sow shoulder lesions in various stages of development. It was determined that the YOLOv5s and YOLOv5m performed the best with a mAP@0.5 of 0.91 and 0.92, respectively. YOLOv8s and YOLOv8m models performed the second best with a mAP@0.5 of 0.84 and 0.82, respectively. Neither the FRCNN–R50 backbone nor the FRCNN–X-101 Backbone performed well with a mAP@0.5 of 0.26 and 0.56, respectively. It was hypothesized that the dataset was too small for the FRCNN models.

After identifying the lesions, this study tested traditional image processing and deep learning-based binarization techniques to estimate lesion size. The shoulder lesions observed in this study changed color as the lesions progressed. This color change resulted in changes in model performance over time. Early in lesion development, the deep-learning algorithms performed slightly better. Overall, Otsu's method performed the best, although this model had overestimated lesion size in the early developing lesions.

This research highlights the effectiveness of using RGB images for detecting and monitoring sow shoulder lesions. However, more work needs to be completed to make this a viable Precision Livestock Farming management technique. In addition, the cost of cameras and computers may not make this approach cost-effective on its own; however, if cameras were added to the farrowing system, this technique could be one of a suite of parameters that could be monitored.

Author Contributions: Conceptualization, S.B., B.T.J., T.M.B.-B., G.A.R. and S.R.S.; methodology, S.B., T.M.B.-B. and S.R.S.; software, S.B.; validation S.B. and T.M.B.-B.; formal analysis, S.B.; resources, S.B., T.M.B.-B. and S.R.S.; data curation, S.B. and S.R.S.; writing—original draft preparation, S.B.; writing—review and editing, T.M.B.-B., B.T.J., G.A.R. and S.R.S.; supervision, T.M.B.-B. All authors have read and agreed to the published version of the manuscript.

Funding: This research was funded in part by the USDA Agricultural Research Service and the USDA NIFA Award No. 2016-67015-24. The USDA is an equal-opportunity employer.

Institutional Review Board Statement: All animal husbandry protocols were performed in compliance with federal and institutional regulations regarding proper animal care practices and were approved by the USMARC Institutional Animal Care and Use Committee (2015–2021). Approval number 5438-32630-006-004.

Informed Consent Statement: Not applicable.

Data Availability Statement: Data are presented in this article in the form of figures and tables.

Acknowledgments: The authors would like to express their deepest gratitude to the dedicated swine area employees at the USDA USMARC (U.S. Meat Animal Research Center, Clay Center, NE, USA) for all their support in managing livestock subjects during data collection and hardware installation.

Conflicts of Interest: The mention of trade names or commercial products in this publication is solely for the purpose of providing specific information and does not imply recommendation or endorsement by the USDA. The USDA prohibits discrimination in all its programs and activities on the basis of race, color, national origin, age, disability, and where applicable, sex, marital status, familial status, parental status, religion, sexual orientation, genetic information, political beliefs, reprisal, or because all or part of an individual's income is derived from any public assistance program (not all prohibited bases apply to all programs). USDA is an equal opportunity employer.

References

1. Herskin, M.S.; Bonde, M.K.; Jørgensen, E.; Jensen, K.H. Decubital Shoulder Ulcers in Sows: A Review of Classification, Pain and Welfare Consequences. *Animal* **2011**, *5*, 757–766. [CrossRef] [PubMed]
2. Nola, G.T.; Vistnes, L.M. Differential Response of Skin and Muscle in the Experimental Production of Pressure Sores. *Plast. Reconstr. Surg.* **1980**, *66*, 728–733. [CrossRef] [PubMed]
3. Nordbø, Ø.; Gangsei, L.E.; Aasmundstad, T.; Grindflek, E.; Kongsro, J. The Genetic Correlation between Scapula Shape and Shoulder Lesions in Sows. *J. Anim. Sci.* **2018**, *96*, 1237–1245. [CrossRef]
4. The Pig Site Nebraska Swine Report 2005: Shoulder Ulcers in Sows and Their Prevention. Available online: https://www.thepigsite.com/articles/nebraska-swine-report-2005-shoulder-ulcers-in-sows-and-their-prevention (accessed on 13 October 2022).
5. Rioja-Lang, F.C.; Seddon, Y.M.; Brown, J.A. Shoulder Lesions in Sows: A Review of Their Causes, Prevention, and Treatment. *J. Swine Health Prod.* **2018**, *26*, 101–107. [CrossRef]
6. Gaab, T.; Nogay, E.; Pierdon, M. Development and Progression of Shoulder Lesions and Their Influence on Sow Behavior. *Animals* **2022**, *12*, 224. [CrossRef] [PubMed]
7. Paudel, S.; Brown-Brandl, T.; Rohrer, G.; Sharma, S.R. Individual Pigs' Identification Using Deep Learning. In Proceedings of the 2023 ASABE Annual International Meeting, Omaha, NE, USA, 9–12 July 2023; p. 1.
8. Staveley, L.M.; Zemitis, J.E.; Plush, K.J.; D'Souza, D.N. Infrared Thermography for Early Identification and Treatment of Shoulder Sores to Improve Sow and Piglet Welfare. *Animals* **2022**, *12*, 3136. [CrossRef]
9. Shao, H.; Pu, J.; Mu, J. Pig-Posture Recognition Based on Computer Vision: Dataset and Exploration. *Animals* **2021**, *11*, 1295. [CrossRef]
10. Xu, Z.; Tian, F.; Zhou, J.; Zhou, J.; Bromfield, C.; Lim, T.T.; Safranski, T.J.; Yan, Z.; Calyam, P. Posture Identification for Stall-Housed Sows around Estrus Using a Robotic Imaging System. *Comput. Electron. Agric.* **2023**, *211*, 107971. [CrossRef]
11. Wang, R.; Gao, Z.; Li, Q.; Zhao, C.; Gao, R.; Zhang, H.; Li, S.; Feng, L. Detection Method of Cow Estrus Behavior in Natural Scenes Based on Improved YOLOv5. *Agriculture* **2022**, *12*, 1339. [CrossRef]
12. Yap, M.H.; Hachiuma, R.; Alavi, A.; Brüngel, R.; Cassidy, B.; Goyal, M.; Zhu, H.; Rückert, J.; Olshansky, M.; Huang, X.; et al. Deep Learning in Diabetic Foot Ulcers Detection: A Comprehensive Evaluation. *Comput. Biol. Med.* **2021**, *135*, 104596. [CrossRef]
13. Wu, J.; Chen, E.Z.; Rong, R.; Li, X.; Xu, D.; Jiang, H. Skin Lesion Segmentation with C-UNet. In Proceedings of the 2019 41st Annual International Conference of the IEEE Engineering in Medicine and Biology Society (EMBC), Berlin, Germany, 23–27 July 2019; pp. 2785–2788. [CrossRef]
14. Kumar, B.N.; Mahesh, T.R.; Geetha, G.; Guluwadi, S. Redefining Retinal Lesion Segmentation: A Quantum Leap With DL-UNet Enhanced Auto Encoder-Decoder for Fundus Image Analysis. *IEEE Access* **2023**, *11*, 70853–70864. [CrossRef]
15. Mosquera-Berrazueta, L.; Perez, N.; Benitez, D.; Grijalva, F.; Camacho, O.; Herrera, M.; Marrero-Ponce, Y. Red-Unet: An Enhanced U-Net Architecture to Segment Tuberculosis Lesions on X-Ray Images. In Proceedings of the 2023 IEEE 13th International Conference on Pattern Recognition Systems (ICPRS), Guayaquil, Ecuador, 4–7 July 2023. [CrossRef]
16. Yuan, Y.; Li, Z.; Tu, W.; Zhu, Y. Computed Tomography Image Segmentation of Irregular Cerebral Hemorrhage Lesions Based on Improved U-Net. *J. Radiat. Res. Appl. Sci.* **2023**, *16*, 100638. [CrossRef]
17. Bagheri, F.; Tarokh, M.J.; Ziaratban, M. Semantic Segmentation of Lesions from Dermoscopic Images Using Yolo-Deeplab Networks. *Int. J. Eng. Trans. B Appl.* **2021**, *34*, 458–469. [CrossRef]
18. Jensen, D.B.; Pedersen, L.J. Automatic Counting and Positioning of Slaughter Pigs within the Pen Using a Convolutional Neural Network and Video Images. *Comput. Electron. Agric.* **2021**, *188*, 106296. [CrossRef]
19. Leonard, S.M.; Xin, H.; Brown-Brandl, T.M.; Ramirez, B.C.; Johnson, A.K.; Dutta, S.; Rohrer, G.A. Effects of Farrowing Stall Layout and Number of Heat Lamps on Sow and Piglet Behavior. *Appl. Anim. Behav. Sci.* **2021**, *239*, 105334. [CrossRef]
20. Ren, H.; Wang, Y.; Dong, X. Binarization Algorithm Based on Side Window Multidimensional Convolution Classification. *Sensors* **2022**, *22*, 5640. [CrossRef] [PubMed]
21. Jocher, G.; Chaurasia, A.; Stoken, A.; Borovec, J.; NanoCode012; Kwon, Y.; Michael, K.; Tao, X.; Fang, J.; Imyhxy; et al. Ultralytics/Yolov5: V7.0—YOLOv5 SOTA Realtime Instance Segmentation. *Zenodo* **2022**. [CrossRef]

22. Ronneberger, O.; Fischer, P.; Brox, T. *U-Net: Convolutional Networks for Biomedical Image Segmentation BT—Medical Image Computing and Computer-Assisted Intervention—MICCAI 2015*; Navab, N., Hornegger, J., Wells, W.M., Frangi, A.F., Eds.; Springer International Publishing: Cham, Switzerland, 2015; pp. 234–241.
23. Oktay, O.; Schlemper, J.; Le Folgoc, L.; Lee, M.; Heinrich, M.; Misawa, K.; Mori, K.; McDonagh, S.; Hammerla, N.Y.; Kainz, B.; et al. Attention U-Net: Learning Where to Look for the Pancreas. *arXiv* **2018**, arXiv:1804.03999.
24. Brüngel, R.; Friedrich, C.M. DETR and YOLOv5: Exploring Performance and Self-Training for Diabetic Foot Ulcer Detection. In Proceedings of the 2021 IEEE 34th International Symposium on Computer-Based Medical Systems (CBMS), Aveiro, Portugal, 7–9 June 2021; pp. 148–153.
25. Teixeira, D.L.; Boyle, L.A.; Enríquez-Hidalgo, D. Skin Temperature of Slaughter Pigs with Tail Lesions. *Front. Vet. Sci.* **2020**, *7*, 198. [CrossRef]
26. Li, R.; Ji, Z.; Hu, S.; Huang, X.; Yang, J.; Li, W. Tomato Maturity Recognition Model Based on Improved YOLOv5 in Greenhouse. *Agronomy* **2023**, *13*, 603. [CrossRef]
27. Dai, M.; Dorjoy, M.M.H.; Miao, H.; Zhang, S. A New Pest Detection Method Based on Improved YOLOv5m. *Insects* **2023**, *14*, 54. [CrossRef] [PubMed]
28. Song, Q.; Li, S.; Bai, Q.; Yang, J.; Zhang, X.; Li, Z.; Duan, Z. Object Detection Method for Grasping Robot Based on Improved Yolov5. *Micromachines* **2021**, *12*, 1273. [CrossRef] [PubMed]
29. Liu, L.; Mou, L.; Zhu, X.X.; Mandal, M. Skin Lesion Segmentation Based on Improved U-Net. In Proceedings of the 2019 IEEE Canadian Conference of Electrical and Computer Engineering (CCECE), Edmonton, AB, Canada, 5–8 May 2019; pp. 1–4.
30. Tong, X.; Wei, J.; Sun, B.; Su, S.; Zuo, Z.; Wu, P. ASCU-Net: Attention Gate, Spatial and Channel Attention U-Net for Skin Lesion Segmentation. *Diagnostics* **2021**, *11*, 501. [CrossRef] [PubMed]
31. Zhao, Y.; Yu, X.; Wu, H.; Zhou, Y.; Sun, X.; Yu, S.; Yu, S.; Liu, H. A Fast 2-D Otsu Lung Tissue Image Segmentation Algorithm Based on Improved PSO. *Microprocess. Microsyst.* **2021**, *80*, 103527. [CrossRef]
32. Yusoff, A.K.M.; Raof, R.A.A.; Mahrom, N.; Noor, S.S.M.; Zakaria, F.F.; Len, P. Enhancement and Segmentation of Ziehl Neelson Sputum Slide Images Using Contrast Enhancement and Otsu Threshold Technique. *J. Adv. Res. Appl. Sci. Eng. Technol.* **2023**, *30*, 282–289. [CrossRef]

animals

MDPI

Article

Deep Learning Models to Predict Finishing Pig Weight Using Point Clouds

Shiva Paudel [1], Rafael Vieira de Sousa [2], Sudhendu Raj Sharma [1] and Tami Brown-Brandl [1,*]

[1] Department of Biological Systems Engineering, University of Nebraska-Lincoln, Lincoln, NE 68583-0726, USA; spaudel6@huskers.unl.edu (S.P.); raj.sharma@unl.edu (S.R.S.)
[2] Department of Biosystems Engineering, University of Sao Paulo, Pirassununga 13635-900, SP, Brazil; rafael.sousa@usp.br
* Correspondence: tami.brownbrandl@unl.edu

Simple Summary: Monitoring the weight of farm pigs is crucial for their well-being. Implementing Cameras and Machine Vision Systems shows promise for automating this process. Traditionally, studies have focused on RGB and depth images for weight prediction, using measurements like volume and body area. However, these methods prove less robust in fluctuating environmental conditions, especially lighting. This study reveals that PointNet, a 3D deep learning architecture trained on point cloud data (3D points), outperforms the conventional approach, and demonstrates stability in varying light conditions due to its ability to learn on spatial information. This finding underscores the potential for PointNet to significantly improve the accuracy and reliability of weight monitoring in farm settings.

Abstract: The selection of animals to be marketed is largely completed by their visual assessment, solely relying on the skill level of the animal caretaker. Real-time monitoring of the weight of farm animals would provide important information for not only marketing, but also for the assessment of health and well-being issues. The objective of this study was to develop and evaluate a method based on 3D Convolutional Neural Network to predict weight from point clouds. Intel Real Sense D435 stereo depth camera placed at 2.7 m height was used to capture the 3D videos of a single finishing pig freely walking in a holding pen ranging in weight between 20–120 kg. The animal weight and 3D videos were collected from 249 Landrace × Large White pigs in farm facilities of the FZEA-USP (Faculty of Animal Science and Food Engineering, University of Sao Paulo) between 5 August and 9 November 2021. Point clouds were manually extracted from the recorded 3D video and applied for modeling. A total of 1186 point clouds were used for model training and validating using PointNet framework in Python with a 9:1 split and 112 randomly selected point clouds were reserved for testing. The volume between the body surface points and a constant plane resembling the ground was calculated and correlated with weight to make a comparison with results from the PointNet method. The coefficient of determination ($R^2 = 0.94$) was achieved with PointNet regression model on test point clouds compared to the coefficient of determination ($R^2 = 0.76$) achieved from the volume of the same animal. The validation RMSE of the model was 6.79 kg with a test RMSE of 6.88 kg. Further, to analyze model performance based on weight range the pigs were divided into three different weight ranges: below 55 kg, between 55 and 90 kg, and above 90 kg. For different weight groups, pigs weighing below 55 kg were best predicted with the model. The results clearly showed that 3D deep learning on point sets has a good potential for accurate weight prediction even with a limited training dataset. Therefore, this study confirms the usability of 3D deep learning on point sets for farm animals' weight prediction, while a larger data set needs to be used to ensure the most accurate predictions.

Keywords: 3D deep learning; PointNet; weight estimation

Citation: Paudel, S.; de Sousa, R.V.; Sharma, S.R.; Brown-Brandl, T. Deep Learning Models to Predict Finishing Pig Weight Using Point Clouds. *Animals* **2024**, *14*, 31. https://doi.org/10.3390/ani14010031

Academic Editor: Emma Fàbrega i Romans

Received: 24 October 2023
Revised: 16 December 2023
Accepted: 19 December 2023
Published: 21 December 2023

1. Introduction

Live weight measurement is an important factor in the management of farm pigs [1,2]. Body weight provides important information regarding the animal's health and well-being [3]. It is noteworthy that in commercial production environments, obtaining accurate records of individual animal weights often encounters challenges [4]. Nevertheless, the systematic capture of individual weights plays a pivotal role in decisions pertaining to swine marketing [5]. Furthermore, continuous weight monitoring can provide the opportunity to promptly identify individuals potentially experiencing digestive and health issues [6].

Automatic weight prediction based on computer vision has been investigated for predicting animal weight in production systems in a non-invasive and economical way [7,8]. Digital image processing is the most widely used vision-based weight prediction method. In digital image processing, images of pigs' bodies are analyzed, and animal dimensions such as area, body length, width, and chest depth are calculated and correlated to animals' weight [9–13]. In the mid-nineties, Brandl et al. [14] analyzed pigs' body areas from digital images to estimate their weights, and were able to predict with less than 6% deviation [14]. Although digital image processing has shown great results, it comes with challenges, such as pigs needing to be under the camera at a somewhat predetermined position, color patterns or dirt on the animals, and lighting conditions needing to be consistent [6,15].

To overcome the challenge faced by digital image analysis, depth images have been used to extract features of pigs' bodies [16,17]. Depth images extract three-dimensional (3D) features of an animal's body such as height, volume, and curvature [18]. Pezzulo et al. [19] used Microsoft Kinect to measure heart girth, length, and height, and used linear and nonlinear models to get the coefficient of determination ($R^2 > 0.95$). Manual feature extraction is insufficient for a precise prediction as a limited number of parameters can be extracted and a small change in posture alters all of the parameters [17]. Okayama et al. [17] tried extracting information about spinal cord bends to make their model stable even with the change in posture. Although their work has shown an improvement in prediction, this approach is strictly restricted to implementation in the predefined close space. Deep learning approaches are presented as a solution to make vision-based weight estimation reliable even with slight changes in the postures of animals.

Artificial Neural Network (ANN) and deep learning have been implemented on digital RGB images of animals to estimate their weights [15,20]. Hidden fully connected or 2D convolution layers have been used to extract the features from the image with an additional regression layer at the end to estimate weight. Jun et al. [21] tried to address the problem of posture and lighting by introducing curvature and deviation, but they could only achieve a coefficient of determination of ($R^2 < 0.79$). In some approaches, depth images were colorized based on the height of the animal before presenting it to ANN [22]. The benefit from the colorization is very minimal and becomes similar to that of using RGB images.

Furthermore, all these approaches are common in using predefined lab setups to collect pigs' RGB or depth images. While studies have shown a correlation up to ($R^2 = 0.99$) with RSME below 1 kg, their constrained experimental design makes their implementation in natural farms impossible [16,22]. There is thus a strong need for a method that can effectively predict animal weight in natural farm settings and conditions. In recent days 3D deep learning methods have been studied for precision livestock farming such as identification [23], posture classification [24], and weight prediction [25,26] as they are robust even with changes in posture and lighting conditions. However, these methods have yet to be tested in commercial farms.

The objectives of this study were to: (1) develop and implement a 3D deep learning method to predict individual pig weight from the point cloud collected with free movement as a typical pen environment without controlled illumination; (2) compare the use of deep learning modeling with a most-studied method of weight prediction using machine vision (in other words, linear regression model between pig weight and volume); and (3) analyze how well these methods perform for different weight groups.

2. Materials and Methods

The animal care and handling techniques were approved by the Animal Use Ethics Committee of the Faculty of Animal Science and Food Engineering at the University of São Paulo (FZEA-USP), under protocol 6526160720. The animals were housed in covered outdoor pens; therefore the lighting was not controlled and varied throughout the day (Figure 1). These natural lighting conditions mimic some commercial conditions in certain parts of the world.

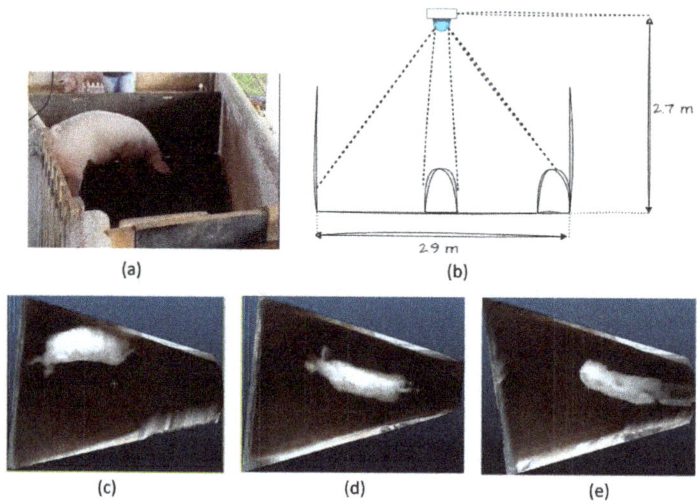

Figure 1. Experimental Setup and collected data; (**a**) pig within the holding pen area while data collection is occurring, (**b**) view of the major axis of data collection holding pen, (**c**–**e**) representative data of pigs in different positions and standing posture within the holding pen area.

A total of 249 unrestrained grow-finish pigs (20–120 kg) were used in an experiment to estimate body weight. This experiment included walking pigs through an individual holding pen and capturing point clouds of these animals. The data was used to estimate the weights of these individual animals. Details of the experiment including the experimental setup, data collection, extraction, and preprocessing of 3D images (point clouds) are included below. In addition, specific methodology employed for volume calculation and the implementation of PointNet are described.

2.1. Data Collection

The collection of 3D videos and weighing of pigs was carried out in experimental farms at FZEA-USP. A total of 249 grow-finish pigs (7 to 20 weeks old) were housed in ten pens, with 25 animals per pen (20–120 kg). The animals had ad libitum access to food and water. On an individual basis, the weights of pigs were collected and 3D videos were capture on five different dates (5 August, 28 October, 4 November, and 9 November).

The 3D video data acquisition was completed using a portable station composed of a 3D camera (Intel Real Sense D435 Stereo Depth Camera© (Intel Corporation, Santa Clara, CA, USA) and a computer. These top-down videos were recorded when pigs were freely moving in a holding pen after the weight was taken. A total of 249 videos of individual pigs were collected (each with a duration of 3–5 s), with the intention of extracting seven point clouds of each animal.

2.2. Experimental Setup

Figure 1 provides a detailed depiction of the experimental setup, focusing on its side view. The camera used in the experiment was positioned at a height of 2.7 m from the floor, and was connected to a computer positioned laterally in the experimental setup via a USB 3.0 cable with a 5 m active extender. Manual triggering was employed for video acquisition utilizing Intel® RealSense™ Viewer software version 2.47.

Camera parameter calibration was autonomously performed on-site using the Intel® Depth Quality Tool software v2.50.0, incorporating plane fit, on-chip self-calibration, and tare calibration methodologies. Video captures of each specimen were conducted at 30 frames per second (fps), employing auto-exposure mode for both the color image sensor (color sensor) and the infrared sensor (depth sensor). The field of view for the color image was 69.4 × 42.5 (H × V) with a resolution of 640 × 480, while the field of view for the depth image was 86 × 57 (H × V) with a resolution of 848 × 480.

The holding pen in which the pigs were observed boasted dimensions with a major axis extending to 2.9 m and a minor axis measuring 1.5 m. Throughout the data collection period, each individual pig was allowed unrestricted movement within the holding pen. The camera captured their dynamic motions within this confined space. The varying positions and standing posture of the pigs within the pen influenced the visibility of different parts of their bodies to the camera as showcased in Figure 1b–e, creating variability in the projected volume of the animal.

2.3. Image Extraction and Preprocessing

The first step of data processing consisted of extracting 3D images (point clouds) from the collected 3D videos. For this, the Intel Realsense Viewer© (Intel Corporation, Santa Clara, CA, USA) application was used (the same application that was used to collect the 3D videos). At this stage, collected 3D videos were locally stored, and five to ten 3D images were manually extracted. The number of extracted images was dependent on the quality of the video.

The entire set of filters and functions developed allowed the 3D images to be read, aligned, filtered, converted, and stored (Figure 2). Points whose values on the *z*-axis were below or above pre-defined limits were removed according to the minimum and maximum average depth found in the 3D images of the pigs. In this way, the points referring to the floor and the side walls were removed. Points referring to structures that are outside the area of interest (that is, any point that is not referring to the swine) were also extracted through "xy" points that were outside of a pre-defined polygon. Further extraction was performed by removing points whose color was different and did not belong to the swine. The color threshold was found automatically by the Otsu method. It should also be noted that the collected data also included point clouds with some unknown noise pattern as demonstrated in Figure 3.

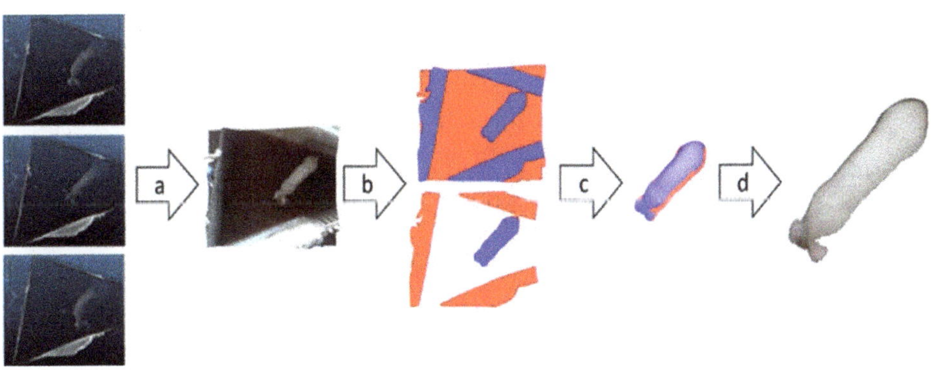

Figure 2. Basic steps for RGB-D image processing to identify and select the animal in the scene (swine): (**a**) image selection; (**b**) thresholding (empty points removal represented by red color); (**c**) color threshold filtering; (**d**) final statistical filtering.

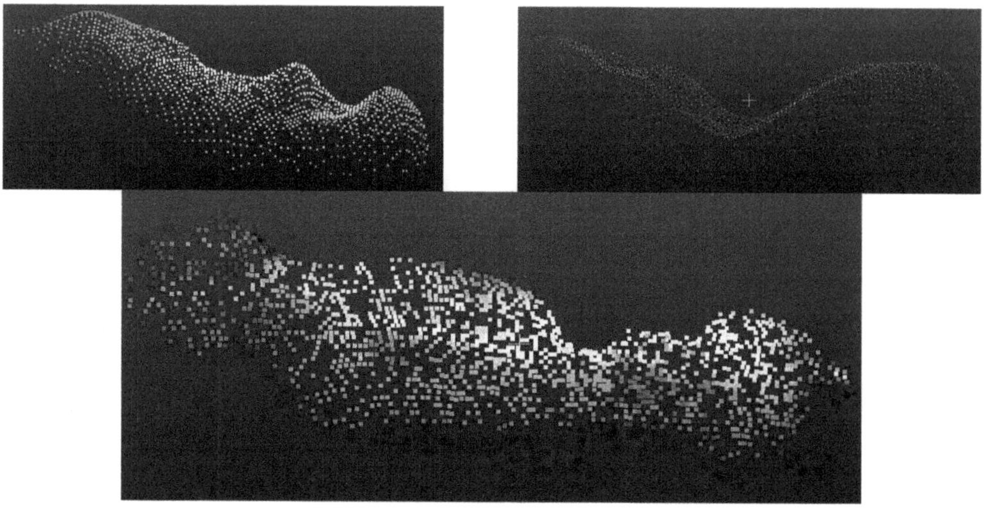

Figure 3. Three representative point cloud data with different quality issues, some unknown wavy pattern, and point clouds with missing large amounts of points.

Finally, the point clouds were randomly subsampled to 1500 points, and removal of outliers took place to obtain the final extracted point cloud of the pig. The weights of the experimented pigs ranged from 20–120 kg. The range was further divided into three different groups: below 55 kg, 55–90 kg, and above 90 kg, in order to analyze whether weight has an effect on correlation as observed in other manuscripts [14]. Correlations were then calculated within each weight range to determine which range could best be explained by volume. The processed data is composed of a total of 1298 point clouds, out of which 644 weigh below 55 kg, 322 are between 55 and 90 kg, and 332 are above 90 kg.

2.4. Volume Calculation

The literature has shown that the volume of the pig's body directly correlates to its mass [16]. The volume was calculated as a projected volume between the pig's top view (Figure 2d) and the pen floor, considered a constant 2.7 m from the camera. The volume was calculated using CloudCompare v2.11.3 software's 2.5D volume function. The grid with step size 0.01 was projected in the 'Z' direction. The maximum height on the body from constant plane 'L' was considered as the height of the point when multiple points fall inside a cell.

The pigs in the study had the freedom to move in any direction within the holding area, including moving their heads up and down. When the pigs lowered their heads, it resulted in limited or no contribution to the overall volume of the pig's body. Conversely, a flat head posture significantly increased the volume, and different ear postures also caused changes in volume. To address the variance in volume caused by head movement, the head of the pigs was manually removed by cropping out any portion above the neck region in all of the point clouds. The postural variations depicted in Figure 1 posed challenges for automating head and neck removal. To ensure consistency across all selected images, a visual inspection was conducted to identify the narrowest width of the neck area. Subsequently, any portions above this defined narrow width were cropped out.

2.5. PointNet

PointNet is a recently proposed algorithm that directly takes points for training [27,28]. PointNet has shown great potential in 3D classification and segmentation [29,30], but it has never been explored for livestock weight prediction. A significant advantage is that it directly takes a set of points as input and extracts features of the point clouds. The PointNet architecture takes a point cloud with n points as input and utilizes input transformation to extract features (Figure 4). The max pooling layer was used to aggregate the point features and finally K-score classes for classification. The input transformation layer is a mini-PointNet network that transforms point clouds into another coordinate system with the same dimension. T-Net is specifically designed to address the challenge of transforming the input point cloud to a canonical or normalized form. It achieves this by learning an optimal linear transformation matrix. This matrix is applied to the local features obtained from the shared MLP, effectively aligning and normalizing the point cloud. Learning and applying these transformations makes the network more robust, ensuring that it can recognize and understand 3D shapes regardless of their orientation or position in space.

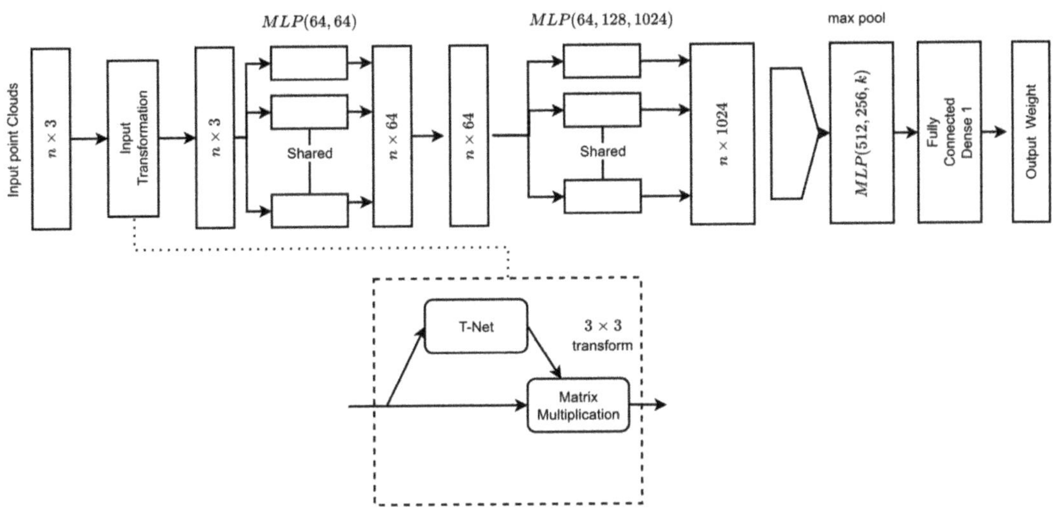

Figure 4. PointNet architecture implemented for weight prediction from a point cloud [15].

Successive multilayered perceptron layers (MLP) were shared by each point on the cloud. The max pooling layer was used to aggregate the global point features vectors. Finally, the global point features were then passed through another MLP with a regression layer on top to output regress weight. The feature transformation layer in between successive shared MLP was removed for computational simplicity. The top layer for K-score classification was replaced with a fully connected dense layer with a RELU activation function that outputs weight. After preprocessing and selection of the point clouds, the PointNet architecture was trained. The PointNet algorithm was implemented on Google Colab GPU Python version 3.7.15 platform with Keras package (Version 2.9.0) using TensorFlow (version 2.9.2) as the backend. Open3d (version 0.16.0) package was used to visualize and read point clouds corresponding to weight levels. PointNet implementation on TensorFlow Keras by David Griffiths was taken as a reference for algorithm implementation. Adam optimizer with a learning rate of 0.01 was used. Root Mean Square Error (RMSE) was used as accuracy metrics and Mean Square Error (MSE) was used as the loss function.

The model was trained for 1000 epochs with early stopping callback with patience of 10. In order to create more variation in the training dataset, data augmentation was used. The points were jittered between −0.005 and +0.005 to introduce variability in the training data while keeping the labels the same. The dataset was split into 9:1 for training and validation. A testing dataset containing 112 different point clouds was randomly selected from the same 249 pigs and was not used for training or validation. The performance of the models was evaluated by comparing the measured values of the target attributes with their respective predicted values based on the determination coefficient (R^2) and RMSE. The iterative fine-tuning process employed in the training phase of the models allowed for the definition of optimal values for the hyperparameters of each supervised learning algorithm. This training and evaluation method ensured the generation of models with the ability to generalize to data not belonging to the database used for training the models.

3. Results and Discussion

A total of 1298 point clouds from 249 pigs were collected for this study. Of these, 644 corresponded to weights less than 55 kg, 323 ranged between 55 and 90 kg, and 333 were above 90 kg. A total of 112 point clouds were manually selected for volume correction. To achieve the highest possible correlation between the volume of animals and the weight-adjusted volume (calculated by removing the head and neck), an additional 112 point clouds from the total of 1298 were reserved specifically for PointNet testing. The remaining 1186 point clouds were then divided into a 9:1 ratio for training and validation. The predicted weight of animals from PointNet was subsequently correlated with the scale weight. The best prediction was achieved with Pointnet, exhibiting a coefficient of determination of 0.94. Following this, the results were further investigated by segmenting the point cloud into three different weight groups.

3.1. Volume to Weight Correlation

Given the quality of data, a subset of point cloud was created for volume correlation. In order to select point clouds for volume correlation, all 1298 point clouds were checked manually and those with minimal quality issues were selected. Out of 246 pigs, 112 best-point clouds, which do not have wavy patterns or missing big chunks of points (Figure 3), were selected for volume correlation. The selected pigs had a uniform weight distribution in the 20–120 kg weight range.

Initially, the correlation between volume and weight was investigated. It was found that an R^2 value of 0.74 was obtained when the complete volume of the pig (including the head) was correlated to the weight (Figure 5). To minimize the effect of free movement, the head of the pigs was removed and volume was calculated. The correlation improved to $R^2 = 0.76$, as shown in Figure 6. This correlation did not come close to that which was reported by Condotta et al. [16] ($R^2 = 0.99$) and Brandl and Jørgensen [14] ($R^2 = 0.98$) in their experiment. It is evident that the point clouds under investigation were taken at different positions and standing postures (Figure 1), with some taken directly under the camera and some towards the edge of the holding barn, which led to variation in the volume of pigs compared to the stationary experiment approach followed in most of the literature. Although the weight seemed to be varying with volume, the changes in the volume itself with the movement of pigs within the barn led to less efficient prediction compared to the previously mentioned studies.

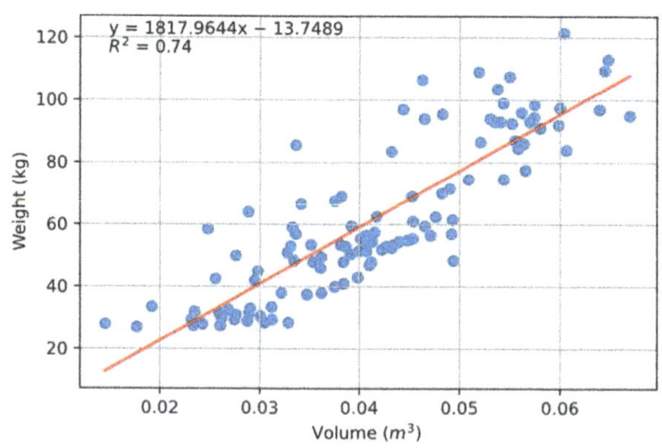

Figure 5. Correlation between the weight of the pig and calculated volume using CloudCompare software's 2.5D function. Blue dots indicate individual pigs, and the middle red line indicates optimal fitting.

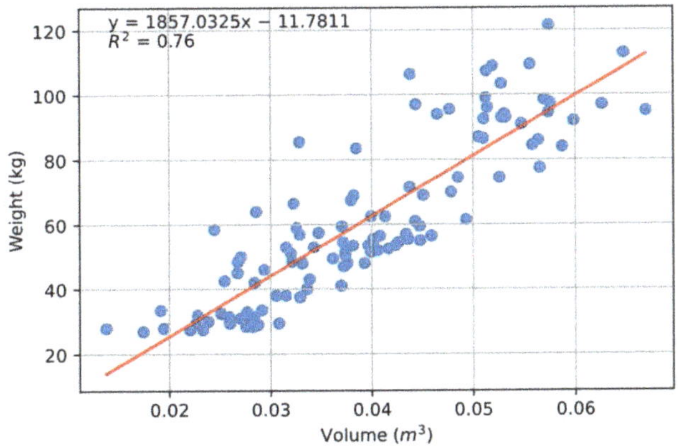

Figure 6. Correlation between adjusted volume as calculated by volume estimation from point cloud after digitally removing head and the neck. Blue dots indicate individual pigs, and the middle red line indicates optimal fitting.

Looking at different weight classes, the weight class below 55 kg was found to be best explained by the volume, with $R^2 = 0.77$, whereas for higher weight classes the correlation was found to be decaying. The correlation between weight and volume for the 55–90 kg weight class was found to be $R^2 = 0.26$, and for weight class above 90 kg, the weight could not be explained by pigs' body volume as the correlation was only $R^2 = 0.01$.

One likely reason for this declining correlation could be attributed to errors occurring during the measurement of scale weight. The larger the animals, the more challenging it was to contain their motion within the scale for accurate weight measurement. Another potential factor contributing to the declining results could be the distribution of data. In weight classes below 55 kg, the average weight of animals was 41.17 kg. For the weight range of 55–90 kg, the average weight of animals was 67.74 kg, and for weight classes above 90 kg, the average weight of animals was 99.31 kg. This indicates that a higher number of animals were concentrated towards the lower range of classes 55–90 kg and above 90 kg, leading to a higher prediction error.

3.2. PointNet for Estimating Weight

One hundred and twelve point clouds were set aside for testing, and the remaining were split into 9:1 for training and validation. Changes in RMSE was recorded throughout the training period, and Figure 7 shows the learning curve of the change in RSME. The minimum RMSE of 6.03 kg was obtained at the 28th epoch of the training. Figure 8 shows point clouds along with true labels and predicted labels. Some of the point clouds as shown in Figure 3 were missing large quantities of points from the body area because of some unknown problem in the experimental setup. Those point clouds were considered good and included in training and testing. While point clouds with prominent noise pattern were not included for the volume correlation, the data set still might have smaller noise leading to poor correlation; however, the deep learning model is strong enough to predict the weight even with such points.

One hundred and twelve random different point clouds were used for the testing. The trained model was presented with the reserved test data for prediction. Figure 9 shows correlation between scale weight and the predicted weight. It achieved an overall R^2 of 0.94 with an RMSE of 6.87 kg. However, although the prediction was better compared to the volume correlation, the performance could have been better with bigger datasets.

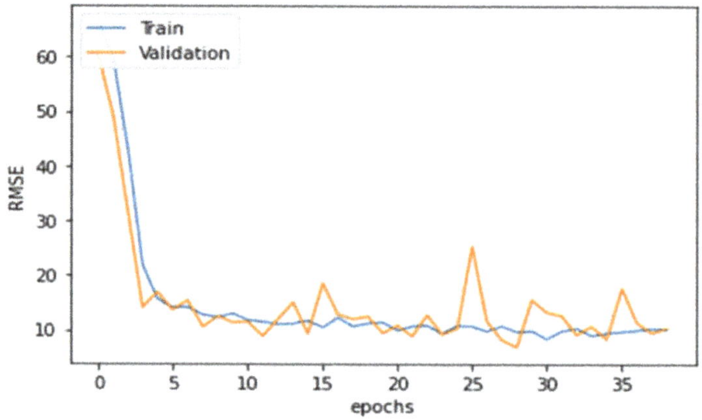

Figure 7. Training history of PointNet architecture over the complete training period.

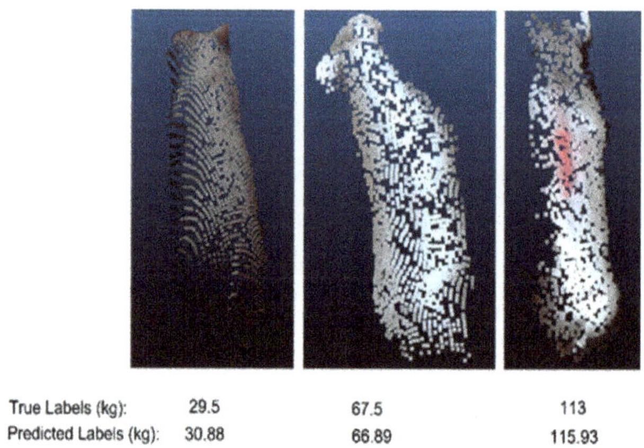

Figure 8. Point clouds, actual and predicted weight values from three different pigs.

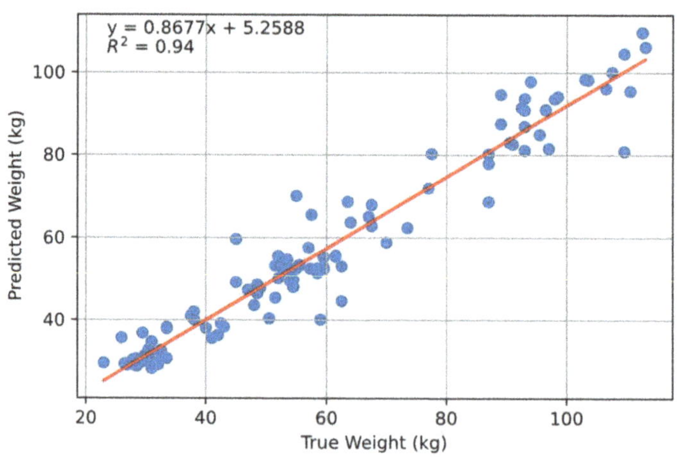

Figure 9. Correlation between predicted weight and scale weight using PointNet model. Blue dots indicate individual pigs, and the middle red line indicates the optimal fitting.

For the weight group below 55 kg, the model achieved an RMSE of 4.50 kg and an R^2 of 0.81. The performance of predictions for the weight class 55–90 kg had an RMSE of 8.20 kg and an R^2 of 0.71. For the weight class above 90 kg, the model achieved an R^2 of 0.44 and an RMSE of 9.95 kg. The trend was similar to the volume correlation, where the correlation worsened with an increase in weight. As mentioned, the training dataset for the below 55 kg class was almost twice as large as the other classes, which likely resulted in significantly better performance. However, for the two remaining classes with a similar number of training data, the lower weight class (between 55–90 kg) performed better than the above 90 kg weight class. This indicates that even with PointNet, the prediction declined with larger pigs.

3.3. *Comparison between Deep Learning and Volume-Based Regreesion*

The PointNet architecture outperformed volume correlation in each class. Table 1 presents the results for all three classes along with overall correlation. It was observed that weight prediction becomes more difficult as the pigs grow larger. It was observed that difficulties in handling animals (constraining their movement during weight measurement leading to errors on ground truth data) might have contributed to higher errors on larger animals. Along with the difficulties in weight measurement, the low accuracy of prediction on large animals can be tied to the nonlinearity of the relationship between the weight and volume of large pigs as suggested by [16].

Table 1. Coefficient of determination and root mean square error (RMSE) of prediction for both the models in each class.

	Volume Correlation		**Deep Learning**	
Weight Class	**Correlation (R^2)**	**RMSE (kg)**	**Correlation (R^2)**	**RMSE (kg)**
Below 55 kg	0.77	9.16	0.81	4.50
55–90 kg	0.26	13.44	0.0.71	8.20
90–120 kg	0.01	17.03	0.44	9.95
Overall	0.76	12.25	0.94	6.87

As discussed in the Section 2, the variability in position within the pen area and variability in standing posture might have been the major cause of poor performance of weight prediction using volume. This variability in positioning affected the projected volume of the pigs' bodies, as highlighted in Figure 1c–e. Looking at Figure 1c, in this instance, one side of the pig's surface was more prominently visible than the other. This observation emphasizes the influence of the pig's spatial orientation within the pen on the visual representation of its anatomy.

The positioning within the pen area is a crucial factor affecting the projected volume of the pig. This is a critical component of this experiment, shedding light on the dynamic nature of their appearances as captured by the camera.

Furthermore, beyond the impact of body visibility to the camera, the variation in the standing posture of the pigs introduces an additional layer of nonuniformity to the perceived volume. Figure 1c–e distinctly illustrates three distinct standing body postures. Notably, these postural variations have a direct and discernible effect on the volume captured by the depth sensor [31].

Examining Figure 1b–d, it becomes evident that changes in posture significantly influence the volume discernible to the depth sensor. The dynamic nature of these postural adjustments presents a challenge when employing computer vision for weight estimation, particularly in scenarios where the pigs are not constrained. The complexity introduced by the interplay of body visibility and postural changes underscores the intricacies involved in accurately estimating the weight of freely moving pigs through computer vision methodologies. This complexity might have led to most experimental setup designs where pigs are confined within a small area. Additionally, it should be noted that in the experimental setup resembling a commercial barn with numerous non-constrained pigs, segmenting individual animals, and calculating the volume of the animal becomes difficult, resulting in inaccurate predictions.

The PointNet approach employed in this study has proven to be effective in addressing the challenges encountered by volume correlation methods. Unlike traditional approaches, PointNet excels in learning features directly from raw, unordered point cloud data. This unique capability enables the model to capture intricate patterns and relationships within the data without the need for predefined structures or order making this method more robust to variation in animals' body projection to the camera.

The PointNet predictions for the worst-performing weight class (above 90 kg) had an RMSE of 9.95 kg. In contrast, the volume-based prediction had a similar RMSE of 9.16 kg in the best-performing weight class (below 55 kg). The results clearly demonstrate that implementing PointNet and other 3D deep learning architectures lead to more accurate weight predictions. Furthermore, it should be noted that the PointNet architecture was trained with only 1186 point clouds, each consisting of only 1500 body surface points. Adding more point clouds for training will certainly improve the prediction accuracy.

Furthermore, it is important to note that the data for this study was collected with only one pig inside the holding pen. However, in farm settings, many pigs are typically housed together in a single pen. Therefore, further studies should be designed to capture and predict the weights of multiple pigs simultaneously. This can be achieved by placing a camera over the drinker or feeder of the pens and segmenting individual pigs in each frame. This approach will allow the algorithm to predict the weights of multiple pigs at once.

The performance of the deep learning algorithm ultimately relies on the size and balance of the database. Therefore, for future studies, it is recommended to collect a higher number of point cloud data. This will ensure a larger and more diverse dataset, leading to improved accuracy and generalizability of the weight prediction model.

4. Conclusions

Weight prediction of individual free roaming pigs from weight to volume correlation was compared to a PointNet-based 3D deep learning model. The point cloud data of individual free roaming pigs in the holding pen was preprocessed to remove the background scene, and then input into the deep learning architecture for training and testing. The adjusted and unadjusted volumes of the same animals were correlated with scale weight. Despite the training data containing noisy point clouds, the deep learning model outperformed the volume correlation. Tested on unseen point clouds, the model achieved a root mean square error of 6.87 kg with a determination coefficient of 0.9421, whereas the best determination coefficient between volume and weight of pigs was found to be 0.7392. Looking at the different weight classes, both methods performed best for pigs weighing below 55 kg. However, the volume correlation was found to rapidly decline for weight prediction as pigs weight grew, while the deep learning method was found to be stable even for bigger pigs. The results show that 3D deep learning can extract features from the point cloud of pigs in general housing conditions to predict weight, despite adverse conditions with the animal in free movement evidenced by the median correlation between animal weight and volume.

In addition, the findings underscore key aspects for enhancing prediction models through the proposed technique. It is noteworthy that the PointNet architecture underwent training with a limited dataset of 1186 point clouds, each comprised of merely 1500 points on the body surface. Augmenting the training set with a greater number of point clouds is poised to significantly enhance prediction accuracy. The efficacy of the deep learning algorithm is ultimately contingent on the scale and equilibrium of the database. Therefore, for prospective investigations, it is advisable to amass a more extensive dataset of point clouds. This approach ensures a broader and more varied dataset, thereby fostering heightened accuracy and generalization within the weight prediction model.

Author Contributions: Conceptualization, S.P., R.V.d.S. and T.B.-B.; methodology, S.P. and R.V.d.S.; software, S.P.; validation, S.P., R.V.d.S. and T.B.-B.; formal analysis, S.P.; resources, T.B.-B. and S.R.S.; data curation, R.V.d.S.; writing—original draft preparation, S.P.; writing—review and editing, T.B.-B. and R.V.d.S.; supervision, T.B.-B. All authors have read and agreed to the published version of the manuscript.

Funding: This study was financed in part by the Coordenação de Aperfeiçoamento de Pessoal de Nível Superior—Brasil (CAPES)—Finance Code 001.

Institutional Review Board Statement: The animal study protocol was approved by the Animal Use Ethics Committee of the Faculty of Animal Science and Food Engineering at the University of São Paulo (FZEA-USP), under protocol 6526160720.

Informed Consent Statement: Not applicable.

Data Availability Statement: The data presented in this study are available on request from the corresponding author. The data are not publicly available due to ethical reasons.

Acknowledgments: The authors would like to express their deepest gratitude to the dedicated animal caretakers at the FZEA-USP (Faculdade de Zootecnia e Engenharia de Alimentos, University de Sao Paulo) experimental farms for their exemplary professionalism and unwavering commitment to the welfare of the research subjects. Their expertise and compassion were indispensable in ensuring the well-being of the animals throughout the duration of this study. The authors also thank the individuals who generously contributed their time and efforts during the data collection phase. Their invaluable assistance played a crucial role in the successful execution of this research endeavor.

Conflicts of Interest: The authors declare no conflict of interest.

References

1. Halachmi, I.; Guarino, M.; Bewley, J.; Pastell, M. Smart Animal Agriculture: Application of Real-Time Sensors to Improve Animal Well-Being and Production. *Annu. Rev. Anim. Biosci.* **2019**, *7*, 403–425. [CrossRef] [PubMed]
2. Kollis, K.; Phang, C.S.; Banhazi, T.M.; Searle, S.J. Weight Estimation Using Image Analysis and Statistical Modelling: A Preliminary Study. *Appl. Eng. Agric.* **2007**, *23*, 91–96. [CrossRef]
3. Dickinson, R.A.; Morton, J.M.; Beggs, D.S.; Anderson, G.A.; Pyman, M.F.; Mansell, P.D.; Blackwood, C.B. An Automated Walk-over Weighing System as a Tool for Measuring Liveweight Change in Lactating Dairy Cows. *J. Dairy Sci.* **2013**, *96*, 4477–4486. [CrossRef] [PubMed]
4. Lee, S.; Ahn, H.; Seo, J.; Chung, Y.; Park, D.; Pan, S. Practical Monitoring of Undergrown Pigs for IoT-Based Large-Scale Smart Farm. *IEEE Access* **2019**, *7*, 173796–173810. [CrossRef]
5. He, Y.; Tiezzi, F.; Howard, J.; Maltecca, C. Predicting Body Weight in Growing Pigs from Feeding Behavior Data Using Machine Learning Algorithms. *Comput. Electron. Agric.* **2021**, *184*, 106085. [CrossRef]
6. Dohmen, R.; Catal, C.; Liu, Q. Computer Vision-Based Weight Estimation of Livestock: A Systematic Literature Review. *New Zeal. J. Agric. Res.* **2022**, *65*, 227–247. [CrossRef]
7. Liu, T.; Teng, G.; Fu, W. Research and Development of Pig Weight Estimation System Based on Image. In Proceedings of the 2011 International Conference on Electronics, Communications and Control (ICECC), Ningbo, China, 9–11 September 2011; pp. 2774–2777. [CrossRef]
8. Pradana, Z.H.; Hidayat, B.; Darana, S. Beef Cattle Weight Determine by Using Digital Image Processing. In Proceedings of the 2016 International Conference on Control, Electronics, Renewable Energy and Communications (ICCEREC), Bandung, Indonesia, 13–15 September 2016; pp. 179–184. [CrossRef]
9. Arulmozhi, E.; Bhujel, A.; Moon, B.E.; Kim, H.T. The Application of Cameras in Precision Pig Farming: An Overview for Swine-Keeping Professionals. *Animals* **2021**, *11*, 2343. [CrossRef]
10. Kaewtapee, C.; Rakangtong, C.; Bunchasak, C. Pig Weight Estimation Using Image Processing and Artificial Neural Networks. *J. Adv. Agric. Technol.* **2019**, *6*, 253–256. [CrossRef]
11. Banhazi, T.M.; Tscharke, M.; Ferdous, W.M.; Saunders, C.; Lee, S.H. Improved Image Analysis Based System to Reliably Predict the Live Weight of Pigs on Farm: Preliminary Results. *Aust. J. Multi-Discip. Eng.* **2011**, *8*, 107–119. [CrossRef]
12. Yu, H.; Lee, K.; Morota, G. Forecasting Dynamic Body Weight of Nonrestrained Pigs from Images Using an RGB-D Sensor Camera. *Transl. Anim. Sci.* **2021**, *5*, 1–9. [CrossRef]
13. Kashiha, M.; Bahr, C.; Ott, S.; Moons, C.P.H.; Niewold, T.A.; Ödberg, F.O.; Berckmans, D. Automatic Weight Estimation of Individual Pigs Using Image Analysis. *Comput. Electron. Agric.* **2014**, *107*, 38–44. [CrossRef]
14. Brandl, N.; Jørgensen, E. Determination of Live Weight of Pigs from Dimensions Measured Using Image Analysis. *Comput. Electron. Agric.* **1996**, *15*, 57–72. [CrossRef]
15. Bhoj, S.; Tarafdar, A.; Chauhan, A.; Singh, M.; Gaur, G.K. Image Processing Strategies for Pig Liveweight Measurement: Updates and Challenges. *Comput. Electron. Agric.* **2022**, *193*, 106693. [CrossRef]

16. Condotta, I.C.F.S.; Brown-Brandl, T.M.; Silva-Miranda, K.O.; Stinn, J.P. Evaluation of a Depth Sensor for Mass Estimation of Growing and Finishing Pigs. *Biosyst. Eng.* **2018**, *173*, 11–18. [CrossRef]
17. Okayama, T.; Kubota, Y.; Toyoda, A.; Kohari, D.; Noguchi, G. Estimating Body Weight of Pigs from Posture Analysis Using a Depth Camera. *Anim. Sci. J.* **2021**, *92*, e13626. [CrossRef] [PubMed]
18. Samperio, E.; Lidón, I.; Rebollar, R.; Castejón-Limas, M.; Álvarez-Aparicio, C. Lambs' Live Weight Estimation Using 3D Images. *Animal* **2021**, *15*, 100212. [CrossRef] [PubMed]
19. Pezzuolo, A.; Guarino, M.; Sartori, L.; González, L.A.; Marinello, F. On-Barn Pig Weight Estimation Based on Body Measurements by a Kinect v1 Depth Camera. *Comput. Electron. Agric.* **2018**, *148*, 29–36. [CrossRef]
20. Cang, Y.; He, H.; Qiao, Y. An Intelligent Pig Weights Estimate Method Based on Deep Learning in Sow Stall Environments. *IEEE Access* **2019**, *7*, 164867–164875. [CrossRef]
21. Jun, K.; Kim, S.J.; Ji, H.W. Estimating Pig Weights from Images without Constraint on Posture and Illumination. *Comput. Electron. Agric.* **2018**, *153*, 169–176. [CrossRef]
22. Meckbach, C.; Tiesmeyer, V.; Traulsen, I. A Promising Approach towards Precise Animal Weight Monitoring Using Convolutional Neural Networks. *Comput. Electron. Agric.* **2021**, *183*, 106056. [CrossRef]
23. Paudel, S.; Brown-Brandl, T.; Rohrer, G.; Sharma, S.R. Individual Pigs' Identification Using Deep Learning. In Proceedings of the 2023 ASABE Annual International Meeting, Omaha, NE, USA, 9–12 July 2023; p. 1.
24. Xu, J.; Zhou, S.; Xu, A.; Ye, J.; Zhao, A. Automatic Scoring of Postures in Grouped Pigs Using Depth Image and CNN-SVM. *Comput. Electron. Agric.* **2022**, *194*, 106746. [CrossRef]
25. He, H.; Qiao, Y.; Li, X.; Chen, C.; Zhang, X. Optimization on Multi-Object Tracking and Segmentation in Pigs' Weight Measurement. *Comput. Electron. Agric.* **2021**, *186*, 106190. [CrossRef]
26. Nguyen, A.H.; Holt, J.P.; Knauer, M.T.; Abner, V.A.; Lobaton, E.J.; Young, S.N. Towards Rapid Weight Assessment of Finishing Pigs Using a Handheld, Mobile RGB-D Camera. *Biosyst. Eng.* **2023**, *226*, 155–168. [CrossRef]
27. Qi, C.R.; Su, H.; Mo, K.; Guibas, L.J. PointNet: Deep Learning on Point Sets for 3D Classification and Segmentation. In Proceedings of the 30th IEEE Conference on Computer Vision and Pattern Recognition CVPR 2017, Honolulu, HI, USA, 21–26 July 2017; pp. 77–85. [CrossRef]
28. Qi, C.R.; Su, H.; Mo, K.; Guibas, L.J. PointNet: Supplementary. In Proceedings of the 30th IEEE Conference on Computer Vision and Pattern Recognition CVPR 2017, Honolulu, HI, USA, 21–26 July 2017; pp. 77–85.
29. Ma, Z.; Dong, Y.; Zi, J.; Xu, F.; Chen, F. Forest-PointNet: A Deep Learning Model for Vertical Structure Segmentation in Complex Forest Scenes. *Remote Sens.* **2023**, *15*, 4793. [CrossRef]
30. Haznedar, B.; Bayraktar, R.; Ozturk, A.E.; Arayici, Y. Implementing PointNet for Point Cloud Segmentation in the Heritage Context. *Herit. Sci.* **2023**, *11*, 2. [CrossRef]
31. Kongsro, J. Estimation of Pig Weight Using a Microsoft Kinect Prototype Imaging System. *Comput. Electron. Agric.* **2014**, *109*, 32–35. [CrossRef]

animals

MDPI

Article

Improving Dry Matter Intake Estimates Using Precision Body Weight on Cattle Grazed on Extensive Rangelands

Hector Manuel Menendez III [1], Jameson Robert Brennan [1,*], Krista Ann Ehlert [2] and Ira Lloyd Parsons [1]

1 Department of Animal Science, South Dakota State University, Rapid City, SD 57703, USA; hector.menendez@sdstate.edu (H.M.M.III); ira.parsons@sdstate.edu (I.L.P.)

2 Department of Natural Resource Management, South Dakota State University, Rapid City, SD 57703, USA; krista.ehlert@sdstate.edu

* Correspondence: jameson.brennan@sdstate.edu

Simple Summary: As precision livestock technology continues to become viable for extensive rangeland systems. it is important to determine which technology has the potential to positively impact grazing management. Utilizing precision weighing systems on rangeland beef cattle provides new and novel insight into individual animal performance throughout grazing periods, which are directly linked to stocking rates (cattle/ha/time), to ensure adequate forage dry matter intake for cattle while avoiding negative environmental impacts. Refining stocking rate estimates using precision body weight measurements and precision system modeling is critical as this management decision is fundamental to rangeland management and livestock productivity across the United States.

Abstract: An essential component required for calculating stocking rates for livestock grazing extensive rangeland is dry matter intake (DMI). Animal unit months are used to simplify this calculation for rangeland systems to determine the rate of forage consumption and the cattle grazing duration. However, there is an opportunity to leverage precision technology deployed on rangeland systems to account for the individual animal variation of DMI and subsequent impacts on herd-level decisions regarding stocking rate. Therefore, the objectives of this study were, first, to build a precision system model (PSM) to predict total DMI (kg) and required pasture area (ha) using precision body weight (BW), and second, to evaluate differences in PSM-predicted stocking rates compared to the traditional herd-level method using initial or estimated mid-season BW. A deterministic model was constructed in both Vensim (version 10.1.2) and Program R (version 4.2.3) to incorporate individual precision BW data into a commonly used rangeland equation using %BW to estimate individual DMI, daily herd DMI, and area (ha) required to meet animal DMI requirements throughout specific grazing periods. Using the PSM, differences in outputs were evaluated using three scenarios: (1) initial BW (business as usual); (2) average mid-season BW; and (3) individual precision BW using data from two precision rangeland experiments conducted at the South Dakota State University Cottonwood Field Station. The data from the two experiments were used to develop PSM case studies. The trial data were collected using precision weight data (SmartScale™) collected from replacement heifers (Case study 1, n = 60) and steers (Case study 2, n = 254) grazing native rangeland. In Case study 1 (heifers), Scenario 1 versus Scenario 3 resulted in an additional 73.41 ha required. Results from Case study 2 indicated an average additional 4.4 ha required per pasture when comparing Scenario 3 versus Scenario 1. Sensitivity analyses resulted in a difference between maximum and minimum simulated values of 27,995 and 4265 kg forage consumed, and 122 and 8.9 pasture ha required for Case studies 1 and 2, respectively. Thus, results from the scenarios indicate an opportunity to identify both under- and over-stocking situations using precision DMI estimates, which helps to identify high-leverage precision tools that have practical applications for enhancing animal and plant productivity and environmental sustainability on extensive rangelands.

Keywords: precision weighing; beef cattle; precision technology; precision system models

Citation: Menendez, H.M., III; Brennan, J.R.; Ehlert, K.A.; Parsons, I.L. Improving Dry Matter Intake Estimates Using Precision Body Weight on Cattle Grazed on Extensive Rangelands. *Animals* **2023**, *13*, 3844. https://doi.org/10.3390/ani13243844

Academic Editors: Brett Ramirez, Janice Siegford, Hao Gan, Yang Zhao, Daniel Berckmans, Robert T. Burns and Lingjuan Wang-Li

Received: 31 October 2023
Revised: 23 November 2023
Accepted: 7 December 2023
Published: 14 December 2023

1. Introduction

Measuring the dry matter intake (DMI) of grazing beef cattle is a significant challenge from basic and applied research and production aspects. For estimating forage utilization and the amount of land area required for cattle grazing extensive rangelands, it is essential to determine the animal unit months (AUMs). For AUMs, the amount of forage needed per animal unit (AU) is defined as a 454 kg cow consuming 2.6% of BW on a dry matter basis, thus equating to 354 kg forage being consumed over one month (1 AUM) [1]. The DMI is required to determine stocking rates and supplementation requirements. The variation in estimated DMI of grazing cattle introduces error at the herd level since estimated DMI is based on full BW multiplied by a percentage of BW. The percentage of BW varies depending on animal class, production phase (e.g., lactating vs. dry), and forage quality [2]. For example, a lactating cow eating low-quality forage (<52% total digestible nutrients) consumes approximately 2.2% of BW on a dry matter basis, whereas a lactating cow eating high-quality forage (>59% total digestible nutrients) consumes 2.7% of BW on a dry matter basis [2,3]. Differences in DMI and BW can also significantly affect stocking rates. For example, a 227 kg heifer consuming 2.5% of BW on rangeland with 700 kg ha^{-1} available for consumption (i.e., 25% of total forage; 2800 kg ha^{-1}) would require 0.24 ha AUM^{-1}, whereas a 381 kg heifer consuming 2.5% of BW would require 0.42 ha AUM^{-1}.

Traditionally, stocking rate estimates have been based on knowledge of rangeland stocking capacity from land manager experience, initial animal BW (if known), peak biomass (kg ha^{-1}), and forage utilization measurements. Frequently, initial herd level average BW is used to calculate AUMs and, subsequently, stocking rate; however, herd level averages may not adequately account for the variability in initial individual animal BW, changes over time due to daily cattle growth (BW + Δ kg d^{-1} (daily gain)), differences in % BW of DMI, or changes in BW due to environmental conditions. Not accounting for this individual animal variability can have a proportional scaling effect from individual livestock operations to landscape scale estimates of AUMs required. The exclusion of individual animal BW data can result in potentially overgrazing forage resources and subsequently have negative impacts on natural resources and animal production.

Range and animal scientists have envisioned in-pasture weighing systems since the 1960s [4,5]. However, the advent of modern precision livestock technology has made in -pen and pasture weighing systems a viable option for research and production [6,7]. The advent of precision data collection for rangeland cattle, including the imminent development of camera-based weighing systems [7], has made it possible to weigh cattle on pasture in real time and provides new insight into individual animal weights throughout the grazing season [8,9]. These data provide the potential to match more closely the stocking rate and carrying capacity of rangeland for economically and environmentally sustainable grazing livestock production. As the role of precision technology grows in extensive rangeland systems, a critical question is how previously unattainable data can be leveraged in precision system models (PSM) [10]. A PSM model is specifically designed to incorporate precision livestock data to help inform management. Using PSMs will help evaluate complex tradeoffs relative to ranch management objectives such as animal efficiency, managing variability in forage resources (surplus and shortfall), environmental impact, and mental models. Deploying a PSM to estimate individual DMI using precision weighing technology may help to identify performance gaps in stocking rates, maximize forage utilization, and prevent overgrazing. Therefore, the objectives of this study were to (1) build a PSM to predict forage DMI and stocking rate in extensively managed cattle using precision BW; and (2) conduct two case studies to compare stocking rate predictions between PSM and traditional methodologies.

2. Materials and Methods

2.1. Study Site

The SDSU Institutional Animal Care and Use Committee approved all procedures involving animals (Approval #2109-054E and #2104-021E). Both case study projects were conducted at the South Dakota State University (SDSU) Cottonwood Field Station (CFS; Cottonwood, SD, USA, GIS cords: 43.989107 N, −101.857228 E), located in the Northern Great Plains and consisting of mixed grass prairie, where dominant forage species included western wheatgrass (*Pascopyrum smithii* (Rydb.) A. Love), crested wheatgrass (*Agropyron cristatum* (L.) Gaertn.), green needlegrass (*Nassella viridula* Trin.), and needle-and-thread (*Hesperostipa Comata* Trin. and Rupr), with the inclusion of sedges (*Carex* spp.), buffalograss (*Bouteloua dactyloides* (Nutt.) J.T.Columbus), and blue grama (*Bouteloua gracilis* (Willd. Ex Kunth.) Lag. Ex Griffiths) [11]. There are also recent introductions of non-native grasses, such as Kentucky bluegrass (*Poa pratensis* Boivin and Love) and Japanese brome (*Bromus japonicus* Thunb.). Elevation at the CFS ranges from 710–784 m, and the climate is an arid cold steppe [12]. The long-term average annual precipitation for the area is 468 mm (1992–2022) [13].

2.2. Precision System Model Development

Mathematical models are becoming increasingly critical for the application of precision livestock technology as they allow the users to ask "what if" questions, such as the question in the current study of how individual BW across grazing periods impacts DMI estimates [14,15]. As the granularity of data increases, it is essential that models are used not only to generate results but that the results are meaningful, providing insight that translates into management interventions [16]. A critical step in fully leveraging mathematical models is to start with a simple modeling approach which ensures that the model achieves its intended purpose. In the current study, we use a simple deterministic model (in terms of mathematical complexity and number of parameters and equations) to assess differences in DMI, grazing area required (ha), and forage consumed (kg), setting the foundation for more complex integration of grassland dynamics [17]. Thus, as the amount of livestock, forage, soil, and climate data (amongst many other sources) become available through precision technology, the most successful models will likely be the ones that are based on a clear understanding of the production system. This understanding will enable the development of meaningful and scalable decision support tools that are based on a clear (i.e., simple) foundation of science such as the basic principles of rangeland management for grazing livestock.

A PSM model evaluating forage consumption (AUM) was constructed in Vensim DSS™, a dynamic visually based modeling software, utilizing equations described in the online SDSU Grazing Calculator [3]. A "precision data" AUM model (AUM_{PSM}) component was built in Vensim DSS to integrate daily individual BW data. The model was also reconstructed in Program R [18] to facilitate open-source PSM development and use. Model results matched both programs [Vensim DSS (version 10.1.2) and Program R (version 4.2.3)]. A 3D smoothing function was applied to raw weight data to minimize variation due to rumen-fill. The AUM_{PSM} included a discrete (daily time step, delta time = 1) first-order differential equation to aggregate the daily estimated DMI and hectares needed for each animal into monthly herd level values. Thus, the model output was total hectares needed per month to meet cattle nutrient requirements and total forage consumed at a herd level. Fixed parameters for the model can be found in Tables 1 and 2. Animal BW data were used from two grazing studies. Case Study 1 utilized data collected from a replacement heifer development study managed on dormant forage. Case Study 2 utilized data collected from summer grazed yearling steers.

Table 1. Model parameters used in the three scenarios for the heifer and steer stocking rate calculations based on the "take-half-leave-half" method suggested by the USDA Natural Resources Conservation Service (NRCS). The "take-half" consists of 50% of forage being taken (25% is consumed; 25% is trampled, urinated, and defecated on), while the "leave-half" leaves 50% of total biomass behind to regenerate plant growth.

Model Parameters	Heifer	Steer 1	Steer 2	Unit
Harvest Efficiency	0.25	0.25	0.25	Dimensionless
Days per Month	30	30	30	Day
Number of Cattle	60	127	135	Head
Percent BW	2.5	2.5	2.5	%

Table 2. Average initial forage biomass of winter heifer pasture and estimated peak forage biomass within each summer steer pasture (kg per ha^{-1}).

Case Study	Pasture	2021	2022
Case Study 1: Replacement heifers	–	–	917.78
Case Study 2: Steers	1	971.78	712.86
	2	935.91	1032.30
	3	1113.00	1305.79
	4	958.33	631.04
	5	1568.07	1064.81
	6	1713.78	1217.24

2.3. *Case Study 1*

Replacement Angus heifers (n = 60; initial BW = 237.6 ± 15.5 kg) grazed dormant native pastures (n = 2; 115.1 ha and 93.4 ha) from November 2021 to May 2022, as part of a broader project to integrate precision feeding technology to precisely manage heifer development [19] (Figure 1). Individual animals were tagged with a radio frequency identification device (RFID; Allflex Inc., Dallas, TX, USA) to pair weight measurements to each animal. Both treatment groups were supplemented with 2.27 kg hd^{-1} d^{-1} of pelleted dried distiller's grains with solubles (DDGs). The supplement was delivered to the control group via a traditional bunk fed method, and the precision group supplement was offered via a Super Smartfeed Producer™ (C-Lock Inc., Rapid City, SD, USA). Individual BWs were calculated daily from partial BWs measured using a front-end scale (Smartscale™, C-Lock Inc. Rapid City, SD, USA; Figure 1) positioned at the water source. Forage samples were collected at the beginning and end of the trial using a grid sampling technique for each pasture (n = 10 per pasture; 0.25 m^2 quadrat) to provide initial available forage values (kg dry matter ha^{-1}). In Case study 1, only the initial forage value was used to estimate stocking rate.

Figure 1. Dormant winter pasture and heifers (**left**) and steers using the SmartScale™ (**right**).

2.4. Case Study 2

Steers were managed on native pastures as part of a long-term (>80 year) grazing study evaluating the effects of stocking density on plant communities and animal performance. The current project's broader objectives are to evaluate opportunities for precision livestock technology to improve animal performance, monitor energy expenditure, and mitigate environmental impacts in extensively grazed cattle. More detailed methods are described by Vandermark [20]; briefly, crossbred Angus steers (n = 254) were managed on native pastures from June to August of 2021 and 2022, respectively. Each steer was weighed, fitted with an RFID tag, and allocated to one of six pastures equipped with a precision scale system (SmartScale™, C-Lock Inc., Rapid City, SD USA) positioned at the water source. Each pasture was assigned either a rotational (RG) or continuous (CG) grazing strategy and one of three stocking rates—low (pasture area = ~70.31 ha), medium (pasture area = ~53.15 ha), and high (pasture area = ~30.77 ha); 0.3, 0.42, and 0.7 AUM's, respectively—in a 2 × 3 factorial design. The pasture sizes and stocking rates (above) were not used in the PSM_{AUM} simulations; rather, only the precision BW and initial forage values were utilized. The purpose of describing the differences in stocking rates is to emphasize that we utilized diverse sources of BW data compared to only a single stocking rate, as different stocking rates have been shown to impact rates of BW gain [11]. Forage samples were collected using random clippings (n = 5, 0.25 m^2 quadrat) in each pasture at bi-weekly intervals for both grazing trials (2021 and 2022) to determine peak biomass.

2.5. Precision System Model Application

Individual animal BW data for both case studies were downloaded through an Automated Programming Interface (API, C-Lock Inc.) into Program R [18]. Weight data were assigned to individual animal records using the data.table package [21], filtered for spurious weight data points utilizing robust regression [22], and organized into longitudinal data frames. Static values (i.e., the same BW for each animal each day) were used to compare the AUM model outputs and the SDSU Grazing Calculator outputs to ensure mathematical accuracy and that double-accounting was avoided. It is important to note that the size or nutritional value of individual pastures for Case study 1 (n = 2) and Case study 2 (n = 6 per year) were not used in the PSM_{AUM}; rather, only the initial (Case study 1) or estimated peak standing forage (Case study 2) and individual precision BW were used to estimate total grazing area needed and the total forage consumed for each of the three scenarios. Using initial dormant forage or estimated peak forage production are commonly employed by rangeland managers for setting stocking rates before turning livestock out to graze.

Three scenarios (described below) were simulated to determine differences in calculated total hectares of pasture needed over each grazing period. The first two scenarios were designed to represent typical calculations used to determine stocking rate, while the third utilized precision BW data. All simulations were based on the DMI estimation method using %BW described above. Model parameters were the same in each scenario (Table 1). Scenario 1 applied traditional rangeland AUM estimation methods based on average initial BW (Table 3) at the beginning of each grazing period. Scenario 2 consisted of calculating a mid-season average BW for the heifers and the steers. Case study 1 was calculated using the desired BW weight at time of breeding in May of 381 kg minus the initial average herd BW (Table 3). Unlike the heifers, the steers in Case study 2 did not have a specific desired end BW, as one of the full experiment's goals was to assess differences in BW across stocking rates, not necessarily to maximize performance [20]. Thus, the steer mid-season BW was calculated using an average of the average initial herd BW (day 1) and ending herd BW (day 60) (Table 3). The impetus behind this scenario was to account for changing BW at a herd level at mid-season in an attempt to capture the same expected variation in BW throughout the grazing period. Finally, Scenario 3 utilized daily BW measurements from the precision scales for each animal over the total length of the grazing period (Table 3) with the AUM_{PSM}.

Table 3. Scenario parameters for body weight (BW).

Study	Pasture	Scenario					
		1: Average Initial BW, kg hd^{-1}		2: Mid-Season BW, kg hd^{-1}		3: Individual Precision BW, kg hd^{-1} d^{-1}	
Case Study 1: Replacement heifers	–	243.45		301.53		Individual weights	
Case Study 2: Steers		2021	2022	2021	2022	2021	2022
	1	352.33	273.66	375.71	300.89	–	–
	2	338.92	274.82	360.82	299.28	–	–
	3	345.80	271.23	366.92	302.58	–	–
	4	348.89	270.50	363.75	298.65	–	–
	5	352.91	271.68	378.12	300.47	–	–
	6	341.20	271.54	367.06	298.41	–	–

Finally, we conducted a Monte Carlo analysis (10,000 runs) to determine the percentiles of potential variation from %BW on AUM_{PSM} calculations on a daily basis impacting total hectares required and total forage used for each grazing period. The BW parameters ranged from 1.8 to 2.7% using a univariate normal distribution.

3. Results

3.1. Case Study 1

The initial BW of replacement heifers was 243.45 kg, with a mid-season BW of 301.53 kg per head. It had a 175 d grazing period; 8 d were removed due to missing BW due to system malfunction, resulting in a 167 d period used for this analysis. The greatest difference in total area needed was 73.41 ha between Scenarios 1 and 3; however, using an estimated mid-season average (Scenario 2) only resulted in 10.03 ha less than the precision estimate (Scenario 3; Table 4; Figure 2). Simulations were deterministic and did not have a probability distribution. The sensitivity analysis results of forage consumption from variation in %BW ranged from 56,019 to 84,014 kg on day 167, increasing as the grazing trial progressed (Figure 3). Like estimated forage consumption, the sensitivity analysis of total hectares required, from variation in %BW, ranged from 244 to 366 ha on day 167 and increased as the trial progressed (Figure 3).

Table 4. Total hectares required for each case study (replacement heifers and grazing steers) estimated according to total forage production (initial or peak), and intake estimated as a function of BW measured according to each of the three scenarios, initial BW, estimated mid-season BW, and individual precision BW.

Study	Pasture	Scenario					
		1: Average Initial BW, Hectares		2: Mid-Season BW, Hectares		3: Individual Precision BW, Hectares	
Case Study 1: Replacement heifers	–	265.77		329.17		334.07	
Case Study 2: Steers		2021	2022	2021	2022	2021	2022
	1	52.21	44.60	50.85	49.04	50.85	44.43
	2	52.73	48.78	56.14	48.78	52.89	41.74
	3	53.80	44.21	57.08	49.32	54.60	54.80
	4	51.70	42.08	53.89	48.97	48.55	42.02
	5	54.90	44.28	58.83	48.97	46.58	34.91
	6	55.61	46.27	59.83	50.85	56.33	42.96

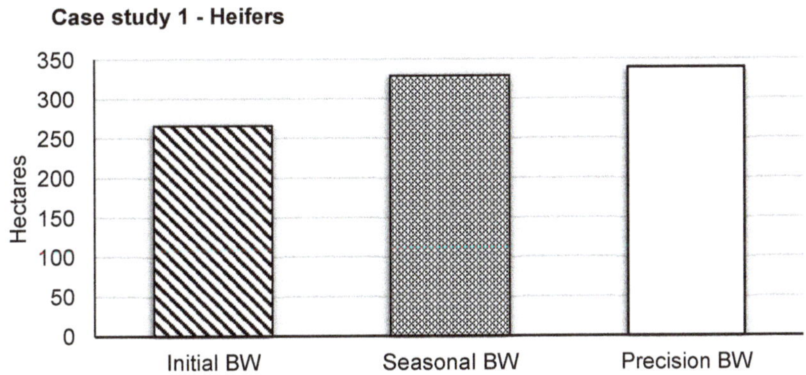

Figure 2. Simulated area (ha) to graze 60 heifers for 167 d based on body weight (BW) dry matter intake estimation in Scenario 1 (initial herd BW), Scenario 2 (herd mid-season BW), and Scenario 3 (precision BW).

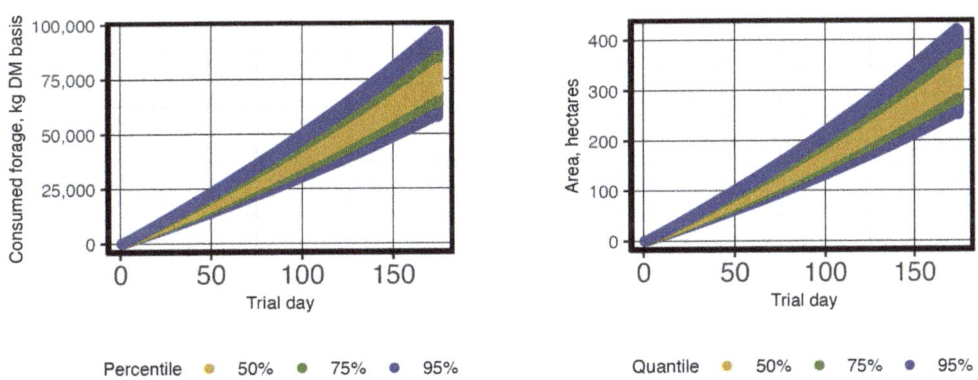

Figure 3. The percentiles of potential variation of daily total forage consumed (kg DM basis) from changes to % body weight (BW) (**left** panel) and required pasture size (**right** panel) over a 167 d period for replacement heifers. Initial animal unit equivalents were 0.54 based on average herd BW.

3.2. Case Study 2

Average initial and mid-season BW are reported in Table 3, with an average initial BW of 293 ± 49.6 and 369 ± 37.1 kg across six pastures for the year 2021 and 2022, respectively. Study length was 68 and 76 d for trial year 1 and 2, respectively; however, due to equipment failures and poor visitation rates, 8 and 16 d were removed from the end of each trial period to obtain a 60 d period for the analysis of each. The average difference was 4.4 ha per pasture between Scenarios 3 and 1; however, using estimated mid-season BW only resulted in a 0.9 ha difference between Scenarios 3 and 2 (Figure 4; Table 4). Sensitivity of average DMI due to variation in %BW ranged from 8530 to 12,795 and 7182 to 10,774 kg per pasture on trial day 60 for steer trials 1 and 2, respectively. This directly resulted in the range of total required ha per pasture of 17.6 to 26.3 and 17.7 to 26.6 ha on day 60 (Figure 5) for steer trials 1 and 2, respectively.

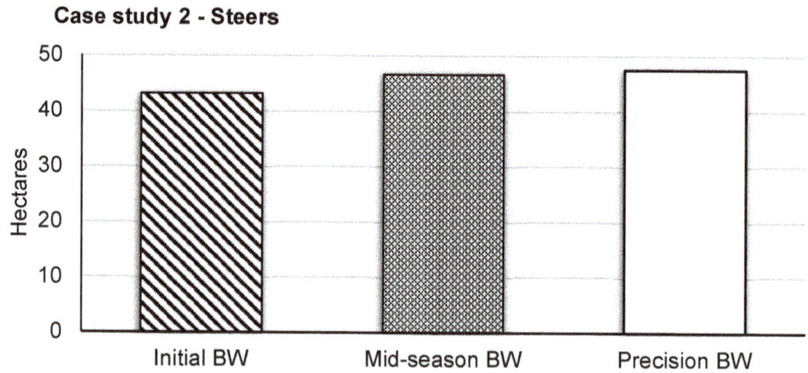

Figure 4. Simulated area (ha) to graze steers (average of 2021 and 2022 BWs; n = 254) in a typical pasture containing ~20 steers) for 60 d based on body weight (BW) dry matter intake estimation in Scenario 1 (initial herd BW), Scenario 2 (herd mid-season BW), and Scenario 3 (precision BW). Overall required grazing ha were calculated for each individual animal (~20) within each group (n = 6) per year (2021, 2022), and all groups (n = 12, i.e., six per year) were averaged to represent the total variation captured by individual steers within both trials (n = 254).

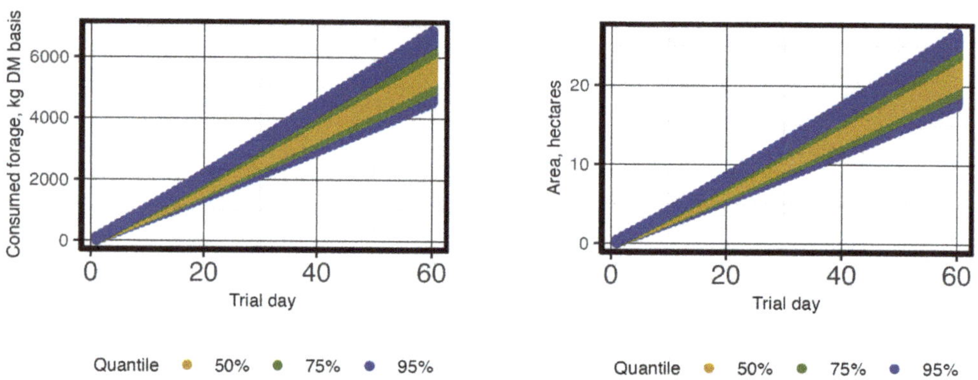

Figure 5. The percentiles of potential variation of daily total forage consumed (kg DM basis) from changes to % body weight (BW) (left panel) and required pasture size (right panel) over a 60 d period for steers. The sensitivity analysis represents the total number of steers (n = 254), similar to Figure 4 above. Initial animal unit equivalents ranged from 0.59 to 0.78 based on average herd initial BW of each pasture group (n = 6) of steers per each year (2021 and 2022; n = 2).

3.3. Considerations for Range Beef Cattle

As expected, considerable differences existed between the traditional AUM method and the precision-informed AUM_{PSM} method, caused by individual animal BW. Differences between Scenarios 1 and 3 were amplified over time as BW became increasingly influential to AUM estimates. This indicates that obtaining initial BW before turn-out for grazing is a critical management factor and that precision weighing can aid both traditional and precision AUM estimates [23]. A mid-season average (Scenario 2) captured similar results as the precision-informed AUM estimate (e.g., ~3% difference for heifers). However, this agreement was found using expected daily gain based on a well-developed nutrition plan (grazing and supplement) and from a sample population that was culled to represent a similar weight range. It is unlikely that such cattle uniformity exists for the typical rancher; therefore, we would expect more variability. Thus, an opportunity exists to evaluate how many ranchers use initial or mid-season BW and the different qualities of BW data available. It is essential to evaluate the usefulness of tools like precision weighing for effective adoption and avoidance of technologies that do not yield returns.

Precision system modeling focuses on identifying high-leverage precision measurement or management tools to minimize a performance gap [5]. Grazing models like APEX use growth functions based on mean animal weights [24]. Although models, like APEX, can incorporate probabilistic functions, the variation applied is random for each animal at any given time-point, which does not reflect grazing behavior or environmental responses affecting performance, while precision BW data is representative of such variations. Thus, the next step in PSM is to identify feedback mechanisms that more adequately represent individual BW variation. These mechanisms will likely include other data streams (e.g., climate data effects on heat/cold stress) to model this individual variation. Thus, PSM will help to further separate rangeland animals into different groups of efficiency/quality. For example, the variation of total forage consumed, and hectares needed caused by the sensitivity analysis of %BW for each heifer or steer (Figures 3 and 5) indicates that individual animal consumption rates (kg dry matter d^{-1}) are impactful. As more information is collected about feed-efficient grazing cattle, there is the potential to select animals with similar efficiency levels to minimize this variation.

Applying precision data and derived model coefficients provides more precise DMI estimates and shows the advantages and disadvantages of herd level estimates, especially in the context of changing cattle demands and forage nutrient composition and availability throughout the grazing season. For instance, unlike the data used in the current study, where forage availability was a fixed resource (i.e., dormant forage), most livestock grazing capitalizes on seasonal changes in forage availability and quality [25]. Thus, the next step is to account for seasonal changes of biomass availability and nutrient composition of the heifer and steer case studies instead of only using initial dormant forage availability or peak biomass estimates. Evaluating precision informed stocking rates for continuous versus rotational grazing is another opportunity for further research using a more complex version of the PSM_{AUM}. Precision BW estimates can help producers fine-tune stocking rates for future pastures (dynamic pasture adjustment using AUM_{PSM}). Further, virtual fencing has made these potential AUM_{PSM} informed rotations feasible because physical labor and infrastructure (e.g., water) requirements are minimized [5].

Grazing livestock research and production are focused on the "margins" like performance and costs, with the aim to maximize rangeland resources [11]. It has been estimated that 20–40% of US grasslands are overgrazed [26]. Heavy grazing has been shown to be equally as profitable as more lightly grazed systems in years with average precipitation; however, limitations in forage productivity and water infiltration are extreme during drought within this type of grazing management system. Precision system models help to answer complex questions regarding the impact of precision livestock technologies on marginal increases in animal productivity at a local and supply chain level [27]. While a comparison of Scenarios 1 and 3 indicates local ranch-level benefits, especially for extensive systems (Figures 2 and 4), viewing supply chain impacts supports efforts for marginal improvements at the local level (e.g., climate-smart commodities). For example, the current number of heifers in South Dakota (US) is 375,000 [28]. We extrapolated our results to a state level using the following Equation (1):

$$Hectares\ Overgrazed = \frac{Heifer_s}{Heifer_l} * Difference\ in\ Hectares, \quad (1)$$

where hectares overgrazed is the number of additional hectares required for a specific grazing period, $Heifers_s$ is the current number of heifers in the state, $Heifers_l$ is the number of Angus heifers used in the current study (n = 60), and the difference in hectares (Δ ha) is the difference between the modeled results of Scenarios 1 and 3 (Table 5).

Table 5. Extrapolation of Scenarios 1 and 3 to estimate hectares required for a 167 d period at a state level based on January 2023 state heifer numbers.

Scenario	ha	Δ ha	State ha Required	Area Overgrazed ha
1	265.55	-	1,659,687	-
3	338.96	73.41	2,118,500	458,812

It is important to note that the estimates provided in Table 5 are based on a specific grazing management system of "take-half-leave-half", which is designed to avoid overgrazing and promote plant regrowth. The current modeled example (Equation (1); Table 5) is based off static values and does not account for potential forage regrowth after dormancy during the spring. Therefore, the overgrazing of 458,812 acres is relative to the "take-half-leave-half" grazing strategy and should not be applied as an indication of potential rangeland degradation from overgrazing for other production settings without accounting for grazing management and potential plant regrowth details. Rather, this estimate of overgrazed areas is only to highlight potential differences that may occur at scale, and which may potentially be improved through the use of precision-derived stocking rate coefficients for different classes of grazing beef livestock.

4. Conclusions

Using technologies like SmartScales™ to provide precision informed stocking rate coefficients is likely to provide a high-leverage management opportunity for rangeland managers. As data increase for animal classes and grazing periods (e.g., winter, summer), more precise stocking estimates can be obtained to maximize pasture use relative to production goals (harvest efficiency) [29]. However, research derived precision AUM coefficients may be sufficient for ranch managers as the implementation of scale-based precision technology is limited by cost and infrastructure within extensive systems (i.e., precision weighing is a means to an end). More investigation is needed for non-growing animal classes such as mature cows that are likely to have much less variation in DMI over the grazing season. Practical steps are needed to evaluate how precision data can be integrated into PSM to guide precision data collection efforts [30]. The next step is using precision-informed forage nutrient composition data from forecast or remote sensing (e.g., near real-time forage production and nutrient composition estimates) and DMI equations that include net energy for maintenance [2]. For example, in some years, grass may come out of dormancy before May, depending on the climate, plant root storage, and soil moisture conditions, which could alter available biomass and its nutrient composition [31] (e.g., increased protein and lower fiber, which was not the case for our heifer case study). Additionally, using PSM to evaluate local-ranch- and supply-chain-level impacts will provide quantitative justification for specific precision livestock tool use or future development.

Author Contributions: Conceptualization, writing, and analysis H.M.M.III; writing and data curation J.R.B.; writing, editing and conceptualization K.A.E.; writing, editing, and analysis I.L.P. All authors have read and agreed to the published version of the manuscript.

Funding: This research was made possible with an equipment grant from C-Lock Inc. (Rapid City, SD, USA), a donation of DDGS pellets from POET Nutrition (Sioux Falls, SD, USA), USDA Hatch grant (#SD00H724-21; SD00H729-21), and the Inter-Disciplinary Engagement in Animal Systems (IDEAS) program (2022-69014-36670) from the USDA National Institute of Food and Agriculture (NIFA) and the SDSU Agriculture Experimental Station.

Institutional Review Board Statement: The SDSU Institutional Animal Care and Use Committee approved all procedures involving animals approval #2109-054E and #2104-021E, approved on 7 October 2021 and 15 April 2021).

Informed Consent Statement: Not applicable.

Data Availability Statement: The data presented in this study are available on request from the corresponding author.

Acknowledgments: The authors would like to thank Logan Vandermark, Lily McFadden, and Anna Dagel for their assistance in data collection and Roger N. Gates for evaluating the initial concept.

Conflicts of Interest: The authors declare no conflict of interest.

References

1. Smart, A.; Walker, J.; Gates, R.N. Setting the Stocking Rate. In *South Dakota State University Extension Beef*; Smart, A., Walker, J., Eds.; South Dakota State University Extension: Brookings, SD, USA, 2020; pp. 1–6.
2. Galyean, M.L.; Beauchemin, K.A.; Caton, J.; Eisemann, J.H.; Engle, T.; Erickson, G.E.; Krehbiel, C.R.; Lemenager, R.P.; Tedeschi, L.O. *Nutrient Requirements of Beef Cattle*, 8th ed.; The National Academies Press: Washington, DC, USA, 2016; pp. 171–186.
3. South Dakota State University Extension: Grazing Calculator. Available online: https://extension.sdstate.edu/grazing-calculator (accessed on 24 October 2022).
4. Martin, S.C.; Barnes, K.K.; Bashford, L. A step toward automatic weighing of range cattle. *J. Range Manag.* **1967**, *20*, 4. [CrossRef]
5. Adams, D.C.; Currie, P.O.; Knapp, B.W.; Mauney, T.; Richardson, D. An automated range-animal data acquisition system. *J. Range Manag.* **1987**, *40*, 256. [CrossRef]
6. MacNeil, M.D.; Berry, D.P.; Clark, S.A.; Crowley, J.J.; Scholtz, M.M. Evaluation of partial body weight for predicting body weight and average daily gain in growing beef cattle. *Trans. Anim. Sci.* **2021**, *5*, txab126. [CrossRef] [PubMed]
7. Wells, R.S.; Interrante, S.M.; Sakkuma, S.S.; Walker, R.S.; Butler, T.J. Accuracy of the VYTELLE SENSE in-pen weighing positions. *Appl. Animal Sci.* **2021**, *37*, 626–634. [CrossRef]
8. Xiong, Y.; Condotta, I.C.; Musgrave, J.A.; Brown-Brandl, T.M.; Mulliniks, J.T. Estimating body weight and body condition score of mature beef cows using depth images. *Transl. Anim. Sci.* **2023**, *7*, txad085. [CrossRef] [PubMed]
9. Currie, P.O.; Volesky, J.D.; Adams, D.C.; Knapp, B.W. Growth patterns of yearling steers determined from daily live weights. *J. Range Manag.* **1989**, *42*, 393–396. [CrossRef]
10. Menendez, H.M., III; Brennan, J.R.; Gaillard, C.; Ehlert, K.; Quintana, J.; Neethirajan, S.; Remus, A.; Jacobs, M.; Teixeira, I.A.M.A.; Turner, B.L.; et al. ASAS–NANP symposium: Mathematical modeling in animal nutrition: Opportunities and challenges of confined and extensive precision livestock production. *J. Anim. Sci.* **2022**, *100*, skac160. [CrossRef] [PubMed]
11. Dunn, B.H.; Smart, A.J.; Gates, R.N.; Johnson, P.S.; Beutler, M.K.; Diersen, M.A.; Janssen, L.L. Long-term production and profitability from grazing cattle in the northern mixed grass prairie. *Rangel. Ecol. Manag.* **2010**, *63*, 233–242. [CrossRef]
12. Beck, H.; Zimmermann, N.; McVicar, T.; Vergopolan, N.; Berg, A.; Wood, E.F. Present and future Köppen-Geiger climate classification maps at 1-km resolution. *Sci Data* **2018**, *5*, 180214. [CrossRef] [PubMed]
13. NOAA. National Oceanic and Atmospheric Administration National Centers for Environmental Information. 2023. Available online: https://www.ncei.noaa.gov/access/monitoring/climate-at-a-glance/divisional/time-series/3905/pcp/all/10/1992-2022normals (accessed on 15 November 2023).
14. Tedeschi, L.O.; Fox, D.G. *The Ruminant Nutrition System: Volume I—An Applied Model for Predicting Nutrient Requirements and Feed Utilization in Ruminants*, 3rd ed.; XanEdu: Ann Arbor, MI, USA, 2020; pp. 1–7.
15. Ford, A. *Modeling the Environment: An Introduction to System Dynamics Models of Environmental Systems*, 2nd ed.; Island Press: Washington, DC, USA, 2010; pp. 3–17.
16. Sterman, J.D. *Business Dynamics: Systems Thinking and Modeling for a Complex World*; Irwin McGraw-Hill: New York, NY, USA, 2000; pp. 1–134.
17. Thornley, J.H.M. *Grassland Dynamics: An Ecosystem Simulation Model*; CAB International: Cambridge, MA, USA, 2001; pp. 1–9, 65–70.
18. R Core Team. R: A Language and Environment for Statistical Computing. In *R Foundation for Statistical Computing*; R Core Team: Vienna, Austria, 2023; Available online: https://www.r-project.org/ (accessed on 7 July 2023).
19. Dagel, A. Evaluating Precision Supplementation Technology and Mathematical Nutrition Models for the Development of Heifers Grazing Dormant Rangeland in Western South Dakota. Master's Thesis, South Dakota State University, Brookings, SD, USA, 21 July 2023.
20. Vandermark, L.R. Impact of Virtual Fence Technology on Yearling Steer Behavior, Performance, and Energetic Expenditure. Master's Thesis, South Dakota State University, Brookings, SD, USA, 5 May 2023.
21. Data.Table: Extension of 'Data.Frame'. Available online: https://cran.r-project.org/web/packages/data.table/index.html (accessed on 1 August 2023).
22. Parsons, I.L.; Norman, D.A.; Karisch, B.B.; Webb, S.L.; Stone, A.E.; Proctor, M.D.; Street, G.M. Automated walk-over-weigh system to track daily body mass and growth in grazing steers. *Comput. Electron. Agric.* **2023**, *212*, 108113. [CrossRef]
23. González-García, E.; Golini, P.D.O.; Hassoun, P.; Bocquier, F.; Hazard, D.; González, L.A.; Ingham, A.B.; Bishop-Hurley, G.J.; Greenwood, P.L. An assessment of walkover-weighing to estimate short-term individual forage intake in sheep. *Animal* **2018**, *12*, 1174–1181. [CrossRef] [PubMed]

24. Fang, Q.X.; Harmel, R.D.; Ma, L.; Bartling, P.N.S.; Derner, J.D.; Jeong, J.; Williams, J.R.; Boone, R.B. Evaluating the APEX model for alternative cow-calf grazing management strategies in Central Texas. *Agric. Syst.* **2022**, *195*, 103287. [CrossRef]
25. Jansen, V.S.; Kolden, C.A.; Schmalz, H.J.; Karl, J.W.; Taylor, R.V. Using satellite based vegetation data for short-term grazing monitoring to inform adaptive management. *Rangel. Ecol. Manag.* **2021**, *76*, 30–42. [CrossRef]
26. Piipponen, J.; Jalava, M.; Leeuw, J.; Rizayeva, A.; Godde, C.; Cramer, G.; Herrero, M.; Kummu, M. Global trends in grassland carrying capacity and relative stocking density of livestock. *Glob. Change Biol.* **2022**, *28*, 3902–3919. [CrossRef]
27. Menendez, H.M.; Tedeschi, L.O. The characterization of the cow-calf, stocker and feedlot cattle industry water footprint to assess the impact of livestock water use sustainability. *J. Agric. Sci.* **2020**, *158*, 416–430. [CrossRef]
28. United States Department of Agriculture National Agricultural Statistical Service: South Dakota Heifer Inventory. Available online: https://quickstats.nass.usda.gov/results/BCFA9CC3-C24C-34D7-BC34-541D1275C3EA (accessed on 1 December 2022).
29. Smart, A.J.; Derner, J.D.; Hendrickson, J.R.; Gillen, R.L.; Dunn, B.H.; Mousel, E.M.; Johnson, P.S.; Gates, R.N.; Sedivec, K.K.; Harmoney, K.R.; et al. Effects of grazing pressure on efficiency of grazing on North American Great Plains rangelands. *Rangel. Ecol. Manag.* **2010**, *63*, 397–406. [CrossRef]
30. Brennan, J.R.; Menendez, H.M., III; Ehlert, K.; Tedeschi, L.O. ASAS-NANP symposium: Mathematical modeling in animal nutrition—Making sense of big data and machine learning: How open-source code can advance training of animal scientists. *J. Anim. Sci.* **2023**, *101*, 1–11. [CrossRef] [PubMed]
31. Díaz-Solís, H.; Grant, W.E.; Kothmann, M.M.; Teague, W.R.; Díaz-García, J.A. Adaptive management of stocking rates to reduce effects of drought on cow-calf production systems in semi-arid rangelands. *Agric. Syst.* **2009**, *100*, 43–50.

animals MDPI

Communication

In-Line Detection of Clinical Mastitis by Identifying Clots in Milk Using Images and a Neural Network Approach

Glenn Van Steenkiste *, Igor Van Den Brulle, Sofie Piepers and Sarne De Vliegher

Department of Internal Medicine, Reproduction and Population Medicine, Faculty of Veterinary Medicine, Ghent University, 9820 Merelbeke, Belgium; sofie.piepers@ugent.be (S.P.); sarne.devliegher@ugent.be (S.D.V.)
* Correspondence: glenn.vansteenkiste@ugent.be

Simple Summary: The study focused on improving the detection of clinical bovine mastitis, the inflammation of the udder in cows as a response to intramammary infection, which can be identified by the presence of clots in the milk. Currently, automated milking systems do not detect this important disease very accurately. To address this, we developed a clots detection program using a neural network. This neural network was trained to recognize clots in milk samples from dairy cows by using a large number of pictures of milk filter socks, some with and some without clots. These pictures were divided into different sets for training, validating, and testing the program, respectively. The settings of the neural network were optimized using a genetic algorithm. The program's interpretations were explained using a method called integrated gradients. The program was found to be 100% accurate in identifying clots in the test pictures. This suggests that the method could be very useful for automatically checking for clinical mastitis on dairy farms, although further field validation through integration into the existing systems is needed.

Abstract: Automated milking systems (AMSs) already incorporate a variety of milk monitoring and sensing equipment, but the sensitivity, specificity, and positive predictive value of clinical mastitis (CM) detection remain low. A typical symptom of CM is the presence of clots in the milk during fore-stripping. The objective of this study was the development and evaluation of a deep learning model with image recognition capabilities, specifically a convolutional neural network (NN), capable of detecting such clots on pictures of the milk filter socks of the milking system, after the phase in which the first streams of milk have been discarded. In total, 696 pictures were taken with clots and 586 pictures without. These were randomly divided into 60/20/20 training, validation, and testing datasets, respectively, for the training and validation of the NN. A convolutional NN with residual connections was trained, and the hyperparameters were optimized based on the validation dataset using a genetic algorithm. The integrated gradients were calculated to explain the interpretation of the NN. The accuracy of the NN on the testing dataset was 100%. The integrated gradients showed that the NN identified the clots. Further field validation through integration into AMS is necessary, but the proposed deep learning method is very promising for the inline detection of CM on AMS farms.

Keywords: deep learning; clinical mastitis; automated milking systems; image recognition; clots in milk

Citation: Van Steenkiste, G.; Van Den Brulle, I.; Piepers, S.; De Vliegher, S. In-Line Detection of Clinical Mastitis by Identifying Clots in Milk Using Images and a Neural Network Approach. *Animals* **2023**, *13*, 3783. https://doi.org/10.3390/ani13243783

Academic Editors: Brett Ramirez, Janice Siegford, Hao Gan, Yang Zhao, Daniel Berckmans, Robert T. Burns and Lingjuan Wang-Li

Received: 24 October 2023
Revised: 30 November 2023
Accepted: 6 December 2023
Published: 8 December 2023

1. Introduction

From an economic viewpoint, mastitis is one of the most important diseases in dairy cows [1–5], due to its effects on animal health and the subsequent losses in milk production, as well as the need to discard abnormal milk or milk from diseased cows (European Union Directive EC/853/2004 and US Food and Drug Administration Grade A pasteurized milk ordinance). Depending on the study, the cost of each clinical mastitis (CM) varies between USD65 and 930 [2–4]. The early detection of CM can reduce both the economic impact and the long-term impact on cow health and welfare [6,7].

Many dairy farms are transitioning to automated milking systems (AMSs), with around 38,000 units installed worldwide in 2017 [8]. When using AMSs, there are fewer opportunities for the farmer to detect CM in individual animals. The current AMS incorporates a variety of milk monitoring and sensing equipment, but the sensitivity and specificity of its CM detection capabilities remain relatively low, with most systems having a sensitivity between 47 and 90% and a specificity between 56 and 99% [9,10]. For reference, the International Standards Organization (ISO) describes a standard target of 90% sensitivity and 99% specificity for the detection of abnormal milk (ISO/FDIS (Final Draft International Standard) 20966 [11]), Annex C (Automatic Milking Installations—Requirements and Testing). Most of these sensor systems try to detect mastitis by measuring and analyzing indirect parameters, such as (but not limited to) electrical conductivity, somatic cell count, milk flow rate, changes in milk color, milk yield per hour or quarter, and cow activity [7,10,12–15].

A typical symptom of CM is the presence of clots in the milk during pre-milking, which has been proposed as the gold standard for the detection of CM [16,17]. Therefore, we propose to use an in-line camera to detect such clots in the filter after the pre-milking phase. A similar sensor has been proposed in the past but was limited in its capabilities of detecting the clots on the filter, as it was developed to score the quality of the milk and needed to be adapted for instances of different detriments occurring in the milk [18]. Another study proposed to measure clot density in quarter milk samples, which could be useful in monitoring milk quality and clinical mastitis [19]. The researchers used in-line filters to collect quarter milk samples and visually scored the clot density, based on the coverage of the filter area. They showed that high scores clustered within certain cows and periods, suggesting a potential threshold for detecting abnormal milk. The objective of the present study was the development and evaluation of a neural network (NN) capable of detecting such clots on pictures of the filters of the milking system after the pre- and/or milking phase at the cow level.

2. Materials and Methods

2.1. Experimental Data

The data for this study were generated by adding debris (including straw, hay, manure, bedding material, mud, teat sealer, calcium, and/or flies) and/or clots from used milk filters of AMSs to milk, before passing this milk through a circular milk filter (Universal Hygia Favorit filters, Universal dairy equipment) mounted in a PVC tube. Debris and clots were collected from 40 filters, half of which had clots and half of which did not. These samples were gathered from multiple AMSs and various cows. A vacuum pump provided suction for pulling the milk through the filter. The filters were painted blue for better visualization of the clots. An iPhone 6s was mounted in the PVC pipe to take a photo with the flashlight after each pass of milk. In total, 696 pictures were taken from filters with clots, and 586 pictures from filters without clots.

2.2. Image Analysis

For the training dataset, the images without clots were randomly resampled using the built-in python random.uniform function to obtain an even number of images with and without clots for balancing the NN weights. In total, 1676 images were used for training, validating, and testing. During the training of the NN, the images were augmented by random rotations, flipping, rescaling, zooming, and shearing (Table 1), using the Keras ImageDataGenerator function.

Table 1. Image augmentation parameters used by the Keras ImageDataGenerator function during training of the neural network.

Augmentation Type	Value
Random rotation	60°
Random shift in width	10%
Random shift in height	10%
Random zoom	100–130%
Horizontal flip probability	50%
Vertical flip probability	50%
Shear range	20°
Fill mode	nearest

To avoid the NN learning features from outside of the filter, e.g., milk spatters on the PVC pipe, the PVC pipe was removed from the picture using OpenCV v4.1. The image was then rescaled to 500 × 500 pixels as the NN input (Figure 1).

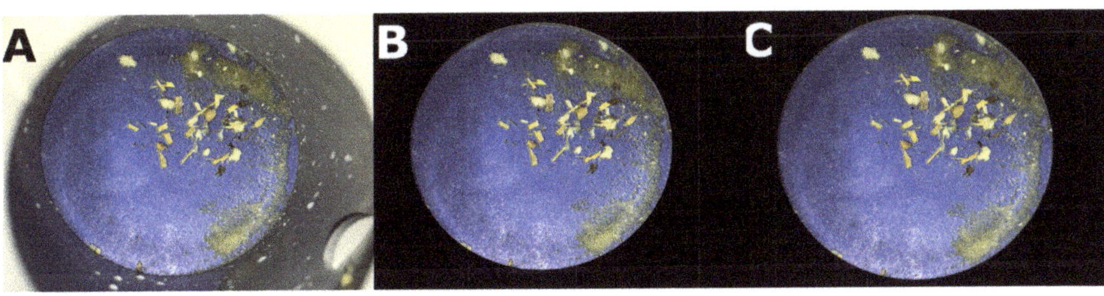

Figure 1. Image pre-processing steps on filter image after the passage of milk with clots. Panel (**A**): original image taken with a resolution of 4032 × 3034 pixels. Panel (**B**): image after applying a black mask on the region of the PVC pipe. Panel (**C**): resulting image used for the neural network after rescaling to 500 × 500 pixels.

A genetic algorithm was used to optimize the hyperparameters of the NN (Figure 2) based on the validation dataset. A non-dominated sorting genetic algorithm II (NSGA-II) was used, with a population size of 20 and 10 optimization generations, using the accuracy as the fitness value [20]. The training of each child NN of the algorithm was ended after 50 epochs or when the training stopped improving for three epochs. Optimization was used for the following hyperparameters: the number of filters, the width of convolution and subsampling for each convolutional layer, the number of neurons for each fully connected layer, L2 regularization, and the dropout used for training (Table 2).

For training, the SoftMax cross entropy was used to calculate the loss with the Adam optimizer and with the default parameters at a learning rate of 0.0001 for updating the weights of the network [21]. The network was trained using parameters optimized by the genetic algorithm, with each epoch evaluated using a validation dataset. This process was repeated for 100 epochs, selecting the best network weights based on validation results, to avoid overfitting the training dataset. The testing dataset was employed for statistical analysis. The network was built in Keras with a TensorFlow v2.0.1 backend [22]. The batch size was set to 16. Afterwards, the integrated gradients of the NN were calculated and compared to a completely black baseline image to obtain an insight into the input–output behavior of the neural network. The attribution of the input pixels to the output labels was projected as a mask over the input image using the OpenCV toolbox (Figure 3).

Table 2. Hyperparameters selected by the genetic algorithm after 10 optimization generations. For the convolutional layers, the first number indicates the residual block and the second number indicates the convolutional layer within the block. Abbreviations: Conv: convolutional layer.

Layer	Hyper Parameter	Value
All	Dropout	0.2
Conv 1.1	Filters	8
	Kernel size	4
	Subsampling	2
Conv 1.2	Filters	8
	Kernel size	1
	Subsampling	1
Conv 2.1	Filters	16
	Kernel size	4
	Subsampling	1
Conv 2.2	Filters	16
	Kernel size	1
	Subsampling	1
Conv 3.1	Filters	24
	Kernel size	4
	Subsampling	1
Conv 3.2	Filters	24
	Kernel size	1
	Subsampling	1
Conv 4.1	Filters	32
	Kernel size	4
	Subsampling	1
Conv 4.2	Filters	32
	Kernel size	1
	Subsampling	1
Conv 5.1	Filters	40
	Kernel size	4
	Subsampling	1
Conv 5.2	Filters	40
	Kernel size	1
	Subsampling	1
Conv 6.1	Filters	48
	Kernel size	4
	Subsampling	1
Conv 6.2	Filters	48
	Kernel size	1
	Subsampling	1
Conv 7.1	Filters	48
	Kernel size	4
	Subsampling	2
Conv 7.2	Filters	48
	Kernel size	1
	Subsampling	1
Conv 8.1	Filters	48
	Kernel size	4
	Subsampling	1
Conv 8.2	Filters	48
	Kernel size	1
	Subsampling	1
Dense 1	Number of neurons	64
	L2 regularization	0.1
Dense 2	Number of neurons	32
	L2 regularization	0.01

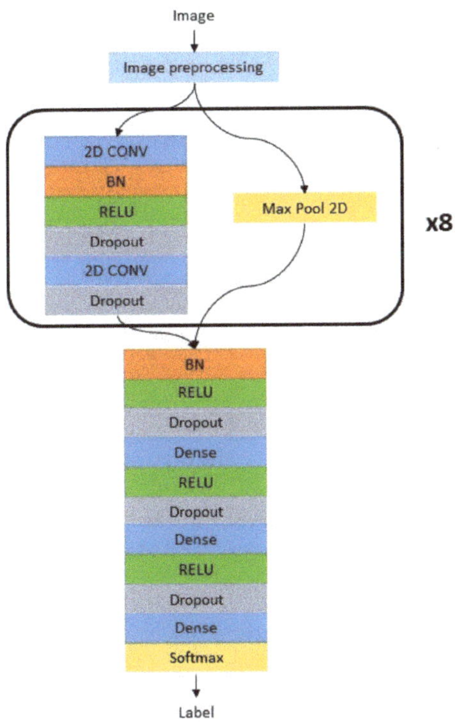

Figure 2. High-level architecture of the neural network. CONV: convolutional block; BN: batch normalization; RELU: rectified linear activation unit; Max Pool 2D: Max pooling operation; Dense: fully connected layer; Softmax: Softmax activation block with the two different output classes.

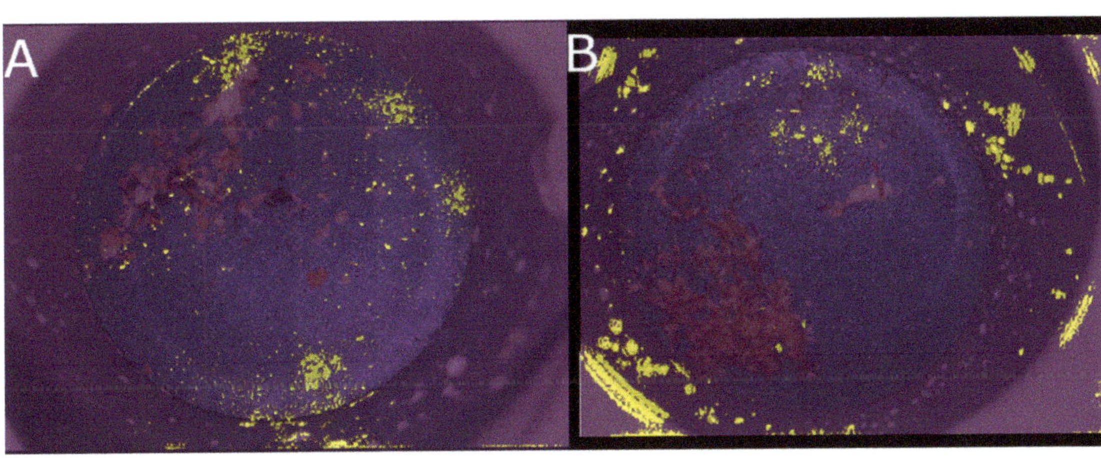

Figure 3. Visualization of the integrated gradients by an attribution mask over the original input image. Yellow indicates a high attribution of the indicated pixels to the output label of the neural network (NN). Panel (**A**) shows the integrated gradients of the currently used NN. Here the pixels (around) the clots attribute the most to the output label. Panel (**B**) shows the integrated gradients of a NN which was trained on images where the PVC pipe was not removed from the image. This 'cheating' NN used the milk spats on the PVC pipe to identify if this was an image with clots.

2.3. Statistical Analysis

The dataset was randomly divided into a training subset containing 60% of the data, a validation dataset containing 20%, and a holdout subset containing 20%, which was not used for tuning the model; this resulted in 1006 training, 335 validation, and 335 holdout images. The following metrics were calculated on the holdout dataset: accuracy, positive and negative predictive values, specificity, and sensitivity.

3. Results

The accuracy, specificity, positive and negative predictive values, and sensitivity results of the NN on the testing dataset were 100% (Table 3). The integrated gradients showed that the NN identified the clots and accurately distinguished the clots from other materials including straw, hay, manure, bedding material, mud, teat sealer, calcium, and flies.

Table 3. Coherence matrix of results.

	Predicted Negative	Predicted Positive
True negative	182	0
True positive	0	153

4. Discussion

The accuracy, positive- and negative predictive value, specificity, and sensitivity of the NN on the holdout dataset are a clear improvement, in comparison with other milk-based CM detection methods. Since the proposed method can reliably detect clots in foremilk, this method could be a feasible approach to detect CM in AMSs. Since clots during foremilking are considered the gold standard for detecting CM, the current approach has the potential to be a more reliable CM detection implementation for AMSs, in comparison with current CM detection sensors, which try to detect CM with other parameters, including electrical conductivity, L-lactate dehydrogenase, milk color, and somatic cell counting [9,16]. Many different sensors and algorithms have been proposed for the detection of CM, but, thus far, none of the published CM detection methods achieved the ISO target of at least 80% sensitivity and 99% specificity [23].

One of the main benefits of the current approach is the extremely high accuracy. The main frustration of dairy farmers is the current high number of false alarms made by the available CM detection sensors in AMSs [24]. Due to the low prevalence of (severe) CM, the majority of alerts will indeed be false positives, leading to a potential underreporting of CM cases, as farmers may stop investigating all alerts [23–25]. With the proposed sensor, the detection and management of severe mastitis on AMS farms could be significantly improved, reducing the number of false-positives and ensuring that all cases of severe CM are accurately identified and treated and that milk is separated. Even if the practical implementation of the current sensor would not have an accuracy/precision of 100%, a NN, as we have proposed, can be adapted in order to maximize the accuracy, by penalizing false positive results during the training process or by calculating the receiver operating characteristic curve and setting a manual threshold for the minimal required specificity [26]. An additional benefit of using a NN is their robustness for the presence of a variety of detriments (such as straw, manure, udder or tail hair, sawdust, sand, and the remainders of internal teat sealants) on the image, without the need to retrain the algorithm for every possible detriment. This is in contrast with the previous proposed work, in which the fuzzy logic algorithm had to be adapted to recognize the different detriments [18]. If environmental changes (e.g., change in filter type) would overwhelm the robustness of the NN, the weights of the NN could be updated on-site using transfer learning to adapt to the new environment [27]. If the farmer receives multiple false positive results from the sensor, he could initiate an update of the algorithm remotely based on the incorrectly classified images without the need to re-engineer the algorithm. In addition, the proposed NN approach could also use an optional reference image to differentiate and track animals with varying mastitis cases. The algorithm identifies changes in clots by comparing new images with the reference image, which would not be possible with the fuzzy logic algorithm.

If a 3D representation of the filter could be created, e.g., by adding a second camera for 3D stereovision, the NN could also be adapted to calculate the volume of the clots. Calculation of the animal's clot volume allows us to monitor diseased animals with CM over time and to estimate the severity of the disease and clinical recovery, and, hypothetically, even the likelihood of a bacteriological cure. For example, if the clot volume of a diseased animal is decreasing between consecutive milkings, the animal is likely to be recovering. If few clots are present, increasing milking frequency may suffice as treatment, reducing antibiotic use. On the other hand, treatment can be necessary when many clots in milk are detected. Monitoring the dynamics, as well as the gradual increase in the densities, of the clots on the filter over a period of time could also be a valuable tool to identify cows with chronic mastitis [19].

The integrated gradients showed that the NN identified the region of the clots as an input feature (Figure 3). Neural networks are notorious for having a "black box" approach. It is difficult to attribute the prediction of an NN to its input features and, thus, to know why an NN tells us which pixels of an image are responsible for picking a certain label [28]. The aim of explainable artificial intelligence (AI) is to understand the input–output behavior of NNs. One such explainable AI method is the integrated gradients approach, in which the attribution of each pixel is calculated by summing the gradients (a partial derivative of each variable, while all others are held constant) of the network on different points at the path between a baseline image (e.g., a black image) to the actual input image (e.g., the image of the filter with clots). If the PVC pipe had not been removed from the input images, the integrated gradients would have clearly shown that the NN had learned to recognize the different milk spatter patterns instead of recognizing the clots.

The current study encountered limitations due to the relatively small dataset, which lacked diversity. Although image augmentation techniques were applied to introduce variability, such approaches do not compare to the larger, more complex datasets typically utilized in deep learning studies [29]. Furthermore, the consistency of the recording setup throughout the study, i.e., exclusively using an iPhone's flash for illumination, has left the model's robustness to alternate lighting conditions untested; this is a notable concern since, in field conditions, lighting can be inconsistent and obstructed. Consequently, the findings presented here should be interpreted as preliminary, serving as an exploratory investigation into the application of deep learning for mastitis detection. It is recommended that future research be conducted with more extensive datasets gathered from field conditions in AMSs, to thoroughly evaluate the model's performance and practicality.

While our proposed milk sensor can greatly enhance the detection and management of severe mastitis on AMS farms, it is important to remember that, due to the sudden onset of severe CM, sensor information based solely on changes in milk and measurements collected during milking may not be sufficient for all cases [23]. Cows with severe CM may not visit the AMS. Therefore, a combination of several sensor-based (including activity sensors) and AMS-based indicators may have to be incorporated to meet the necessary demands. It is worth noting that other proposed methods already work at the quarter level [10], and by incorporating additional filters and cameras into the AMS at the quarter, we could potentially enhance the performance of our proposed detection system to also achieve this level of monitoring. The performance of a sensor-based detection system may also be enhanced by the combination of sensor-based or automatic milking-based monitoring systems with additional monitoring strategies, such as visual observations. Therefore, while our sensor offers significant advancements, it should be used in conjunction with other tools and strategies for optimal results.

5. Conclusions

The current paper proposed an inline CM detection sensor for AMSs. Further validation and integration on farms with AMSs are necessary, but the proposed method appears to be very promising for the accurate inline detection of CM on AMS farms.

Author Contributions: Conceptualization, I.V.D.B. and S.P.; data curation, G.V.S.; formal analysis, G.V.S.; methodology, G.V.S. and I.V.D.B.; software, G.V.S.; validation, G.V.S.; writing—original draft, G.V.S.; writing—review and editing, G.V.S., I.V.D.B., S.P. and S.D.V. All authors have read and agreed to the published version of the manuscript.

Funding: This research received no external funding.

Institutional Review Board Statement: Not applicable.

Informed Consent Statement: Not applicable.

Data Availability Statement: Data are available upon reasonable request to the corresponding author.

Conflicts of Interest: The authors declare no conflict of interest.

References

1. Halasa, T.; Huijps, K.; Østerås, O.; Hogeveen, H. Economic Effects of Bovine Mastitis and Mastitis Management: A Review. *Vet. Q.* **2007**, *29*, 18–31. [CrossRef] [PubMed]
2. Ibrahim, N. Review on Mastitis and Its Economic Effect. *Can. J. Res.* **2017**, *6*, 13–22. [CrossRef]
3. Getaneh, A.M.; Mekonnen, S.A.; Hogeveen, H. Stochastic Bio—Economic Modeling of Mastitis in Ethiopian Dairy Farms. *Prev. Vet. Med.* **2017**, *138*, 94–103. [CrossRef] [PubMed]
4. Azooz, M.F.; El-Wakeel, S.A.; Yousef, H.M. Financial and Economic Analyses of the Impact of Cattle Mastitis on the Profitability of Egyptian Dairy Farms. *Vet. World* **2020**, *13*, 1750. [CrossRef] [PubMed]
5. Hogeveen, H.; Van Der Voort, M. Assessing the Economic Impact of an Endemic Disease: The Case of Mastitis. *Rev. Sci. Et Tech. Off. Int. Des Epizoot.* **2017**, *36*, 217–226. [CrossRef] [PubMed]
6. Heikkilä, A.M.; Nousiainen, J.I.; Pyörälä, S. Costs of Clinical Mastitis with Special Reference to Premature Culling. *J. Dairy Sci.* **2012**, *95*, 139–150. [CrossRef] [PubMed]
7. Willits, S.; Infrared, R. Infrared Thermography for Screening and Early Detection of Mastitis Infections in Working Dairy Herds. *InfraMation Proc. ITC* **2005**, *42*, 1–5.
8. Sandgren, C.H.; Emanuelson, U. Is There an Ideal Automatic Milking System Cow and How Is She Different from an Ideal Parlor Milked Cow? In Proceedings of the National Mastitis Council 56th Annual Meeting, St. Pete Beach, FL, USA, 28 January 2017; pp. 61–68.
9. Penry, J.F. Mastitis Control in Automatic Milking Systems. *Vet. Clin. North Am. Food Anim. Pract.* **2018**, *34*, 439–456. [CrossRef]
10. Khatun, M.; Thomson, P.C.; Kerrisk, K.L.; Lyons, N.A.; Clark, C.E.F.; Molfino, J.; García, S.C. Development of a New Clinical Mastitis Detection Method for Automatic Milking Systems. *J. Dairy Sci.* **2018**, *101*, 9385–9395. [CrossRef]
11. *ISO 20966:2007*; Automatic Milking Installations—Requirements and Testing. ISO: Geneva, Switzerland, 2007.
12. Fogsgaard, K.K.; Bennedsgaard, T.W.; Herskin, M.S. Behavioral Changes in Freestall-Housed Dairy Cows with Naturally Occurring Clinical Mastitis. *J. Dairy Sci.* **2015**, *98*, 1730–1738. [CrossRef]
13. de Mol, R.M.; Ouweltjes, W. Detection Model for Mastitis in Cows Milked in an Automatic Milking System. *Prev. Vet. Med.* **2001**, *49*, 71–82. [CrossRef] [PubMed]
14. Naqvi, S.A.; King, M.T.M.; Matson, R.D.; DeVries, T.J.; Deardon, R.; Barkema, H.W. Mastitis Detection with Recurrent Neural Networks in Farms Using Automated Milking Systems. *Comput. Electron. Agric.* **2022**, *192*, 106618. [CrossRef]
15. Adkins, P.R.F.; Middleton, J.R. Methods for Diagnosing Mastitis. *Vet. Clin. North Am. Food Anim. Pract.* **2018**, *34*, 479–491. [CrossRef] [PubMed]
16. Mein, G.A.; Rasmussen, M.D. Performance Evaluation of Systems for Automated Monitoring of Udder Health: Would the Real Gold Standard Please Stand Up? Brill Wageningen Academic: Paderborn, Germany, 2008; ISBN 9789086860852.
17. Claycomb, R.W.; Johnstone, P.T.; Mein, G.A.; Sherlock, R.A. An Automated In-Line Clinical Mastitis Detection System Using Measurement of Conductivity from Foremilk of Individual Udder Quarters. *New Zealand Vet. J.* **2009**, *57*, 208–214. [CrossRef] [PubMed]
18. Wiethoff, M.; Suhr, O. Method and Device for Determining the Quality of Milk Produced by Machine Milking. US Patent, US8261597B2 2007.
19. Hallén Sandgren, C.; Anglart, D.; Klaas, I.C.; Rönnegård, L.; Emanuelson, U. Homogeneity Density Scores of Quarter Milk in Automatic Milking Systems. *J. Dairy Sci.* **2021**, *104*, 10121–10130. [CrossRef] [PubMed]
20. Deb, K.; Pratap, A.; Agarwal, S.; Meyarivan, T. A Fast and Elitist Multiobjective Genetic Algorithm: NSGA-II. *IEEE Trans. Evol. Comput.* **2002**, *6*, 182–197. [CrossRef]

21. Kingma, D.P.; Ba, J. Adam: A Method for Stochastic Optimization. In Proceedings of the ICLR 2015, San Diego, CA, USA, 7–9 May 2015; pp. 1–15.
22. Chollet, F. Keras. Available online: https://keras.io (accessed on 22 September 2019).
23. Hogeveen, H.; Klaas, I.C.; Dalen, G.; Honig, H.; Zecconi, A.; Kelton, D.F.; Sánchez Mainar, M. Novel Ways to Use Sensor Data to Improve Mastitis Management. *J. Dairy Sci.* **2021**, *104*, 11317–11332. [CrossRef]
24. Mollenhorst, H.; Rijkaart, L.J.; Hogeveen, H. Mastitis Alert Preferences of Farmers Milking with Automatic Milking Systems. *J. Dairy Sci.* **2012**, *95*, 2523–2530. [CrossRef]
25. Deng, Z.; Koop, G.; Lam, T.J.G.M.; van der Lans, I.A.; Vernooij, J.C.M.; Hogeveen, H. Farm-Level Risk Factors for Bovine Mastitis in Dutch Automatic Milking Dairy Herds. *J. Dairy Sci.* **2019**, *102*, 4522–4535. [CrossRef]
26. Fawcett, T. An Introduction to ROC Analysis. *Pattern Recognit. Lett.* **2006**, *27*, 861–874. [CrossRef]
27. van Steenkiste, G.; van Loon, G.; Crevecoeur, G. Transfer Learning in ECG Classification from Human to Horse Using a Novel Parallel Neural Network Architecture. *Sci. Rep.* **2020**, *10*, 186. [CrossRef] [PubMed]
28. Sundararajan, M.; Taly, A.; Yan, Q. Axiomatic Attribution for Deep Networks. In Proceedings of the 34th International Conference on Machine Learning, ICML 2017, Sydney, NSW, Australia, 6–11 August 2017; Volume 7, pp. 5109–5118.
29. Razzak, M.I.; Naz, S.; Zaib, A. *Deep Learning for Medical Image Processing: Overview, Challenges and the Future*; Springer: Berlin/Heidelberg, Germany, 2018; Volume 26, pp. 323–350.

animals

MDPI

Article

Behavioral Adaptations of Nursing Brangus Cows to Virtual Fencing: Insights from a Training Deployment Phase

Shelemia Nyamuryekung'e [1,*], Andrew Cox [2], Andres Perea [2], Richard Estell [3], Andres F. Cibils [4], John P. Holland [5], Tony Waterhouse [5], Glenn Duff [2], Micah Funk [2], Matthew M. McIntosh [3], Sheri Spiegal [3], Brandon Bestelmeyer [3] and Santiago Utsumi [2,*]

[1] Division of Food Production and Society, Norwegian Institute of Bioeconomy Research (NIBIO), PB 115, N-1431 Ås, Norway
[2] Department of Animal and Range Sciences, New Mexico State University, Las Cruces, NM 88003, USA; arcox@nmsu.edu (A.C.); arperea@nmsu.edu (A.P.); glennd@nmsu.edu (G.D.); funkm@nmsu.edu (M.F.)
[3] United States Department of Agriculture-Agriculture Research Service, Jornada Experimental Range, Las Cruces, NM 88003, USA; rick.estell@usda.gov (R.E.); mattmac@nmsu.edu (M.M.M.); sheri.spiegal@usda.gov (S.S.); brandon.bestelmeyer@usda.gov (B.B.)
[4] United States Department of Agriculture Southern Plains Climate Hub, United States Department of Agriculture-Agriculture Research Service, Oklahoma and Central Plains Agricultural Research Center, El Reno, OK 73036, USA; andres.cibils@usda.gov
[5] SRUC Hill and Mountain Research Centre, Scotland's Rural College, Kirkton Farm, Crianlarich, Perthshire FK20 8RU, UK; john.holland@sruc.ac.uk (J.P.H.); tony.waterhouse@sruc.ac.uk (T.W.)
* Correspondence: shelemia@nibio.no (S.N.); sutsumi@nmsu.edu (S.U.); Tel.: +47-406-04-100 (S.N.); +1-575-646-2514 (S.U.)

Simple Summary: The study explores the use of virtual fencing technology for managing livestock distribution, focusing on nursing Brangus cows. The study investigates how these animals learn to avoid restricted areas and increase their reliance on auditory cues over time. The findings support the effectiveness of virtual fencing in controlling cow spatial behavior and highlight their ability to adapt to virtual boundaries rapidly. The research also presents a safe and efficient training protocol for implementing virtual fence systems.

Abstract: Virtual fencing systems have emerged as a promising technology for managing the distribution of livestock in extensive grazing environments. This study provides comprehensive documentation of the learning process involving two conditional behavioral mechanisms and the documentation of efficient, effective, and safe animal training for virtual fence applications on nursing Brangus cows. Two hypotheses were examined: (1) animals would learn to avoid restricted zones by increasing their use of containment zones within a virtual fence polygon, and (2) animals would progressively receive fewer audio-electric cues over time and increasingly rely on auditory cues for behavioral modification. Data from GPS coordinates, behavioral metrics derived from the collar data, and cueing events were analyzed to evaluate these hypotheses. The results supported hypothesis 1, revealing that virtual fence activation significantly increased the time spent in containment zones and reduced time in restricted zones compared to when the virtual fence was deactivated. Concurrently, behavioral metrics mirrored these findings, with cows adjusting their daily travel distances, exploration area, and cumulative activity counts in response to the allocation of areas with different virtual fence configurations. Hypothesis 2 was also supported by the results, with a decrease in cueing events over time and increased reliance with animals on audio cueing to avert receiving the mild electric pulse. These outcomes underscore the rapid learning capabilities of groups of nursing cows in responding to virtual fence boundaries.

Keywords: virtual fence; conditional behavior; animal tracking; precision livestock farming; GPS location; accelerometer and activity

Citation: Nyamuryekung'e, S.; Cox, A.; Perea, A.; Estell, R.; Cibils, A.F.; Holland, J.P.; Waterhouse, T.; Duff, G.; Funk, M.; McIntosh, M.M.; et al. Behavioral Adaptations of Nursing Brangus Cows to Virtual Fencing: Insights from a Training Deployment Phase. *Animals* **2023**, *13*, 3558. https://doi.org/10.3390/ani13223558

Academic Editors: Yang Zhao, Daniel Berckmans, Hao Gan, Brett Ramirez, Janice Siegford, Lingjuan Wang-Li and Robert T. Burns

Received: 16 October 2023
Revised: 13 November 2023
Accepted: 15 November 2023
Published: 17 November 2023

1. Introduction

Virtual fencing has emerged as a promising commercialized solution for efficiently managing livestock distribution, introducing an innovative paradigm for controlling their movements [1,2]. This method employs collars equipped with auditory-electric cues, designed to deter animals from crossing predetermined virtual fence boundaries [3,4]. These smart collars, integrated with online embedded micro-controllers, calculate the animal's GPS position, travel direction, and speed in relation to the virtual fence-designated polygon boundary. As an animal approaches this boundary, the collar triggers a gradually intensifying audio pitch that indicates proximity to the virtual boundary (https://www.nofence.no/en/, accessed on 20 February 2022). If the animal stops moving or changes direction, the audio signal ceases, and no electric pulse is administered. Conversely, if the animal persists toward the virtual fence boundary, the audio tone is succeeded by a mild electric pulse to deter further encroachment [3–5].

One of the key advantages of virtual fencing technology is its flexibility in defining land and forage resource allocations using customizable polygons in terms of shape, size, and duration. This flexibility empowers land managers with unprecedented control over grazing practices, addressing the challenge of managing heterogeneous resources across both space and time [6–8]. The potential applications of virtual fencing are expansive, encompassing immediate benefits such as the prevention of unwanted access in ecologically sensitive areas, including riparian zones, while concurrently reducing labor requirements [1,4].

Although concerns about animal welfare have arisen regarding virtual fence applications, many of these worries can be attributed to a lack of understanding of the technology's operational principles. Notably, the core similarity between electric fencing and virtual fencing lies in their use of visual (electric fence) or acoustic (virtual fence) cues to signal the deterrent electric pulse [9,10]. In addition, animals readily adapt to these cues, mitigating potential negative effects on their comfort, well-being, and overall welfare [9,11,12].

Successful implementation of virtual fencing technology hinges on a suitable training phase [4,13–15] for the animals to anticipate and move away from the virtual fence boundaries to avoid the electric cue that acts as a physical deterrent. This learning process involves two conditional behavioral mechanisms. The first mechanism, known as the skin defense system, enables animals to steer clear of landscape locations associated with prior exposure to an electric pulse [12,16]. The second mechanism involves associative learning, where animals associate an audio tone with signaling an impending mild electric pulse [8,17,18]. Consistent timing of the device's delivery of an electric pulse following the audio tone is critical to providing the animal with the necessary predictability to alter foraging trajectories. The gradually intensifying audio pitch as the collar progresses towards the virtual fence boundary serves as a cue for animals to anticipate the onset of the electric pulse. This associative learning also enables animals to adapt to periodic changes in virtual fence boundaries.

Despite prior research efforts, the scientific literature highlights significant knowledge gaps in the development of efficient, effective, and safe animal training protocols for virtual fence applications. Moreover, establishing standard training protocols can serve as benchmarks for future research on virtual fencing, facilitating cross study comparisons and enabling validation of the application across varying equipment and animal physiological and behavioral differences.

The primary aim of this study was to investigate the virtual fence application, with an emphasis on the learning process associated with the two conditional behavioral mechanisms in a group of nursing Brangus cows (n = 28). These cows were equipped with commercial Nofence virtual fence collars (https://www.nofence.no/en/, accessed on 20 February 2022) and utilized an existing ranch infrastructure. In addition, the research outlines a detailed training protocol for the implementation of a virtual fence system.

Initially, our investigation focused on hypothesis 1, which postulated that animals would acquire the ability to avoid restricted zones located outside the virtual fence-designated polygon area by increasing their utilization of the containment zones within the virtual fence-designated polygon area. To explore this hypothesis, we analyzed the GPS coordinates collected using the collars to assess how the animals utilized the spatial dimensions of the arena. Furthermore, we utilized behavioral metrics derived from the collar data to validate the animal's responses (daily travel distances, exploration of areas, and activity levels) to changes in the allocated area with the virtual fence application.

Subsequently, hypothesis 2 was examined to determine if animals received progressively fewer audio-electric cues over time (Period) and increasingly relied on auditory cues to modify their behavior, in accordance with the associative learning mechanism. This facet of the study involved the evaluation of the number of audio and electric cues received by the animals. Two computations were used to evaluate the relationship between these cues: (1) the ratio of electric to audio cueing and (2) the percentage of electric cues relative to the total number of cues emitted by the collar. Also, the duration of audio cues emitted by the collars was examined to gauge the total length of the audio tone played and to track variations in the length of the emitted audio tone per cue.

2. Materials and Methods

2.1. Animal and Virtual Fence Collars

All procedures were approved by the New Mexico State University Institutional Animal Care and Use Committee (Protocol 2021–010). Cows with calves over 1.5 months of age were used in the training phase to mitigate challenges associated with younger calves, as recommended by Nyamuryekung'e et al. [19]. Two groups of cows were used in the training phase (n = 11 and 17, respectively). These groups were organized based on the calving dates (ranging from 21/2 to 11 April 2022 for Group 1 and from 14/3 to 20 May 2022 for Group 2) to ensure the average initial calf age was approximately 2 months (1.96 vs. 2.25 months for Groups 1 and 2, respectively) at the beginning of the deployment (11 May 2022 and 6 July 2022 for Groups 1 and 2, respectively).

Within their respective groups, nursing Brangus cows that were naïve to the virtual fence technology were selected. These cows had approximate mean weights of 490 kg in Group 1 and 402 kg in Group 2. Cows were equipped with C2 Nofence virtual fence collars (Batnfjordsøra, Norway), each weighing 1.45 kg (https://www.nofence.no/en/, accessed on 20 February 2022) two weeks before the virtual fence training. Collars operated on solar and battery energy, using LTE Cat-M1 or 2G network (cellular) to register virtual fence polygons and communicate real-time animal positions and activity at 15 and 30 min intervals, respectively, with reliable cellular coverage and in near real-time when cellular coverage was suboptimal. During this two-week period of acclimatization, the cows wore the collars while grazing freely alongside their uncollared offspring in pastures near the New Mexico State University's Chihuahuan Desert Rangeland Research Center headquarters (32°31′53.2″ N 106°48′15.6″ W). It is important to note that no virtual fence cueing was initiated during this two-week adaptation Period, and electric fence lines were not employed.

The virtual fence polygons generated using the Nofence device delineated three distinct spatial areas:

- Containment Zones: these areas resided within the virtual fence polygon, granting collared animals the freedom to move without restrictions;
- Cueing Zone: Positioned approximately 5–10 m wide from the containment zone's edge or virtual fence boundary, this area progressively subjected collared animals to three pairings of audio-electric cueing. Following each cue, an alert was sent to the manager;
- Restricted Zone: Also referred to as the escape zone, these areas were located outside the virtual fence polygon. While collars continued to record data in this zone, no cueing was initiated. In cases of breaches, escape notifications were dispatched to the manager.

Within the cueing zone, the pitch of the audio signal increased progressively, culminating in the highest pitch signal at 82 dB. The complete audio ramp cycle lasted between 5 and 20 s, signaling the administration of the electric pulse upon cycle completion. The actual duration of the audio cueing cycle varied depending on the speed of the animal's movement as it approached the virtual fence boundary.

Virtual fence boundaries, unlike fixed physical fence lines, exhibit variability due to inherent GPS position errors. The Nofence collars exclusively accept GPS positions with an accuracy of 3.5 m or better for cue initiation. The collars were deactivated safely in case of failure or after emitting three electric pulses in close proximity. Additionally, the Nofence collars featured directional cueing, meaning that animals received cues only when leaving the virtual fence containment zone. Hence, animals were permitted to re-enter the virtual fence polygon without encountering any cues. However, once inside, the collars were reactivated to deter any further attempts to exit the containment zone.

2.2. Training Deployment Design

At the onset of the training phase, the first group of virtual fence-collared cows (n = 11) with uncollared calves (11 May 2022) were moved to a holding pen situated at the New Mexico State University's Chihuahuan Desert Rangeland Research Center headquarters. The pen (0.2 hectares and devoid of vegetation) contained three permanent feeding stations spaced evenly within each third of the area. Baled, beardless wheat hay (approximately 590 kg) was provided *ad libitum*. Additionally, a single water trough was placed on one side of the pen (Figure 1). Throughout the deployment, to facilitate replenishing of the feeding stations when near depletion, the animals were temporarily moved to the southwestern pen area, and access to the arena was restricted for approximately one hour.

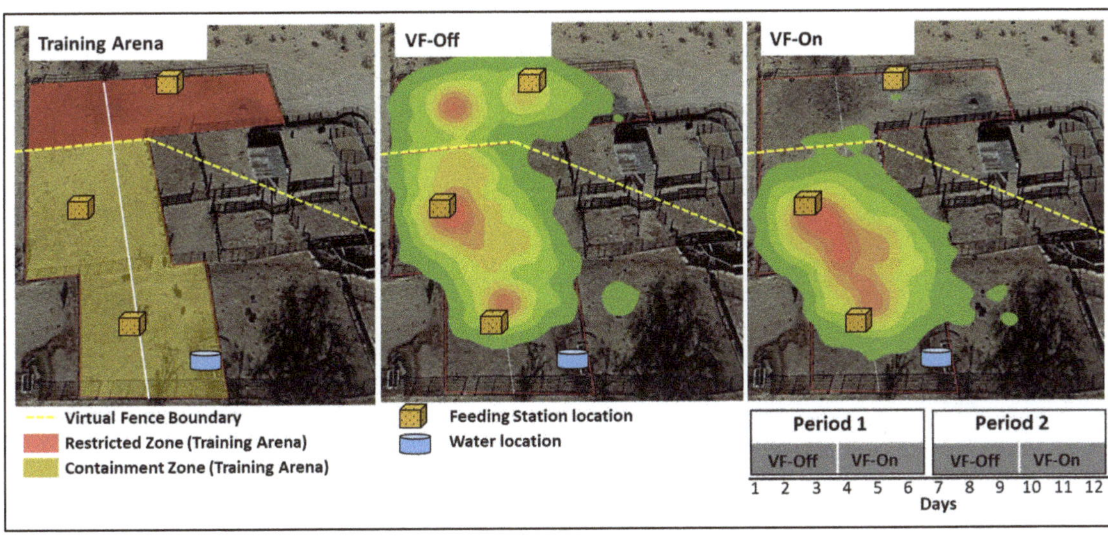

Figure 1. Top left: Training arena (0.2 ha) devoid of vegetation, with three permanent feeding stations spaced along the major axis (white line, 65.80 m), with a single water trough location. The yellow dashed line indicates the virtual fence boundary encapsulating the southern portion of the training arena, delineating it into a containment zone (amber area, southern portion of the training arena, 0.12 ha) and a restricted zone (vermilion area, northern portion of the training arena, 0.08 ha). Heat maps, generated using the kernel density tool with 7 quantile classification categories, depict the spatial distribution (ranging from green as low to red as high) derived from GPS locations of Brangus cows from both groups (n = 11 and 17) within the training arena wearing virtual fence-deactivated (middle: VF-Off) or virtual fence-activated (right: VF-On) collars. Bottom right: training deployment design across the two Periods.

Collared cows were initially allowed to feed freely at the three feeding stations for three consecutive days while wearing virtual fence-deactivated collars (VF-Off). On the fourth day, all collars were configured with a virtual fence polygon drawn to restrict the collared cows from accessing the farthest 40% of the pen area, designating it as the restricted zone (0.08 hectares), while allowing access to the two remaining feeding stations and the water trough location, defined as the containment zone (0.12 hectares). This virtual fence configuration was maintained for three consecutive days (VF-On). The activation or deactivation of the virtual fence configuration occurred at approximately 07:00 am during the initiation of the 3-day interval when all animals were situated within the containment zones. This sequence of three days with virtual fence-deactivated collars (VF-Off), followed by three days with virtual fence-activated collar (VF-On) configuration, was repeated for a second 6-day Period (total duration of the training phase = 12 days) (Figure 1). The same pens and training protocol described above were replicated with a second group of naïve collared cows ($n = 17$) with uncollared calves (6/7/2022).

2.3. Data Processing

At the end of the deployment (24 May 2022 for Group 1 and 19 July 2022 for Group 2), data from the Nofence collars were retrieved as CSV files using an online portal from the Nofence website service. The data included time stamps for variables processed by each collar registering the animal positions (latitude, longitude, horizontal accuracy, horizontal dilution of precision, and number of satellites), accelerometer-based activity count, ambient environmental condition (humidity, pressure, and temperature), communication variables (global system for mobile communication operator and communication strength), and collar status (battery voltage, solar charge, VF status, and polygon ID). In addition, the collars registered a time stamp and position for each cueing event, either audio (including duration of the signal) or electric pulse, and escape notifications.

Data were screened for duplicates and filtered to include only data from the training phase for both groups, which occurred from 11 May 2022 to 24 May 2022 for Group 1 and from 6 July 2022 to 19 July 2022 for Group 2. Furthermore, data collected an hour before and after the virtual fence configuration status changed (06:00 am–08:00 am) and from when the animals were temporarily relocated to an adjacent pen area (~one hour) for feeding station refill were excluded from the analysis.

To facilitate the analysis, a virtual fence status column (VF-On or VF-Off) was created, categorizing the data into segments representing the three consecutive days. The virtual fence polygon used during the VF-On status and the position data collected from the collars were imported and projected to the NAD 1983 UTM Zone 13N coordinate system using ArcGIS software (ESRI 2018, ArcMap Desktop v. 10.6). Position data from each collar were overlaid onto the virtual fence polygon layer to extract precise location information pertaining to either being on the containment or restricted zones. Data from each collar were processed by day to include the percentage of time (%Time) GPS location data were within the containment and restricted zones during both virtual fence deactivated (VF-Off) and virtual fence activated (VF-On) status.

Furthermore, from projected GPS coordinate data, the daily distance traveled per collar (Dist) was calculated using the Pythagoras theorem applied to sequential GPS coordinates within a day. To assess the daily area explored by each collared animal (Area), the Minimum Bounding Geometry tool (ESRI 2018, ArcMap Desktop v. 10.6) was employed, which generated a polygon encompassing the minimum area containing all GPS coordinates recorded within a given day [19,20]. In addition, accelerometer-based activity data from the Nofence CSV files, represented as a count of motion intensity using internal threshold values within 30 min intervals, were aggregated into a cumulative count within a day (#Activity).

Variables related to collar cueing events were processed to include the daily count of audio (#Audio) and electric pulses (#Electric). Additionally, two derived variables were computed: one representing the ratio of electric to audio cueing per day (Electric/Audio), and the other indicating the percentage of electric cues relative to the total number of cues emitted by the collar per day (%Electric/Cue; equates to (#Electric/(#Electric + #Audio)) $\times$ 100). As for the duration of the audio signal, two daily variables were generated, including the sum of the total daily audio signal length (Sum_DurAudio) and the average length of a single audio cue (Avg_DurAudio). Notably, both variables related to the duration of the audio signal were computed exclusively when a registered audio warning was present.

2.4. Data Analysis

To prepare the data for statistical analysis, all daily variables related to spatial analysis (%Time, Dist, and Area) and activity (#Activity) of the animals, as well as variables related to cueing events (#Audio, #Electric, Electric/Audio, %Electric/Cue, Sum_DurAudio, and Avg_DurAudio), were averaged for each cow over three-day Periods, aligning with the concurrent virtual fence status (VF-On or VF-Off). This approach not only helped smooth the dataset by mitigating the daily variability inherent in biosensing data but also addressed the issue of an unequal number of sampling days [20]. This disparity was particularly evident in the case of cueing event data, for which there were instances of VF-On status within a day when several animals did not interact with the virtual fence boundary. For instance, when analyzing data on a daily scale, the percentage of days with no cueing events was 66.2% (59.1% in Period 1 and 73.6% in Period 2). However, with Period averaging, the percentage of data points with no cueing events decreased to 25.5%, with 21.4% in Period 1 and 29.6% in Period 2.

The averaged daily variables related to spatial analysis (%Time, Dist, and Area) and activity (#Activity) were subjected to statistical analysis using SAS 9.3 software (SAS Institute, Cary, NC, USA). The PROC MIXED procedure, along with a "covtest" statement, was employed for the analysis. Analysis of Variance with the Kenward–Roger degrees of freedom statement was used to model the effects of virtual fence status (VF-On vs. VF-Off), the Period (1 vs. 2), and their interaction on the averaged daily variables related to spatial analysis (%Time, Dist, and Area) and activity (#Activity) of the animals. Random effects were considered for Groups (1 and 2) and CollarID (n = 11 and n = 17 for Groups 1 and 2, respectively) to account for variations associated with these factors. Least squares means were calculated for the main effects of the virtual fence status and Period individually, with the "pdiff" statement used for pairwise comparisons. Furthermore, using the least squares means computation for the interaction effect, the virtual fence status effect (VF-On vs. VF-Off) within the Period was examined using the "slice" statement.

Descriptive analysis was initially conducted on the raw cueing event data, which included #Audio and #Electric from the collar dataset. Similarly, the averaged daily variables related to cueing events (#Audio, #Electric, Electric/Audio, %Electric/Cue, Sum_DurAudio, and Avg_DurAudio) were analyzed using SAS 9.3 software (SAS Institute, Cary, NC, USA). However, this analysis focused exclusively on the dataset that considered virtual fence-activated status (VF-On). Employing the PROC MIXED procedure with a "covtest" statement, Analysis of Variance was carried out with the Kenward–Roger degrees of freedom statement, focusing on the effects of the Period (1 vs. 2) on the averaged daily variables associated with cueing events. Random effects were incorporated for Groups (1 and 2) and CollarID (n = 11 and n = 17 for Groups 1 and 2, respectively). The main effect of the Period was assessed via computing least squares means, and pairwise comparisons were conducted using the "pdiff" statement. In all analyses, statistical significance was declared at $p < 0.05$.

3. Results

The percentage of time collared animals spent within the containment and restricted zones was significantly influenced by their main effects, namely the virtual fence status ($p < 0.01$) and Period ($p < 0.01$), with no significant interaction effect ($p = 0.10$). When considering the main effect of virtual fence status, VF-On increased the percentage of time collared animals spent within the containment zones (98.0 vs. 70.7 ± 1.2% Time; $p < 0.01$) while reducing the percentage of time within the restricted zones (12.0 vs. 29.3 ± 1.2% Time; $p < 0.01$) compared to VF-Off status (Figure 2). In terms of the main effect of the Period, which encompassed 6 concurrent days (3 days of VF-Off followed by 3 days of VF-On), Period 2 resulted in a higher percentage of time collared animals spent within the containment zones (86.5 vs. 82.2 ± 1.2% Time; $p < 0.01$) and a lower percentage of time within the restricted zones (13.5 vs. 17.8 ± 1.2% Time; $p < 0.01$) compared to Period 1 (Figure 2).

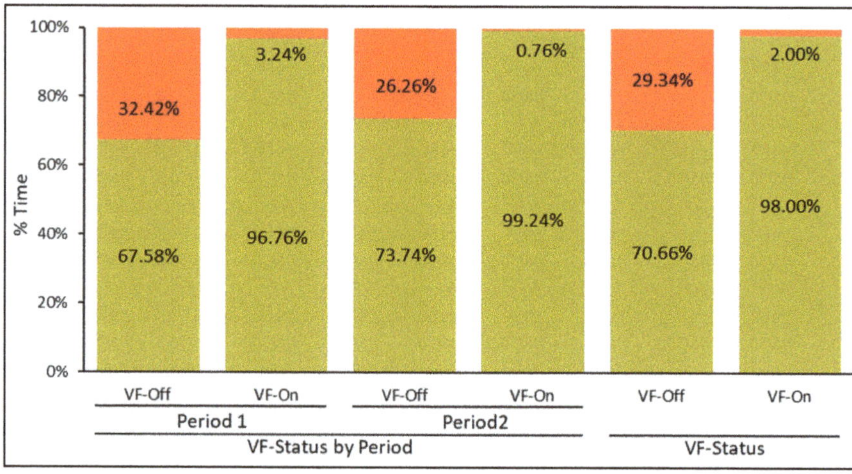

Figure 2. Analysis of the percentage of time spent by virtual-fence-collared nursing Brangus cows within the containment (amber bars) and restricted (vermilion bars) zones across the two different Periods under varying VF-Status (virtual fence deactivated (VF-Off) vs. virtual fence activated (VF-On)). The figure also includes an overall comparison of VF-Status (VF-Off vs. VF-On) as the main effect (represented by the two bars on the right, VF-Status). Significance testing indicates that all pairwise comparisons of VF-Status (VF-Off vs. VF-On) were statistically significant for the percentage of time allocation within either the containment or restricted zones.

Similar trends were observed for the remaining variables related to spatial analysis (Dist, Area) and activity (#Activity) of the animals, where significant main effects were detected for the virtual fence status (Dist; $p < 0.01$, Area; $p < 0.01$, and #Activity; $p < 0.01$) and Period (Dist; $p < 0.01$ and #Activity; $p < 0.01$), except for Area ($p = 0.31$). The interaction effect was not significant for these variables (Dist; $p = 0.15$, Area; $p = 0.06$, and #Activity; $p = 0.19$) (Table 1). For the main effect of virtual fence status, VF-On resulted in a decrease in the daily distance traveled (648.0 vs. 883.8 ± 82.7 m; $p < 0.01$), area explored within the arena (1193.8 vs. 1982.1 ± 121.4 m^2; $p < 0.01$), and the cumulative activity count of the animal (14,988 vs. 17,300 ± 1335.5; $p < 0.01$) compared to the VF-Off status. Regarding the main effects of the Period, Period 2 exhibited lower values for the daily distance traveled (719.0 vs. 812.7 ± 82.7 m; $p < 0.01$) and the cumulative activity count of the animal (15,330 vs. 16,958 ± 1335.5; $p < 0.01$) when compared to Period 1.

Table 1. Least square means estimates and their standard errors (SE) for the interaction term (virtual fence status (VF-Status) × Period) illustrating daily travel distances, area exploration, and activity levels in virtual-fence-collared nursing Brangus cows. The *p*-values indicate the significance of the main effects of VF-Status (virtual fence deactivated collar status (VF-Off) vs. virtual fence activated collar status (VF-On)), Period (representing two 6-day Periods with each composed of 3 days of VF-Off followed by 3 days of VF-On), and the interaction effect.

	Period 1		Period 2		SE	*p*-Value		
	VF-Off	VF-On	VF-Off	VF-On		VF-Status	Period	Interaction
Dist (m)	946.28	679.19	821.31	616.75	84.06	<0.01	<0.01	0.14
Area (m^2)	1956.87	1277.30	2007.40	1110.25	128.01	<0.01	0.31	0.06
Activity	18,406	15,510	16,194	14,465	1371.71	<0.01	<0.01	0.19

The cueing events during the training deployment included 305 audio tones (153 for Group 1 and 152 for Group 2) and 101 electric pulses (54 for Group 1 and 47 for Group 2) in the two groups of nursing Brangus cows (Figure 3). The total number of cueing events per collared cow within the two groups ranged from 8 to 26 for Group 1 and 1 to 30 for Group 2 (Figure 3).

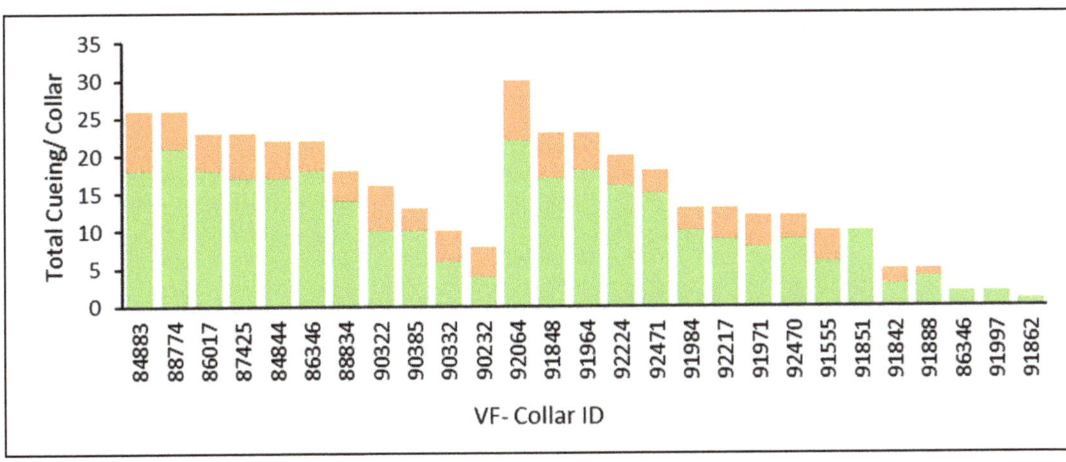

Figure 3. Ranking of total auditory (green) and electrical pulses (orange) emitted per cow during the training phase for naïve nursing Brangus cows from two groups (n = 11 and 17).

The daily number of emitted audio and electric pulses was significantly influenced by the Period (#Audio; $p < 0.01$ and #Electric; $p < 0.01$). During Period 1, when the virtual fence was first activated, individual cows experienced a higher frequency of auditory (1.9 vs. 0.6 ± 0.4; $p < 0.01$) and electrical cues (0.7 vs. 0.2 ± 0.2; $p < 0.01$) compared to Period 2. In terms of variables representing the animals' reliance on audio warnings, the ratio of electric to audio cues per cow (Electric/Audio) was not significant for the Period ($p = 0.10$). However, the percentage of electric cues relative to the total number of cues emitted by the collar differed between Periods (%Electric/Cue; $p = 0.04$). During Period 1, exposure to the virtual fence activated status resulted in a higher percentage of electric cues relative to the total cues per cow (23.0 vs. 13.3 ± 3.8%; $p = 0.04$) compared to Period 2. Lastly, regarding the duration of the audio signal, the length of the sum of the total daily audio signals (Sum_DurAudio) differed between Periods ($p = 0.02$), while the average length of a single audio cue (Avg_DurAudio) did not differ between Periods ($p = 0.91$). Animals in Period 1 were exposed to longer durations within the day with collars emitting the audio warning (58.2 vs. 26.7 ± 8.7 sec; $p = 0.02$) compared to Period 2 (Table 2).

Table 2. Least square means estimates with standard errors (SE) for Period, illustrating the daily count of audio and electric pulses, the ratio of electric to audio cueing (Electric/Audio), the percentage of electric cues relative to the total number of cues (%Electric/Cue), the sum of the total daily audio signal length (Sum_DurAudio), and the average length of a single audio cue (Avg_DurAudio) emitted per virtual fence-collared nursing Brangus cows. The *p*-values denote the significance across Periods, representing 3 days of virtual fence-activated status.

	Period		SE	*p*-Value
	1	2		
#Audio	1.88	0.64	0.45	<0.01
#Electric	0.70	0.17	0.17	<0.01
Electric/Audio	0.36	0.22	0.08	0.10
%Electric/Cue	23.02	13.31	3.76	0.04
Sum_DurAudio (sec.)	58.16	26.70	8.74	0.02
Avg_DurAudio (sec/audio.)	11.87	11.66	1.38	0.91

4. Discussion

In this study, we set out to investigate hypothesis 1, that animals would develop the ability to avoid restricted zones outside the virtual fence-designated polygon area by increasing their use of the containment zones (skin defense mechanism). Our findings support this hypothesis. Virtual fence activation significantly reduced the percentage of time animals spent within the restricted zones while increasing the time spent within the containment zones. In essence, during the training phase, virtual fence activation successfully contained the cows within their designated zones approximately 98.0% of the time, consistent with other findings [9–13,21]. Additionally, the significant main effect of the Period further suggests an ongoing learning process among the cows regarding the virtual fence polygon's configuration. Notably, with regard to the escape notifications derived from collar data, which indicated instances when animals breached the containment zone, we observed 10 such notifications, all occurring in Period 1, involving three animals in Group 1 and a single animal in Group 2, while Period 2 had no recorded escape events.

Furthermore, our study leveraged behavioral metrics derived from collar data, specifically focusing on daily travel distance, area exploration, and accelerometer-based activity levels. These metrics provided valuable insights into how the animals responded to changes in the allocated area imposed by the virtual fence application. Activation of the virtual fence led to a reduction in all these behavioral metrics, as the animals were confined to 60% of their initial designated pen area (0.12 hectares out of 0.2 hectares). However, the reduction in the activity variable during virtual fence-activated status may also suggest some level of adjustments in the animal's time budget, as observed in other studies [9,21,22]. This aspect of the result warrants further investigation.

In addition, the behavior metrics displayed Period-dependent effects, with the second Period showing lower values compared to the initial Period. This adjustment can be attributed to the following rationale: as the second Period began with VF-Off status for three days, the animals displayed a delayed response from the previous three days of VF-On status in the first Period. This delayed response of the animals when exploring previously restricted zones might be associated with animals taking time to become familiar with the arena area or waiting for their peers to venture into the restricted zones before following suit [23]. However, during the VF-On status of the second Period, the animals drew upon their prior learning experiences from the first period to correctly adjust their behavior, resulting in lower behavior metrics compared to the VF-On status of the first Period.

The real-time data collected from sensors integrated into virtual fencing devices hold great promise for economical and effective animal monitoring [1,2,24,25]. These data can be tailored to address questions related to spatial use (animal position) [20,26], activity patterns (motion sensors) [27,28], and even ambient environmental conditions (humidity, pressure, and temperature). This enables a detailed and accurate analysis of the animals' operational status. The integration of such information, harnessed from virtual fence sensor data, provides invaluable support for the surveillance and monitoring of animals, enhancing data-driven decision-making processes for land managers in the context of precision livestock farming within rangeland management [24,25]. The sensitivity of the behavioral metrics derived from our dataset further underscores the potential of virtual fence collar data for animal surveillance and monitoring within precision livestock farming applications.

Our study provides empirical support for Hypothesis 2, which proposed that animals would gradually receive fewer audio-electric cues over time and would increasingly depend on auditory cues to modify their behavior through associative learning. We indeed observed a clear reduction in the frequency of audio-electric cues over time, corroborating this hypothesis. Specifically, during Period 2, when the virtual fence was activated, cows received notably fewer auditory and electrical cues compared to Period 1. This observed reduction is consistent with findings reported in previous studies on virtual fence applications [10–12,21,29]. The reduction during Period 2, as quantified using the equation ((1 − (Period 2/Period 1)) × 100), was 66% for auditory and 76% for electric cues. Additionally, the observed numerical difference in the reduction rates during Period 2 lends support to the notion that animals increasingly rely on auditory over electric cues as they become more familiar with the virtual fence system.

In parallel with the cueing results, the variables assessing animals' reliance on audio warnings, specifically the ratio of electric to audio cueing per cow, were not significant for Period. This is in contrast to Lomax et al. [11], who reported a decrease in the ratio of electric to audio cueing. However, a significant reduction in the percentage of electric cues relative to the total number of cues was evident, with animals receiving a smaller proportion of electric cues relative to the total in the second Period. The inconsistency between the results of these two variables could be attributed to their differing sensitivity levels as well as the averaging of the variables across Periods before statistical analysis. Lomax et al. [11] reported differences in the ratio of electric to audio cueing on a daily scale, with statistical differences only present on the first day of animals encountering the virtual fence configuration.

Lastly, regarding the duration of the audio signal, animals in Period 1 experienced longer durations during the day when collars emitted audio warnings compared to Period 2. This finding corroborates the total number of audio warnings emitted across the two Periods. However, the lack of significance in the average length of a single audio warning was unexpected. The collars were programmed to deliver audio warnings progressively with varying pitches, ultimately reaching the highest pitch signal at 82 dB, designed to help animals learn when the electric pulse would be emitted. However, our results suggest that this progressive pitch strategy might not have had the expected impact. This finding suggests the need for further investigation in future research to decipher the most suitable audio signal or whether animals respond consistently as long as there is an audio cue.

The individual variation among animals in total cueing events, as revealed in the descriptive analysis, presents an intriguing opportunity for further investigation of virtual fence applications through individual herd selection to address animal welfare concerns. However, breed and within-breed individual differences related to both social- and asocial-learning mechanisms could potentially influence an individual's ability to adapt to the technology [11,13,14,21,29].

The concept of selectively using virtual fence collars on specific individuals has sparked interest in its potential applications for large herds. However, empirical support for this idea has been limited, with only a single study conducted on sheep suggesting that virtual fence collaring the entire herd, or 66% of it, would yield comparable results [22]. In our study, we employed mother–offspring pairings during the training phase, with only the cows wearing the virtual fence collars. While the recommended approach for effective virtual fencing is to collar all individuals within a herd continuously, regardless of their social rank, as advised by Keshavarzi et al. [13], our research demonstrates the effective containment and learning capabilities of the dams, regardless of the behavior of their offspring. Informal observations revealed that, on occasion, some calves ventured into the restricted zones during virtual fence-activated status. However, this behavior was not systematically monitored, presenting an opportunity for further research investigations.

Cattle are gregarious in nature; hence, learning about novel situations can be facilitated through their conspecifics [13]. In the group deployment, this response was noted through the anecdotical observation of the reaction of a neighbor cow to the cues emitted from a peer's collar and was also corroborated by the other literature [29]. In addition, the gregarious nature of cows may encourage cued individuals to respond appropriately during training by re-joining herd members inside the containment zone when challenged with the tactile stimulus, imitating a predatory response (skin defense mechanism). As a precaution to the cued individuals' reactions during the training phases, adjacent non-collared cows outside the training arena at the headquarters pen location were restricted to areas behind the containment zone. This training approach extends to pasture deployment, as cautioned by Verdon et al. [23], who recommended avoiding the use of virtual fence applications for managing cattle with close visual contact with other herds.

However, to facilitate proper animal training for the virtual fence application, it is advised that all collared individuals receive at least a single cueing event [14]. Our training deployment design could facilitate this by incorporating additional Periods within the deployment with the alternative virtual fence-deactivated (VF-Off) status configuration followed by the virtual fence-activated (VF-On) status configuration. Moreover, our choice of a 3-day sequence for the virtual fence configuration demonstrated effectiveness in our training deployment. However, factors such as animal density ($n = 11$ and $n = 17$ for Groups 1 and 2, respectively) within the containment zone (0.12 ha) and the ratio of the virtual fence boundary (23.87 m) to the physical fence boundary (139.45 m) within the containment zone perimeter are noteworthy variables that should be thoroughly investigated to determine the optimal duration for the sequencing of virtual fence configurations.

The deactivation status of the virtual fence plays a crucial role in establishing a baseline for the virtual fence application. The initial period, commencing with virtual fence deactivated status (VF-Off), serves as a control to assess the effectiveness of the virtual fence application. In our training phase setup, with three feed stations providing baled hay evenly distributed across each third of the arena area, the animals spent approximately 67.6% of their time in the containment zone, which was roughly 60% (0.12 ha) of the area, and 32.4% of their time in the restricted zone, equivalent to the remaining 40% (0.08 ha) of the arena area. The assessment of the selectivity of the containment versus restricted zone using Ivlev's electivity index ($E = ((r - p)/(r + p))$, where r is the proportion of time allocated within a zone and p is the proportion of the zone area within the arena, yielded a value of 0.06 for the containment zone and −0.10 for the restricted zone [20,30]. This indicates that, at the start of the training phase, due to the elective index being closer to 0, animals had indifferent selectivity for the containment and restricted zones, substantiating the effectiveness of the virtual fence application. Furthermore, establishing this baseline can provide insights into variations in cueing rates across different virtual fence studies by defining the preference for the restricted area before application, hypothesizing that a higher preference may result in more cueing. Additionally, the second period, commencing with VF-Off status, serves as a control to ensure that animals do not associate spatial location with cueing but rather the presence of the audio cue as a warning of an impending mild electric pulse. The inclusion of a single water trough location within the containment zone further enhanced training success, as animals that ventured beyond the containment zones could be recaptured with the system when prompted by thirst. Consequently, the training protocol presented here offers an efficient, effective, and safe approach to animal training for virtual fence applications.

5. Conclusions

Our results indicate that groups of cows learn rapidly to respond to VF boundaries by reducing time spent within the restricted areas and minimizing the frequency of cueing events to alter behavior. Assuming animals complete the training phase successfully, adverse effects are minimized as animals learn to respond to the audio and avoid the mild electric cue, which minimizes adverse effects on animal comfort, well-being, and welfare. In addition, the associative learning process enables animals to adjust readily and respond to periodic changes in virtual fence boundaries, providing managers with the flexibility to allocate land and forage resources with polygons of configurable shape, size, and duration.

Author Contributions: Conceptualization, S.N., S.U. and A.F.C.; methodology, S.N. and S.U. software, S.N.; formal analysis, S.N.; investigation, S.N., A.C., A.P. and M.F.; resources, G.D., S.U., R.E., A.C., S.S. and A.F.C.; data curation, S.N.; writing—original draft preparation, S.N.; writing—review and editing, A.C., R.E., A.F.C., J.P.H., T.W., G.D., M.F., M.M.M., S.S., B.B. and S.U.; visualization, S.N. and M.M.M.; supervision, A.F.C. and S.U. All authors have read and agreed to the published version of the manuscript.

Funding: This research was funded by the Long-Term Agroecosystem Research network (LTAR) of the United States Department of Agriculture (USDA), the USDA National Institute of Food and Agriculture, and the USDA National Institute of Food and Agriculture Sustainable Agriculture Systems Coordinated Agricultural Project grant # 12726269. S.N. received partial funding from the Norwegian Institute of Bioeconomy Research (NIBIO), while J.P.H was supported by the Scottish Government's Rural and Environment Science and Analytical Services Division (SRUC-B2-2).

Institutional Review Board Statement: The animal study protocol was approved by the Institutional Review Board (or Ethics Committee) of New Mexico State University Institutional Animal Care and Use Committee (Protocol 2021–010, 7 May 2021).

Informed Consent Statement: Not applicable.

Data Availability Statement: The raw data concerning the virtual fence sensors showcased in this study are available upon request from the corresponding authors.

Acknowledgments: The authors would like to express their sincere appreciation to the management team of the Chihuahuan Desert Rangeland Research Center (CDRRC) for their unwavering support during the experiment. Special gratitude is extended to the U.S. Precision Livestock Farming 2023 committee members for their generous scholarship and for selecting the research to be highlighted in the Special Issue. Furthermore, the authors acknowledge the editors and anonymous reviewers for their invaluable feedback, which significantly contributed to the improvement of the final version of the manuscript.

Conflicts of Interest: The authors declare no conflict of interest. The funders had no role in the design of the study; in the collection, analyses, or interpretation of data; in the writing of the manuscript; or in the decision to publish the results. Furthermore, references to specific trade names, proprietary products, or vendors do not imply any endorsement or warranty of the product by the USDA, nor does it indicate an exclusive approval over other products or vendors that could be equally suitable.

References

1. Waterhouse, T. Virtual Fencing Systems: Balancing Production and Welfare Outcomes. *Livestock* **2023**, *28*, 227–234. [CrossRef]
2. Goliński, P.; Sobolewska, P.; Stefańska, B.; Golińska, B. Virtual Fencing Technology for Cattle Management in the Pasture Feeding System—A Review. *Agriculture* **2023**, *13*, 91. [CrossRef]
3. Anderson, D.M.; Estell, R.E.; Holechek, J.L.; Ivey, S.; Smith, G.B. Virtual Herding for Flexible Livestock Management—A Review. *Rangel. J.* **2014**, *36*, 205–221. [CrossRef]
4. Umstatter, C. The Evolution of Virtual Fences: A Review. *Comput. Electron. Agric.* **2011**, *75*, 10–22. [CrossRef]
5. Muminov, A.; Na, D.; Lee, C.; Kang, H.K.; Jeon, H.S. Modern Virtual Fencing Application: Monitoring and Controlling Behavior of Goats Using GPS Collars and Warning Signals. *Sensors* **2019**, *19*, 1598. [CrossRef] [PubMed]
6. Campbell, D.L.M.; Haynes, S.J.; Lea, J.M.; Farrer, W.J.; Lee, C. Temporary Exclusion of Cattle from a Riparian Zone Using Virtual Fencing Technology. *Animals* **2019**, *9*, 5. [CrossRef] [PubMed]
7. Boyd, C.S.; O'Connor, R.; Ranches, J.; Bohnert, D.W.; Bates, J.D.; Johnson, D.D.; Davies, K.W.; Parker, T.; Doherty, K.E. Virtual Fencing Effectively Excludes Cattle from Burned Sagebrush Steppe. *Rangel. Ecol. Manag.* **2022**, *81*, 55–62. [CrossRef]

8. Lee, C.; Henshall, J.M.; Wark, T.J.; Crossman, C.C.; Reed, M.T.; Brewer, H.G.; O'Grady, J.; Fisher, A.D. Associative Learning by Cattle to Enable Effective and Ethical Virtual Fences. *Appl. Anim. Behav. Sci.* **2009**, *119*, 15–22. [CrossRef]
9. Hamidi, D.; Grinnell, N.A.; Komainda, M.; Riesch, F.; Horn, J.; Ammer, S.; Traulsen, I.; Palme, R.; Hamidi, M.; Isselstein, J. Heifers Don't Care: No Evidence of Negative Impact on Animal Welfare of Growing Heifers When Using Virtual Fences Compared to Physical Fences for Grazing. *Animal* **2022**, *16*, 100614. [CrossRef]
10. Langworthy, A.D.; Verdon, M.; Freeman, M.J.; Corkrey, R.; Hills, J.L.; Rawnsley, R.P. Virtual Fencing Technology to Intensively Graze Lactating Dairy Cattle. I: Technology Efficacy and Pasture Utilization. *J. Dairy Sci.* **2021**, *104*, 7071–7083. [CrossRef]
11. Lomax, S.; Colusso, P.; Clark, C.E.F. Does Virtual Fencing Work for Grazing Dairy Cattle? *Animals* **2019**, *9*, 429. [CrossRef] [PubMed]
12. Nyamuryekung'e, S.; Hansen, I.; Grøva, L.; Karlo, M.; Jørgensen, G.H.M. *Behaviour and Welfare of Cattle Wearing Dagros Virtual Fencing System*; NIBIO Report vol 9, nr 102; NIBIO: Osaka, Japan, 2023; 38p, ISBN 978-82-17-03337-0.
13. Keshavarzi, H.; Lee, C.; Lea, J.M.; Campbell, D.L.M. Virtual Fence Responses Are Socially Facilitated in Beef Cattle. *Front. Vet. Sci.* **2020**, *7*, 543158. [CrossRef] [PubMed]
14. Colusso, P.I.; Clark, C.E.F.; Lomax, S. Should Dairy Cattle Be Trained to a Virtual Fence System as Individuals or in Groups? *Animals* **2020**, *10*, 1767. [CrossRef]
15. Tiedemann, A.R.; Quigley, T.M.; White, L.D.; Lauritzen, W.S.; Thomas, J.W.; McInnis, M.L. Electronic (Fenceless) Control of Livestock. In *USDA Forest Service Research Papers RMRS*; US Department of Agriculture, Forest Service, Pacific Northwest Research Station: Washington, DC, USA, 1999; pp. 1–23.
16. Bishop-Hurley, G.J.; Swain, D.L.; Anderson, D.M.; Sikka, P.; Crossman, C.; Corke, P. Virtual Fencing Applications: Implementing and Testing an Automated Cattle Control System. *Comput. Electron. Agric.* **2007**, *56*, 14–22. [CrossRef]
17. Kearton, T.; Marini, D.; Cowley, F.; Belson, S.; Keshavarzi, H.; Mayes, B.; Lee, C. The Influence of Predictability and Controllability on Stress Responses to the Aversive Component of a Virtual Fence. *Front. Vet. Sci.* **2020**, *7*, 580523. [CrossRef] [PubMed]
18. Campbell, D.L.M.; Lea, J.M.; Haynes, S.J.; Farrer, W.J.; Leigh-Lancaster, C.J.; Lee, C. Virtual Fencing of Cattle Using an Automated Collar in a Feed Attractant Trial. *Appl. Anim. Behav. Sci.* **2018**, *200*, 71–77. [CrossRef]
19. Nyamuryekung'e, S.; Cibils, A.F.; Estell, R.; VanLeeuwen, D.; Steele, C.; Roacho-Estrada, J.O.; Rodriguez Almeida, F.; Gonzalez, A.L.; Spiegal, S.; Nyamuryekung'e, S.; et al. Do Young Calves Constrain Movement Patterns of Nursing Raramuri Criollo Cows on Rangeland? *Rangel. Ecol. Manag.* **2020**, *73*, 84–92. [CrossRef]
20. Nyamuryekung'e, S.; Cibils, A.F.; Estell, R.E.; Vanleeuwen, D.; Spiegal, S.; Steele, C.; Gonz, A.L.; Mcintosh, M.M.; Gong, Q. Movement, Activity, and Landscape Use Patterns of Heritage and Commercial Beef Cows Grazing Chihuahuan Desert Rangeland. *J. Arid. Environ.* **2022**, *199*, 104704. [CrossRef]
21. Campbell, D.L.M.; Lea, J.M.; Keshavarzi, H.; Lee, C. Virtual Fencing Is Comparable to Electric Tape Fencing for Cattle Behavior and Welfare. *Front. Vet. Sci.* **2019**, *6*, 445. [CrossRef]
22. Marini, D.; Kearton, T.; Ouzman, J.; Llewellyn, R.; Belson, S.; Lee, C. Social Influence on the Effectiveness of Virtual Fencing in Sheep. *PeerJ.* **2020**, *8*, e10066. [CrossRef]
23. Verdon, M.; Horton, B.; Rawnsley, R. A Case Study on the Use of Virtual Fencing to Intensively Graze Angus Heifers Using Moving Front and Back-Fences. *Front. Anim. Sci.* **2021**, *2*, 663963. [CrossRef]
24. Nyamuryekung'e, S. Transforming Ranching: Precision Livestock Management in the IoT Era. *Rangelands* 2023, *in press*.
25. Stevens, D.R.; Thompson, B.R.; Johnson, P.; Welten, B.; Meenken, E.; Bryant, J. Integrating Digital Technologies to Aid Grassland Productivity and Sustainability. *Front. Sustain. Food Syst.* **2021**, *5*, 602350. [CrossRef]
26. Nyamuryekung'e, S.; Duff, G.; Utsumi, S.; Estell, R.; McIntosh, M.M.; Funk, M.; Cox, A.; Cao, H.; Spiegal, S.; Perea, A.; et al. Real-Time Monitoring of Grazing Cattle Using LORA-WAN Sensors to Improve Precision in Detecting Animal Welfare Implications via Daily Distance Walked Metrics. *Animals* **2023**, *13*, 2641. [CrossRef]
27. Versluijs, E.; Niccolai, L.J.; Spedener, M.; Zimmermann, B.; Hessle, A.; Tofastrud, M.; Devineau, O.; Evans, A.L. Classification of Behaviors of Free-Ranging Cattle Using Accelerometry Signatures Collected by Virtual Fence Collars. *Front. Anim. Sci.* **2023**, *4*, 1083272. [CrossRef]
28. Fogarty, E.S.; Swain, D.L.; Cronin, G.M.; Moraes, L.E.; Trotter, M. Behaviour Classification of Extensively Grazed Sheep Using Machine Learning. *Comput. Electron. Agric.* **2020**, *169*, 105175. [CrossRef]
29. Aaser, M.F.; Staahltoft, S.K.; Korsgaard, A.H.; Trige-Esbensen, A.; Alstrup, A.K.O.; Sonne, C.; Pertoldi, C.; Bruhn, D.; Frikke, J.; Linder, A.C. Is Virtual Fencing an Effective Way of Enclosing Cattle? Personality, Herd Behaviour and Welfare. *Animals* **2022**, *12*, 842. [CrossRef]
30. Jacobs, J. International Association for Ecology Quantitative Measurement of Food Selection: A Modification of the Forage Ratio and Ivlev 's Electivity Index. *Oecologia* **1974**, *14*, 413–417. [CrossRef]

animals

MDPI

Article

Using Sound Location to Monitor Farrowing in Sows

Elaine van Erp-van der Kooij [1,*], Lois F. de Graaf [1], Dennis A. de Kruijff [1], Daphne Pellegrom [1], Renilda de Rooij [1], Nian I. T. Welters [2] and Jeroen van Poppel [1]

[1] Department of Animal Husbandry, HAS Green Academy, University of Applied Sciences, P.O. Box 90108, 5200 MA 's-Hertogenbosch, The Netherlands; j.vanpoppel@has.nl (J.v.P.)

[2] Department of Applied Biology, HAS Green Academy, University of Applied Sciences, P.O. Box 90108, 5200 MA 's-Hertogenbosch, The Netherlands

* Correspondence: l.verp@has.nl; Tel.: +316-2160-0396

Simple Summary: Automated monitoring can help farmers care during the farrowing of sows and piglets. Five sows were monitored in two field studies. A sound camera with small microphones showing sounds as coloured spots and a security camera was used to record the farrowing of sows and piglets. First, sound spots were compared with audible sounds, analysing video data. This gave many false positives, including visible sound spots but no audible sounds. Of piglet births, 23 of 50 piglet births were visible, but none were audible. The sow's behaviour changed when farrowing started. One piglet was silently crushed. Secondly, data were analysed at a slower speed, and sound spots were compared with sounds and behaviour separately. This resulted in better results, but again, many sound spots showed without audible sound. When adding up audible sounds and visible sow behaviour and comparing sound spots with the combination of sound and behaviour, results were much improved, with an accuracy of 91.2%, an error of 8.8%, a sensitivity of 99.6%, and a specificity of 69.7%. We conclude that sound cameras are promising tools, detecting sounds more accurately than humans. The most promising application for the sound camera is detecting the onset of farrowing.

Citation: van Erp-van der Kooij, E.; de Graaf, L.F.; de Kruijff, D.A.; Pellegrom, D.; de Rooij, R.; Welters, N.I.T.; van Poppel, J. Using Sound Location to Monitor Farrowing in Sows. *Animals* **2023**, *13*, 3538. https://doi.org/10.3390/ani13223538

Academic Editors: Roy Neville Kirkwood, Brett Ramirez, Janice Siegford, Hao Gan, Yang Zhao, Daniel Berckmans, Robert T. Burns and Lingjuan Wang-Li

Received: 2 October 2023
Revised: 11 November 2023
Accepted: 14 November 2023
Published: 16 November 2023

Abstract: Precision Livestock Farming systems can help pig farmers prevent health and welfare issues around farrowing. Five sows were monitored in two field studies. A Sorama L642V sound camera, visualising sound sources as coloured spots using a 64-microphone array, and a Bascom XD10-4 security camera with a built-in microphone were used in a farrowing unit. Firstly, sound spots were compared with audible sounds, using the Observer XT (Noldus Information Technology), analysing video data at normal speed. This gave many false positives, including visible sound spots without audible sounds. In total, 23 of 50 piglet births were visible, but none were audible. The sow's behaviour changed when farrowing started. One piglet was silently crushed. Secondly, data were analysed at a 10-fold slower speed when comparing sound spots with audible sounds and sow behaviour. This improved results, but accuracy and specificity were still low. When combining audible sound with visible sow behaviour and comparing sound spots with combined sound and behaviour, the accuracy was 91.2%, the error was 8.8%, the sensitivity was 99.6%, and the specificity was 69.7%. We conclude that sound cameras are promising tools, detecting sound more accurately than the human ear. There is potential to use sound cameras to detect the onset of farrowing, but more research is needed to detect piglet births or crushing.

Keywords: pigs; animal welfare; sound camera; behaviour; piglet crushing

1. Introduction

It is estimated that by 2050, the human world population could reach >9 billion, consuming 50–60% more food than at present. The majority of people still prefer animal proteins over plant-based food, and the demand for livestock products continues to grow, as does global food insecurity [1]. Sustainable intensification is one of the solutions [2]. With the intensification of food production and the industrialisation of animal production systems comes the fear of decreased animal welfare [3]. While meat production increases, society expects that animals used for meat are treated humanely and individually. Precision Livestock Farming (PLF) can improve or monitor animal welfare on farms if properly implemented [4,5]. PLF can be defined as managing livestock production using the principles of process engineering. Smart sensors are used to measure and monitor animal health and welfare [6–10]. Several sensors have been developed for the livestock sector. For pigs, the main focus is on the health and productivity of pigs, with sensors such as cameras, microphones, thermometers, and accelerometers being developed and applied [1,11–14]. A new development is the application of a sound camera, providing sound source localisation through an array of 64 microphones and visualising sound sources as coloured spots. Sound cameras are presently used for crowd control under outdoor conditions [15] and have been introduced in ecology [16] and agriculture, specifically in poultry and pigs [17–19]. Potentially, these cameras can be of use in monitoring welfare in pig farms since automated behaviour monitoring via sound and vision could help farmers to prevent health and welfare issues [20–24].

The farrowing phase in pig production and breeding is a crucial moment that has a great impact on the economics of the farm but also on the welfare of the sow and piglets. Piglet mortality is a big problem in modern sow farming [25]. For sow farms, relevant welfare issues around farrowing are a prolonged birthing process, possibly resulting in stillborn piglets and the crushing of piglets after farrowing [26–28]. Globally, approximately one piglet per litter is stillborn [29]. More stillbirths occur in sows that are confined in a crate during farrowing than in sows in open pens [30]. Stillbirths are often due to asphyxiation during the farrowing process. Piglets born later in the farrowing process have a higher chance of asphyxiation [29]. Asphyxiation is related to dystocia in sows due to prolonged farrowing or weak uterine contractions. This requires the stockman to assist the sow during farrowing. Improved farrowing management and human supervision during farrowing might decrease piglet mortality [25]. The automated monitoring of sows and alerting the farmer when the farrowing process stagnates can aid the farmer in optimising the health and welfare of his animals [8]. Therefore, in this study, we focus on the automatic monitoring of the farrowing process. Sows show specific behaviours and behavioural patterns before, during, and after farrowing. If we can monitor these behaviours, we might detect the onset of farrowing and follow the process of farrowing. Before farrowing, sows show a natural pattern of nest-building behaviour, such as foraging, rooting, and pawing, which are motivated by the desire to build a shelter to protect their offspring against predators and cold weather. Domestic sows, when given nest-building materials, still perform nest-building behaviours [31]. In farming systems with farrowing crates, nest-building behaviour is reflected in 'playing' with the available material, which can be a jute sack or a handful of straw. During farrowing, sows show pain-related behaviour such as tail flicks, straining, and pushing the back leg forward [30]. Approximately 50% of postnatal deaths in piglets are caused by the crushing of the sow [28]. Piglet crushing happens mostly within the first four days after farrowing. There is a large individual variation in piglet crushing between sows; sows show a more protective mothering style and crush fewer piglets. Especially during posture changes of the sows, the young piglets are at risk. Piglets vocalise during trapping events but also during other stress-related events. However, detecting these sounds might be used in a monitoring device [32].

In this study, we have used a sound camera together with a security camera to monitor sounds, vision, and sound location around farrowing as the first step in developing a sound-based early warning system for a stagnating birthing process and the prevention of piglet crushing.

2. Materials and Methods

2.1. Sound Camera

The L642V is a camera with an array of 64 microphones. The device uses delay-and-sum beamforming to localise different sound sources and visualises these on an acoustic map. Delay-and-sum beamforming signal processing can be divided into four steps. Since the sound of each source travels to every microphone along a different path (Step 1), the signals captured by the microphones are similar in wave form but differ in their delay and phase. Delay and phase are proportional to the travelled distances. The delays can be determined from the sound speed, the distance between the microphones, and the sound sources (Step 2). The Beamformer targets the point of one of the sound sources, shifting the signal of each microphone via the difference in runtime depending on the focus point. Therefore, the signal components of this one sound source are in phase, and of the other sound sources are out of phase (Step 3). Finally, the signals of the microphones are summed together and normalised by the number of microphones (Step 4). If a certain target point does not contain a source, the signals partly cancel each other out due to destructive interference. At target points with a source, the signals align, and add up due to positive interference. The maximum amplitude is calculated from the time signal and the sound source is visualised on the acoustic map. Due to the positive interference, target points with a source have a higher magnitude than those that do not, and thus, source locations can be found. In this study, we denoised sound by selecting a frequency band for which the sounds were visualised. In a pilot study, we tested different frequency bands and found that in a range from 39,000 to 49,000 Hz, the sounds of the sow and her environment were best visible, with the lowest influence of background noise. In the present study, each camera was manually tuned to a specific optimal frequency band of 2000 Hz within these limits, according to the noise in the farrowing unit in that period. The camera has a built-in spatial filter, which means it only shows sound sources within the selected area, which, in our study, was the farrowing pen for each sow. Finally, the camera also denoised automatically by not visualising a sound when no clear source could be found.

2.2. Study Design

The study was performed at a commercial pig farm with two farrowing units for 64 sows each. Sows were housed in farrowing crates within a farrowing pen, with access to a jute sack in the pen but no straw and a solid concrete floor. Sows were monitored around farrowing, staying in a farrowing pen of 2.80 × 1.75 m with a farrowing crate of 2.1 × 1.0 m. Two Bascom XD10-4 security cameras that showed sound and vision were placed above the pens to record audible sounds and the behaviour of the sow and piglets, with each camera viewing two pens. Three Sorama L642V sound cameras were placed directly above three farrowing pens, with each camera viewing one pen (Figure 1).

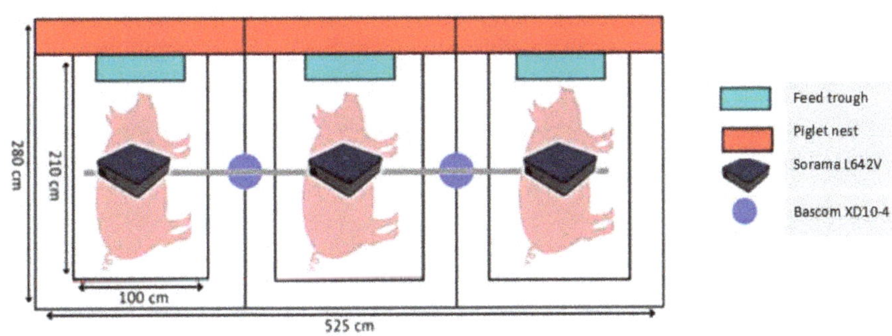

Figure 1. Experimental set-up for three sows with two Bascom XD10-4 security cameras and three L642V sound cameras.

Data from the cameras could not directly be recorded due to the safety settings. We, therefore, streamed the data to three laptops in the office of the farm. The screen of the laptops showed the image of the security camera and the sound camera side by side, as well as a clock, to synchronise the images if necessary (Figure 2). We recorded data using the screen recorder software Open Broadcast Software version 27.2.1 (OBS-studio), resulting in video files in the mp3 format. Laptops were remotely controlled using TeamViewer.

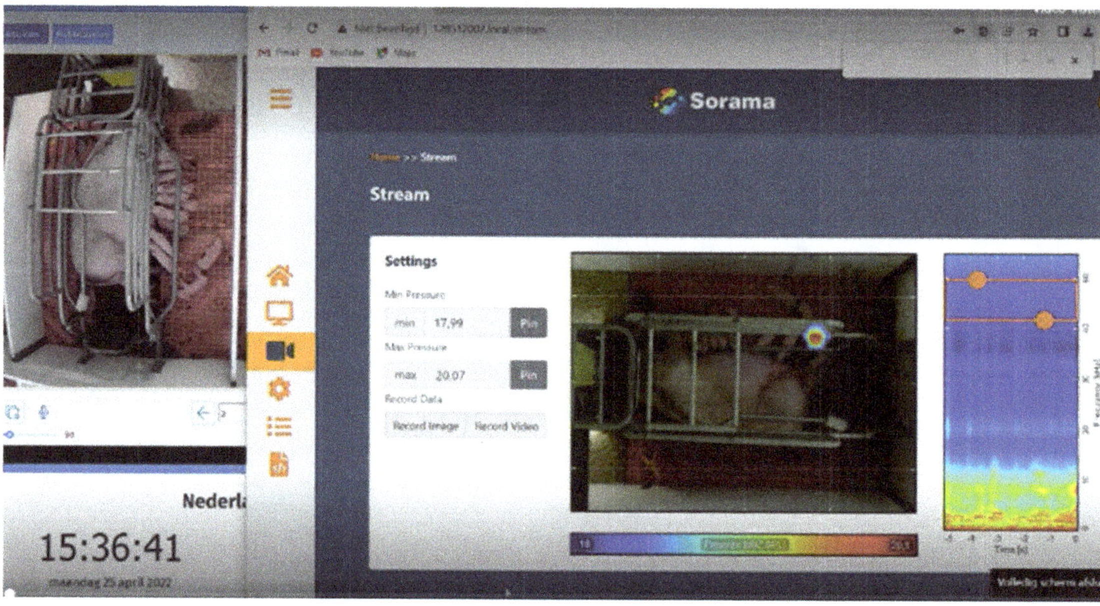

Figure 2. The screen capture that was used for the analysis of behaviour and sound in the farrowing pen.

Approximately 45 min of video data from each of the four sows were analysed after the first field study. Birthing events and crushing events were recorded from the video footage using the Observer XT (Noldus Information Technology). Four hours before farrowing and during farrowing, sow behaviours were recorded that were possibly associated with the birthing process. We used a simplified ethogram with three behavioural categories: lying (inactive), playing with the jute sack/rooting (snout on the concrete floor), and sitting/standing (inactive). Audible sounds were recorded from the recorded video of the security camera. Visible sounds were recorded from the sound camera. When a sound spot was visible at roughly the right location within ± 1 s from the audible sound, the spot was considered correct and positive. Audible sounds with no corresponding sound spot were considered false negatives. Sound spots with no audible sounds were considered false positives. Finally, when no sound was visible or audible for 2 s, this was considered correct negative. In Table 1, the connected audible and visible sounds and sound locations are shown.

Table 1. Sound spots and corresponding audible sounds that were considered correct for field study 1.

Sound Spot	Audible Sound
Head of the sow	Sow in crate
Rump of the sow	Sow in crate
Fence	Metal fence
Head of the sow	Trough
Trough	Trough
Outside pen	Neighbour sow
Cloud of spots	Neighbour sow

In Table 2, the connected audible and visible sounds and behaviours are shown.

Table 2. Sound spots and corresponding audible sounds and corresponding behaviours that were considered correct for field study 2.

Sound Spot	Audible Sound	Sound Spot	Behaviour
		Head of the sow	Playing with sack
Head of the sow	Front of the pen	Head of the sow	Eating
Rump of the sow	Metal fence	Trough	Eating
Fence	Metal fence	Rump of the sow	Standing
Head of the sow	Trough	Fence	Standing
Trough	Trough	Fence	Moving leg lying
Outside pen	Neighbour sow	Outside pen	None
Cloud of spots	Neighbour sow	Cloud of spots	None
		None	Lying down

Data from three sows were analysed after the second field study. From the security cameras, sow behaviours, and audible sounds were recorded using the Observer XT, and from the sound camera, sound spots were recorded for a short period during farrowing. When a sound spot was visible at roughly the right location within ± 1 s from the audible sound, the spot was considered correct and positive (CP). Audible sounds with no corresponding sound spot were considered false negatives (FN). Sound spots with no audible sounds were considered false positives (FP). Finally, when no sound was visible or audible for 2 s, this was considered correct negative (CN). For validation purposes, accuracy, error%, sensitivity and specificity were calculated as follows:

$$accuracy = \left(\frac{(CP + CN)}{total}\right) \times 100\%$$

$$error\ \% = \left(\frac{(FP + FN)}{total}\right) \times 100\%$$

$$sensitivity = \left(\frac{CP}{FN + CP}\right) \times 100\%$$

$$specificity = \left(\frac{CN}{(FN + CP)}\right) \times 100\%$$

3. Results

Data from five sows in two field studies were gathered and analysed for visible sounds (sound spots) and audible sounds, and, in field test 2, visible sow behaviour as well. There was a minor time lag in the recording of the visible sounds of approximately 1.5 s, which was corrected by adding 1.5 s to the recorded times of the sound spots. We adjusted the frequency settings for each camera per sow manually by testing which frequency band showed the best visualising of sound sources with the least noise at that time and place. This resulted in frequency bands of 39,570–41,570 Hz (camera 1), 46,060–48,060 Hz (camera 2) and 41,730–43,730 Hz (camera 3) in the first study and an adjustment to 45,630–47,630 Hz for camera 1 in the second study.

3.1. Field Study 1

In the first field study, we compared the audible and visible sounds of the sows before farrowing and recorded 13,351 sound spots and 981 audible sounds in 177 min of video data. We found a low agreement between the sound and vision data (Table 3).

Table 3. Results of visible and audible sounds before farrowing in the first field study (N = 4 sows, 177 min); sound spots without audible sound are false positives (FP), sound spots with audible sound are correct positive (CP); audible sounds without sound spot are false negative (FN) and 2 s of no audible sound and no sound spot indicates a correct negative (CN). If a sound spot was in the wrong location but at the correct time for an audible sound, it was considered either as a false negative (1st column) or as a correct positive (2nd column).

	Wrong Location Considered False Negative	Wrong Location Considered Correct Positive
FP	10,751	10,751
FN	1823	76
CP	1509	3256
CN	1582	1582
Accuracy	19.7	30.9
Error%	80.3	69.1
Sensitivity	45.3	97.7
Specificity	12.8	12.8

3.2. Field Study 2

In the second field study, we analysed 6 min (360 s) of video from two sows during farrowing. Video data were analysed at a 10-fold slower speed, and audible sounds, sow behaviour, and sound spots were recorded. This resulted in a somewhat higher but still unsatisfactory agreement between the sound and vision data when sound spots were compared to either audible sounds or visible behaviour (Table 4). However, when comparing sound spots with the combination of audible sounds and visible behaviour, results were much improved, with an accuracy of 91.2, an error% of 8.8, and a specificity of 69.7 (Table 4). In this analysis, we added up sounds and behaviours to compare with sound spots, considering an event as a correct positive if either sound or behaviour (or both) were shown at the same time and location as the sound spot.

Table 4. Sound spots compared with audible sound, visible behaviour, or sound and behaviour combined.

	Sound	Behaviour	Sound and Behaviour
# Sows	**4**	**2**	**2**
Accuracy	50.6	71.4	91.2
Error%	49.4	28.6	8.8
% Correct positive	35.4	57.5	71.3
Precision	41.8	66.8	89.2
Sensitivity	99.7	99.9	99.6
Specificity	23.6	32.7	69.7

3.3. Birthing Events and Piglet Crushing

For the detection of birthing events, data from four sows were used. A total of 50 piglets were born from these sows. For 23 of 50 piglet births, a sound spot was visible at the correct time; piglet births were usually not audible, but most were visible on the safety camera. This resulted in an accuracy of 71.4%. Sound spots were visible in the area behind the sow for some time after the piglet was born, but the movement of the piglets was not audible.

One piglet was crushed shortly after birth, but no sound was heard, and no sound spots were visible for this event, which was only visible on the video footage of the Bascom camera.

When we analysed behaviours around parturition, we found that the sows showed specific behaviours at specific times: playing with the jute sack in the farrowing pen was seen more before parturition and especially during the last hours before the piglets were born. Once farrowing started, this behaviour stopped almost completely (Table 5). During farrowing, typical movements of the legs of the sows were seen, seemingly associated with the birthing events.

Table 5. Behaviour of sows (N = 4) four hours before farrowing, during farrowing and four hours after farrowing: percentage of time inactive, either lying or sitting/standing, or being active, playing with the jute sack or rooting on the floor.

Behaviours	Before Farrowing	During Farrowing	After Farrowing
Lying	71.7%	93.5%	100.0%
Sitting/standing	3.3%	6.2%	0%
Jute sack/rooting floor	25%	0.3%	0%

4. Discussion

For this study, we used data from five sows to study the application of a sound camera in pig farming. Further research is needed using more sows and more repetitions to validate these findings. However, in this study, we were able to optimise the methods for analysing the results of sound spots, sound, and behaviour, and we found promising results for the application of the sound camera to monitor farrowing sows.

In this study, the human observer was the gold standard for audible sound. When comparing manual to automated scoring, there were some problems with finding the gold standard. Clinical research has shown that manual scores are usually qualitative or semiquantitative and subjective, even when conducted by a seasoned observer, while automated image analysis is quantitative, reproducible, and objective. Manual image analysis has some drawbacks [33] that can easily be extrapolated to manual sound analysis. The sources of bias include the illusion of size (size being influenced by the context), distinguishing colours, and lateral inhibition (increased response to edges). For sound analysis, these can translate into an illusion of loudness (being influenced by the loudness of other sounds), a distinguishing pitch (depending on the pitch of surrounding sounds), and an increased response to short and sharply defined sounds. General sources of bias include inattentional blindness (i.e., not paying attention) and confirmation bias (i.e., hearing what you expect or want to hear). Labelling audible sounds from videos recorded with a safety camera probably resulted in many false negatives for audible sounds and inaccuracies in labelling since the human observer either hears the sound and reacts too late or does not hear the sound at all, while the sound camera does receive the sound. Furthermore, the labelling of the sound spots was probably not accurate enough since we labelled at a normal speed. This resulted in many 'cloud of sound spots' events, with a cluster of sound spots occurring at once. Playing the videos at a 10-fold slower speed showed that the sound spot clouds were actually a series of sound spots that started with one spot in the correct place, followed by a cluster of spots in the area. Therefore, we adjusted the analysis for the second field study.

Most sows farrowed at night, with low visibility on the cameras, which increased the number of false positives (i.e., sounds visible in a different spot than audible) due to human error. In all tests, we counted many more sound spots than visible behaviours or audible sounds. This may very well be due to human error. A reliability analysis for labelling sound spots between the observers showed an agreement of 82%, but a Kappa value of 0.17 (indicating slight agreement) was obtained. The high number of sound spots and almost no silent periods lead to a high agreement by a chance of 0.78. This lowered the Kappa value [34,35]. In addition, the manual labelling of data as the gold standard is a point of discussion.

In the second field study, adding visible behaviour gave much better results, correctly classifying sound spots for visible but inaudible behaviours. Furthermore, we recorded sound from the Bascon camera and listened to the recordings, which is an indirect way of working with data and probably lowered the audibility for human observers. It seems that the sound camera is much more accurate and precise at detecting sounds than we, as human observers, are. Therefore, we advise combining audible sounds and visible behaviours when validating sound location sensors such as the sound camera we used.

We detected less than half of the piglet birth events with the sound camera. These sounds were low-pitched and soft. Filtering the noise from the sound camera and using the high-frequency bands to visualise sounds might have increased the number of false negatives. One piglet was crushed during the study without an audible sound or a sound spot from the sound camera. We expected to capture high-pitched sounds that accompanied crushing events, but if it happened fast, the piglet had no time to scream. From only one crushing event, we could not validate whether the sound camera could detect piglet crushing. In a study where sound was used to detect piglet distress, it was found that many more piglet stress calls were associated with other stress-related events than associated with trapping events. Although adding context-based event filters increased the results, sound might not be the preferred method for detecting crushing events [32]. Crushing events are mostly associated with posture changes in the sow, such as rolling over during resting. Sows that crush fewer piglets show fewer posture changes [28]. If these posture changes could be detected with a sound camera, this might be used to prevent the crushing of piglets.

The automatic monitoring of sow behaviours can be performed using cameras or activity meters [20,36–38]. To investigate whether the sound camera can be used to detect specific behaviours associated with the onset of farrowing, behaviours were recorded around farrowing. We found that playing with the jute sack and rooting was seen almost only before farrowing. This behaviour is associated with nest-building behaviour [31]. Typical leg movements were seen predominantly during farrowing and seemed to be associated with birthing events. These behaviours resembled the leg behaviours that were reported in previous research, which were classified as pain-related behaviours during farrowing [30,39]. Nest building behaviour was only seen before farrowing, and leg movements during farrowing; after farrowing, these behaviours stopped completely. These behaviours were associated with sound spots in the corresponding, typical locations as follows: playing with the jute sack or rooting on the floor was seen as sound spots near the head of the sow, and leg movements were seen as sound spots near the front or hind legs, detected with the sound camera. This is interesting because the cross-over from showing nest-building behaviour (playing with the jute sack and rooting on the floor) to lying down and showing only some typical leg movements seems to mark the onset of farrowing. If so, we could use the detection of these behaviours with the camera as a signal for the farmer that farrowing has started. Furthermore, the absence of both these behaviours, with sows lying down and being fully inactive, marks the end of the farrowing process. This could be combined with the fact that sound spots behind the sow showed the movement of piglets that were just being born. As long as new spots appeared behind the sow, the farrowing process was still in progress. The combination of these findings could be used to monitor the farrowing process and its duration. However, more research is needed to validate these assumptions.

5. Conclusions

Sound cameras are potentially interesting to apply in pig farming since they can detect sounds and sound locations better than the human observer. The behaviour of the sow and the movement of piglets in the pen could be reliably detected. However, we could not reliably detect piglet births and crushing events in this study. When analysing sound and visual data, it is important that a slower speed must be used to record the order of events and sound spots and that sound data are connected not only to audible sounds but also to behaviours that are inaudible to the human ear. The most interesting application for sound cameras seems to be detecting the onset of farrowing by recording sounds from the prepartum sow as she is preparing for the farrowing process and monitoring the farrowing process by detecting sounds of newly born piglets in the area behind the sow. Further research is needed to test this application, using more sows and more repetitions.

Author Contributions: Conceptualisation, E.v.E.-v.d.K.; methodology, E.v.E.-v.d.K. and J.v.P.; software, E.v.E.-v.d.K.; validation, E.v.E.-v.d.K.; formal analysis, E.v.E.-v.d.K.; investigation, E.v.E.-v.d.K., J.v.P., L.F.d.G., D.A.d.K., D.P., R.d.R., and N.I.T.W.; resources, E.v.E.-v.d.K.; data curation, L.F.d.G., D.A.d.K., D.P., R.d.R., and N.I.T.W.; writing—original draft preparation, E.v.E.-v.d.K.; writing—review and editing, E.v.E.-v.d.K. and J.v.P.; visualisation, D.P., R.d.R., and N.I.T.W.; supervision, E.v.E.-v.d.K. and J.v.P.; project administration, J.v.P.; funding acquisition, E.v.E.-v.d.K. All authors have read and agreed to the published version of the manuscript.

Funding: This project was funded by Holland High Tech with PPS-premium for research and development in the top sector HTSM, project HTSM 21.0382.

Institutional Review Board Statement: Not applicable.

Informed Consent Statement: Animals were observed on a commercial farm with cameras only and no handling took place, nor were the animals disturbed in any way due to the study. The farmer signed an informed consent statement, allowing us to use the data of their sows.

Data Availability Statement: Data in this study are not publicly archived due to privacy reasons since the study was performed at a commercial farm. Datasets are available from the author upon request.

Acknowledgments: We thank Twan Dirks for the opportunity to perform this study at his farm 'Beter Varken' and use the data of his sows. We are thankful for the loan of one sound camera from Noldus Information Technology, and the technical support from Ben Loke of Noldus Information Technology and Koen Diesveld of Sorama.

Conflicts of Interest: The authors declare no conflict of interest.

References

1. Benjamin, M.; Yik, S. Precision Livestock Farming in Swine Welfare: A Review for Swine Practitioners. *Animals* **2019**, *9*, 133. [CrossRef] [PubMed]
2. Charles, H.; Godfray, H.; Garnett, T. Food Security and Sustainable Intensification. *Philos. Trans. R. Soc. B Biol. Sci.* **2014**, *369*, 6–11. [CrossRef]
3. Broom, D.M. Animal Welfare: An Aspect of Care, Sustainability, and Food Quality Required by the Public. *J. Vet. Med. Educ.* **2010**, *37*, 83–88. [CrossRef] [PubMed]
4. Banhazi, T.M.; Lehr, H.; Black, J.L.; Crabtree, H.; Schofield, P.; Tscharke, M.; Berckmans, D. Precision Livestock Farming: An International Review of Scientific and Commercial Aspects. *Int. J. Agric. Biol. Eng.* **2012**, *5*, 1–11. [CrossRef]
5. van Erp-van der Kooij, E.; Rutter, S.M. Using Precision Farming to Improve Animal Welfare. *CAB Rev. Perspect. Agric. Vet. Sci. Nutr. Nat. Resour.* **2020**, *15*, 1–10. [CrossRef]
6. Wathes, C.M.; Kristensen, H.H.; Aerts, J.M.; Berckmans, D. Is Precision Livestock Farming an Engineer's Daydream or Nightmare, an Animal's Friend or Foe, and a Farmer's Panacea or Pitfall? *Comput. Electron. Agric.* **2008**, *64*, 2–10. [CrossRef]
7. Neethirajan, S. Recent Advances in Wearable Sensors for Animal Health Management. *Sens. Bio-Sens. Res.* **2017**, *12*, 15–29. [CrossRef]
8. Berckmans, D. General Introduction to Precision Livestock Farming. *Anim. Front.* **2017**, *7*, 6–11. [CrossRef]
9. Li, N.; Ren, Z.; Li, D.; Zeng, L. Review: Automated Techniques for Monitoring the Behaviour and Welfare of Broilers and Laying Hens: Towards the Goal of Precision Livestock Farming. *Animal* **2020**, *14*, 617–625. [CrossRef]
10. Halachmi, I.; Guarino, M.; Bewley, J.; Pastell, M. Smart Animal Agriculture: Application of Real-Time Sensors to Improve Animal Well-Being and Production. *Annu. Rev. Anim. Biosci.* **2019**, *7*, 403–425. [CrossRef]
11. Chapa, J.M.; Maschat, K.; Iwersen, M.; Baumgartner, J.; Drillich, M. Accelerometer Systems as Tools for Health and Welfare Assessment in Cattle and Pigs—A Review. *Behav. Process.* **2020**, *181*, 104262. [CrossRef] [PubMed]
12. Berckmans, D. Precision Livestock Farming Technologies for Welfare Management in Intensive Livestock Systems. *OIE Rev. Sci. Tech.* **2014**, *33*, 189–196. [CrossRef] [PubMed]
13. Wallenbeck, A.; Keeling, L.J. Using Data from Electronic Feeders on Visit Frequency and Feed Consumption to Indicate Tail Biting Outbreaks in Commercial Pig Production. *J. Anim. Sci.* **2013**, *91*, 2879–2884. [CrossRef] [PubMed]
14. D'Eath, R.B.; Jack, M.; Futro, A.; Talbot, D.; Zhu, Q.; Barclay, D.; Baxter, E.M. Automatic Early Warning of Tail Biting in Pigs: 3D Cameras Can Detect Lowered Tail Posture before an Outbreak. *PLoS ONE* **2018**, *13*, e0194524. [CrossRef]
15. Nguyen, Q.; Shen, G.; Choi, J.S. Sound Detection and Localization in Windy Conditions for Intelligent Outdoor Security Cameras. *Circuits Syst. Signal Process.* **2016**, *35*, 233–251. [CrossRef]

16. Mennill, D.J.; Battiston, M.; Wilson, D.R.; Foote, J.R.; Doucet, S.M. Field Test of an Affordable, Portable, Wireless Microphone Array for Spatial Monitoring of Animal Ecology and Behaviour. *Methods Ecol. Evol.* **2012**, *3*, 704–712. [CrossRef]
17. Exadaktylos, V.; Silva, M.; Ferrari, S.; Guarino, M.; Berckmans, D. Sound Localisation in Practice: An Application in Localisation of Sick Animals in Commercial Piggeries. In *Advances in Sound Localization*; IntechOpen: London, UK, 2011.
18. Du, X.; Lao, F.; Teng, G. A Sound Source Localisation Analytical Method for Monitoring the Abnormal Night Vocalisations of Poultry. *Sensors* **2018**, *18*, 2906. [CrossRef]
19. Adrion, F.; Keller, M.; Bozzolini, G.B.; Umstatter, C. Setup, Test and Validation of a UHF RFID System for Monitoring Feeding Behaviour of Dairy Cows. *Sensors* **2020**, *20*, 7035. [CrossRef]
20. Oczak, M.; Bayer, F.; Vetter, S.; Maschat, K.; Baumgartner, J. Comparison of the Automated Monitoring of the Sow Activity in Farrowing Pens Using Video and Accelerometer Data. *Comput. Electron. Agric.* **2022**, *192*, 106517. [CrossRef]
21. Shao, B.; Xin, H. A Real-Time Computer Vision Assessment and Control of Thermal Comfort for Group-Housed Pigs. *Comput. Electron. Agric.* **2008**, *62*, 15–21. [CrossRef]
22. Nilsson, M.; Herlin, A.H.; Ardö, H.; Guzhva, O.; Aström, K.; Bergsten, C. Development of Automatic Surveillance of Animal Behaviour and Welfare Using Image Analysis and Machine Learned Segmentation Technique. *Animal* **2015**, *9*, 1859–1865. [CrossRef]
23. Chen, C.; Zhu, W.; Liu, D.; Steibel, J.; Siegford, J.; Wurtz, K.; Han, J.; Norton, T. Detection of Aggressive Behaviours in Pigs Using a RealSence Depth Sensor. *Comput. Electron. Agric.* **2019**, *166*, 105003. [CrossRef]
24. Wurtz, K.; Camerlink, I.; D'Eath, R.B.; Fernández, A.P.; Norton, T.; Steibel, J.; Siegford, J. Recording Behaviour of Indoor-Housed Farm Animals Automatically Using Machine Vision Technology: A Systematic Review. *PLoS ONE* **2019**, *14*, e0226669. [CrossRef]
25. Okinda, C.; Lu, M.; Nyalala, I.; Li, J.; Shen, M. Asphyxia Occurrence Detection in Sows during the Farrowing Phase by Inter-Birth Interval Evaluation. *Comput. Electron. Agric.* **2018**, *152*, 221–232. [CrossRef]
26. Skovbo, D.K.F.; Hales, J.; Kristensen, A.R.; Moustsen, V.A. Comparison of Management Strategies for Confinement of Sows around Farrowing in Sow Welfare And Piglet Protection Pens. *Livest. Sci.* **2022**, *263*, 105026. [CrossRef]
27. Singh, C.; Verdon, M.; Cronin, G.M.; Hemsworth, P.H. The Behaviour and Welfare of Sows and Piglets in Farrowing Crates or Lactation Pens. *Animal* **2017**, *11*, 1210–1221. [CrossRef]
28. Andersen, I.L.; Berg, S.; Bøe, K.E. Crushing of Piglets by the Mother Sow (Sus Scrofa)—Purely Accidental or a Poor Mother? *Appl. Anim. Behav. Sci.* **2005**, *93*, 229–243. [CrossRef]
29. Leenhouwers, J.I.; Wissink, P.; van der Lende, T.; Paridaans, H.; Knol, E.F. Stillbirth in the Pig in Relation to Genetic Merit for Farrowing Survivall. *J. Anim. Sci.* **2003**, *81*, 2419–2424. [CrossRef]
30. Nowland, T.L.; van Wettere, W.H.E.J.; Plush, K.J. Allowing Sows to Farrow Unconfined Has Positive Implications for Sow and Piglet Welfare. *Appl. Anim. Behav. Sci.* **2019**, *221*, 104872. [CrossRef]
31. Yun, J.; Swan, K.M.; Vienola, K.; Farmer, C.; Oliviero, C.; Peltoniemi, O.; Valros, A. Nest-Building in Sows: Effects of Farrowing Housing on Hormonal Modulation of Maternal Characteristics. *Appl. Anim. Behav. Sci.* **2013**, *148*, 77–84. [CrossRef]
32. Manteuffel, C.; Hartung, E.; Schmidt, M.; Hoffmann, G.; Schön, P.C. Online Detection and Localisation of Piglet Crushing Using Vocalisation Analysis and Context Data. *Comput. Electron. Agric.* **2017**, *135*, 108–114. [CrossRef]
33. Aeffner, F.; Wilson, K.; Martin, N.T.; Black, J.C.; Hendriks, C.L.L.; Bolon, B.; Rudmann, D.G.; Gianani, R.; Koegler, S.R.; Krueger, J.; et al. The Gold Standard Paradox in Digital Image Analysis: Manual versus Automated Scoring as Ground Truth. *Arch. Pathol. Lab. Med.* **2017**, *141*, 1267–1275. [CrossRef] [PubMed]
34. Banerjee, M.; Fielding, J. Focus on Quantitative Methods Interpreting Kappa Values for Two-Observer Nursing Diagnosis Data. *Res. Nurs. Health* **1997**, *20*, 465–470. [CrossRef]
35. Byrt, T.; Bishop, J.; Carlin, J.B. Bias, Prevalence and Kappa. *J. Clin. Epidemiol.* **1993**, *46*, 423–429. [CrossRef]
36. Oczak, M.; Maschat, K.; Baumgartner, J. Dynamics of Sows' Activity Housed in Farrowing Pens with Possibility of Temporary Crating Might Indicate the Time When Sows Should Be Confined in a Crate before the Onset of Farrowing. *Animals* **2020**, *10*, 6. [CrossRef] [PubMed]
37. Oczak, M.; Maschat, K.; Berckmans, D.; Vranken, E.; Baumgartner, J. Classification of Nest-Building Behaviour in Sows on the Basis of Accelerometer Data. *Biosyst. Eng.* **2015**, *140*, 632–640. [CrossRef]
38. Küster, S.; Kardel, M.; Ammer, S.; Brünger, J.; Koch, R.; Traulsen, I. Usage of Computer Vision Analysis for Automatic Detection of Activity Changes in Sows during Final Gestation. *Comput. Electron. Agric.* **2020**, *169*, 105177. [CrossRef]
39. Mainau, E.; Manteca, X. Pain and Discomfort Caused by Parturition in Cows and Sows. *Appl. Anim. Behav. Sci.* **2011**, *135*, 241–251. [CrossRef]

animals

Article

A Framework for Transparency in Precision Livestock Farming

Kevin C. Elliott [1,*] and Ian Werkheiser [2,*]

1 Lyman Briggs College, Department of Fisheries and Wildlife, and Department of Philosophy, Michigan State University, East Lansing, MI 48825, USA
2 Department of Philosophy, University of Texas Rio Grande Valley, Edinburg, TX 78539, USA
* Correspondence: kce@msu.edu (K.C.E.); ian.werkheiser@utrgv.edu (I.W.)

Simple Summary: The emergence of precision livestock farming (PLF) raises important issues for many different social groups, including farmers, consumers, regulators, and the food industry. This paper explores how those who develop PLF systems can communicate more effectively with different groups about the technologies that they are creating. We suggest that developers reflect on four issues: (1) the different kinds of information that various groups might want to know; (2) the audiences that might care about these different kinds of information; (3) the major difficulties involved in providing the information; and (4) potential strategies for overcoming those difficulties.

Abstract: As precision livestock farming (PLF) technologies emerge, it is important to consider their social and ethical dimensions. Reviews of PLF have highlighted the importance of considering ethical issues related to privacy, security, and welfare. However, little attention has been paid to ethical issues related to *transparency* regarding these technologies. This paper proposes a framework for developing responsible transparency in the context of PLF. It examines the kinds of information that could be ethically important to disclose about these technologies, the different audiences that might care about this information, the challenges involved in achieving transparency for these audiences, and some promising strategies for addressing these challenges. For example, with respect to the information to be disclosed, efforts to foster transparency could focus on: (1) information about the goals and priorities of those developing PLF systems; (2) details about how the systems operate; (3) information about implicit values that could be embedded in the systems; and/or (4) characteristics of the machine learning algorithms often incorporated into these systems. In many cases, this information is likely to be difficult to obtain or communicate meaningfully to relevant audiences (e.g., farmers, consumers, industry, and/or regulators). Some of the potential steps for addressing these challenges include fostering collaborations between the developers and users of PLF systems, developing techniques for identifying and disclosing important forms of information, and pursuing forms of PLF that can be responsibly employed with less transparency. Given the complexity of transparency and its ethical and practical importance, a framework for developing and evaluating transparency will be an important element of ongoing PLF research.

Keywords: precision livestock farming; transparency; philosophy of science; responsible innovation; open science; community epistemic capacity; stakeholder engagement; epistemic justice

Citation: Elliott, K.C.; Werkheiser, I. A Framework for Transparency in Precision Livestock Farming. *Animals* **2023**, *13*, 3358. https://doi.org/10.3390/ani13213358

Academic Editors: Yang Zhao, Daniel Berckmans, Hao Gan, Brett Ramirez, Janice Siegford, Lingjuan Wang-Li and Robert T. Burns

Received: 27 September 2023
Revised: 20 October 2023
Accepted: 25 October 2023
Published: 29 October 2023

1. Background

Precision livestock farming (PLF) is an important developing suite of technologies [1–4]. The goal of PLF is "to manage individual animals through continuous real-time monitoring of health, welfare, production/reproduction, and environmental impact" [2]. It involves collecting information about the well-being of livestock through a range of sensors and analytic tools that can generate data through images, sounds, heart rate monitors, accelerometers, chemical analysis of waste, and a range of other tools. Some of these tools are currently in use, others are being used on a small scale as proof-of-concept, and many others are only in development [5–7]. The data from these systems are typically analyzed using algorithms that turn the low-level data into meaningful information that can guide the decision-making of farmers. PLF also has the potential to be operated via "closed loop" systems, whereby systems can self-correct based on the collected data without depending on human guidance [8–10], though these systems lead to concerns that will be discussed in a later section of this paper. These closed-loop systems can be facilitated by machine learning applied to large datasets of instrument data and outcomes, which can generate both insights into connections between variables that human farmers would not look for or notice as well as faster and more accurate predictions of outcomes for animals based on limited information [11,12].

PLF has the potential to generate a number of benefits for farmers, consumers, and farmed animals. Nevertheless, PLF also generates or exacerbates preexisting ethical and social issues that need to be addressed if it is to be implemented in a socially responsible fashion. While a fully comprehensive review of potential benefits and concerns is beyond the scope of this paper, some examples will be helpful for our later discussion of how to reap the benefits while mitigating some of the concerns. For example, an example of potential benefits is that as the number of animals on farms increases, it can be more difficult to ensure their welfare, but PLF can help farmers keep closer track of their livestock [1,13]. For animals who experience stress in the presence of humans, PLF could also ease their stress by allowing them to be monitored with less direct human interaction [14]. PLF could also ease the workload on farmers by creating automated systems that address potential problems without requiring human intervention [8–10]. PLF can also promote sustainability [15,16], such as by enabling farmers to feed their animals with more precision, thereby avoiding waste [17,18], thus increasing both environmental and economic sustainability for the farm. PLF can even benefit consumers by facilitating more careful tracking of animals through the supply chain, thereby providing greater transparency for consumers who want to know where animals have been and how they have been treated [3].

As an example of potential concerns, PLF could contribute to the general trend toward consolidating smaller farms into larger ones, thereby altering rural communities and eliminating agricultural jobs [14]. One might also worry that PLF could be used as a "cover" to argue that agricultural intensification is compatible with protecting animal welfare, whereas critics might contend that farm animals would actually be better off on smaller, more traditional farms [6]. Privacy is another important ethical issue raised by PLF [3]. Given all the data collected through these systems, it will be essential to develop policies governing the sharing of this data with outside parties. Finally, one might worry about the ways that PLF could change the relationships between farmers and their livestock by eliminating the direct connections that farmers currently have as they assess the well-being of their animals [14,19].

2. Transparency and PLF

Many ethical issues have begun to be discussed in the scholarly literature on PLF; however, the issue of transparency about PLF has received relatively little attention thus far. There has been some discussion about the potential for PLF, in conjunction with technologies like blockchain (the decentralized transaction and data management technology most known for its use in cryptocurrency [20]), to provide transparency for consumers about the paths that animals have taken through the production process [3], but this literature does not focus on the need for transparency about the nature of PLF technologies themselves. Although a few authors have begun to call for this kind of broader transparency about the values embedded in PLF systems [6], there has been little discussion about how best to achieve this form of transparency or about the particular challenges that arise in doing so. This is an important gap because transparency will be crucial for pursuing PLF in a responsible fashion. The open science movement has recently highlighted the important role that transparency plays in promoting reproducible science, accelerating advances, and fostering public engagement with scientific research [21–23]. In the context of PLF, transparency is especially important because there are a number of values at play in this area of research (e.g., animal welfare, profit, sustainability, rural development, and so on).

These values can come into conflict, and they can be interpreted in different ways [14], so transparency is important to enable farmers and consumers to decide what kinds of PLF systems actually help them to achieve their goals.

While it seems clear that transparency is important in this context, transparency is not a simple concept. As one of the authors (Elliott) has put it: "In a very basic sense, something is transparent when one can see through it. Thus, transparency is used as a metaphor in fields like politics and science to express the notion that information or processes have been 'made visible'" [24]. However, he goes on to point out that this basic idea of making information visible can involve a great deal of complexity in practice. He proposes a taxonomy that distinguishes different forms of transparency in terms of their purposes, the audience for the information being provided, the content of the information being provided, and the timeframe, actors, mechanisms, and venues through which the information is provided [25]. He also notes that there are dangers associated with pursuing transparency, and those dangers have to be weighed against the benefits.

Although it is not necessary for our purposes to go into all these dimensions, it is helpful to keep in mind some of the major reasons for pursuing transparency and concerns about doing so. One possible reason for pursuing transparency is the notion that it is either inherently good or that an essential feature of scientific practice is to be open about one's work [26]. But even if one rejects this "intrinsic" argument for transparency, there are a wide variety of instrumental reasons for pursuing it. As intimated above, these include: (1) promoting higher-quality science by facilitating external scrutiny of it; (2) promoting faster innovation; (3) fostering trust; (4) fostering greater diversity and inclusion in the scientific community; and (5) equipping the recipients of information to make better decisions for themselves. These benefits need to be weighed against a variety of potential concerns: (1) using up limited resources, including time, in an effort to provide information; (2) generating confusion on the part of those receiving information; (3) revealing private information or confidential business information; (4) assisting "bad actors" who aim to use the information inappropriately; and (5) creating rigid requirements that detract from the diversity of the scientific community's methods or practices [27]. Elliott emphasizes that one can typically address these concerns without abandoning the pursuit of transparency altogether [25]. One of the benefits of thinking about transparency in terms of a taxonomy with multiple dimensions is that one can explore different forms of transparency that are not as susceptible to concerns. For example, even if it would violate research participants' privacy to provide detailed information about them, it might still be possible to provide some transparency in the form of more limited or de-identified information about the participants. Similarly, if it would threaten confidential business information to provide all the data underlying a study, it might still be possible to provide some transparency in the form of information about the methods and principles used to analyze the data.

This paper draws on these insights in order to propose a framework for pursuing transparency about PLF. Building on the more general taxonomy of transparency developed by Elliott [25], the framework consists of four parts: audience, content, challenges, and strategies (see Figure 1; Table 1). According to this framework, those seeking to promote transparency about PLF should first consider, in specific contexts, the audiences toward which they are striving to provide information. Building on this consideration of audiences, they can consider the specific content that is most relevant to disclose. In order to communicate this content in a meaningful way, it is important to recognize the challenges that can make it difficult to achieve. Finally, drawing on the other elements of the framework, strategies can be developed for achieving meaningful forms of transparency. The following section employs philosophical methods of analysis (especially conceptual clarification) to elaborate on each of the four elements of the figure in the context of PLF. It provides examples of major audiences, content, challenges, and strategies that could be important to consider in a PLF transparency initiative, with the understanding that further empirical investigation of each element would help provide additional guidance for those seeking to implement the framework. Although this framework has been developed specifically for application to PLF, many of its features could also be applicable to other areas of agriculture and biotechnology in general.

Figure 1. Representation of a framework for achieving transparency in the context of PLF.

Table 1. A sample of major audiences, content, challenges, and strategies to be considered as part of transparency efforts.

Audience	Content	Challenges	Strategies
• Other scientists and engineers • Farmers • Consumers • Industry groups • Regulators	• Basic operation, strengths, and weaknesses of PLF system • Data generated by the system • Basic goals of the designers • Implicit values of the system • Operation and nature of underlying ML algorithms	• Difficulty identifying and communicating with relevant audiences • Not knowing the information to be disclosed • Lack of trust • Problematic motivations of the developers or communicators • Opaqueness of ML algorithms	• Collaborations with end users • Design to minimize transparency needs • Acknowledgment of major guiding values • Independent verification • Community epistemic capacities • Explainable AI

3. A Framework for Transparency

3.1. Audience

In order to provide appropriate transparency, it is important to consider the audience toward which information is being directed because different audiences have different informational needs and different ways of obtaining information. For example, the open science movement has generally focused on communicating information in ways that serve other scientists and technologists. While some elements of the open science movement can be helpful to non-specialists (e.g., publishing articles in open access formats), most features of the open science movement (e.g., making raw data available in publicly accessible databases) are geared primarily toward the scientific community. To meet the needs of non-specialists, it is often insufficient merely to make information available; the information generally needs to be interpreted in ways that are meaningful to them [21]. In the context of PLF, we suggest at least five different audiences that could have unique informational needs: scientists and engineers, farmers, consumers, industry groups, and regulators.

Scientists and engineers are likely to be interested in fairly traditional elements of open science, such as open data and open access to research materials [28]. In contrast, farmers are less likely to want this technical information and are more likely to want the "take-home" lessons about what these systems can do, how they work, and what their limitations are [5,29]. Consumers would probably not even care about the working of the systems, but at least some consumers might be interested in the "implicit values" associated with the systems (e.g., whether animal products from farms that employ PLF systems promote particular values concerning animal welfare or sustainability) [30]. Various industry groups, such as meat-packing companies, distributors, wholesalers, grocery stores, and restaurants, are likely to have a mixture of informational needs that could vary depending on how closely they work with farmers, regulators, or consumers. Finally, regulators are likely to be interested in the extent to which PLF systems can be designed to ensure compliance with regulations, whether they could inadvertently violate them, and whether compliance with regulations would become more or less difficult to verify when using the systems. Of course, these five categories do not include all the audiences that could be considered. For example, PLF system designers are often called on to consider "society" or the "public" [5]. However, we contend that this is such a large group with so many different informational needs that it cannot be usefully analyzed as a single audience in this model. Instead, it would need to be broken down into different interest groups. We also acknowledge that we have characterized audiences in terms of their likely informational needs, but it would be important to actually interact with these audiences in order to determine their informational needs in more concrete detail.

3.2. Content

The second component of our framework for transparency about PLF is the content to be disclosed. We have already seen that different audiences are likely to care about different kinds of information. Without providing an exhaustive discussion of all the kinds of content that could be discussed, this section probes more deeply into five major categories of information about PLF that could be disclosed as part of a transparency initiative. The first category of information concerns the basic workings of a PLF system, including its major strengths and limitations. For example, people might want to know what features of the animals the system measures, what outputs the system strives to maintain, the basic features of how the system functions, the conditions under which the system was developed and tested, and the safeguards that are in place to prevent the system from malfunctioning. When new PLF systems are being implemented, audiences will want to know enough about them to feel comfortable that they will work successfully. They may also want to have this information in comparison to other existing or possible alternative PLF systems.

A second category of information focuses not so much on how PLF systems work but on the data generated by them. For example, consumers and regulators might want to receive information about how often animals experience disease or other forms of stress. They might also want to know how often closed-loop PLF systems need to make particular sorts of adjustments and what those adjustments are. In addition, they might want to use the information generated by PLF systems to track the movements of animals through the agricultural system or to ensure that the animals meet criteria for particular kinds of certification (e.g., organic or cage-free). All the groups mentioned in the previous section may well also be interested in comparing data from one operation to others (though in this case some of those groups, most notably farmers and industry groups, would both want that information and possibly want to keep private information about their own operation).

In addition to the basic information about how a PLF system works and the data generated by it, a third major form of content concerns the goals that the system was designed to promote. For example, one might wonder whether the developers were particularly focused on efficiency, animal welfare, or environmental sustainability and how they prioritized those values when they came into conflict. One might also wonder how they defined those concepts, such as what elements of animal welfare or environmental sustainability they focused on. Research has shown that farmers, to pick one audience as an example, often have a few values they prioritize and want to actively maximize, but at the same time see other values as constraints on that maximization rather than something they are also trying to maximize [31].

A fourth category of content about PLF systems concerns the implicit values embedded in them. This is a more difficult form of information to disclose because it is not always obvious to those developing and working with the systems. Implicit values arise when developers make particular choices when designing systems (e.g., measuring particular things, analyzing the data in particular ways), and those choices end up serving some values rather than others (e.g., promoting animal welfare over profit, or vice versa) [32]. This form of content can be very significant because even if the users of PLF systems feel comfortable that their values are generally aligned with those of the system designers (e.g., desiring to promote some particular definition of animal welfare), they might still worry that the system could inadvertently reflect implicit values that they disagree with. For example, when testing the system, the developers might have regarded a 95% success rate at identifying injuries to the animals as adequate. However, some users might have higher standards and would have wanted something like a 99% success rate in order to feel comfortable relying on the PLF system. Along the same lines, developers might be concerned with the overall rate of injuries, while some users might be more concerned about the injuries to mothers, the young, the weak, etc. Developers of a PLF system have to make numerous decisions about how accurate all their sensors need to be, what endpoints they need to measure, and how to handle tradeoffs between optimizing different features of the system. All of these choices can make the system as a whole more prone to various sorts of errors, and the users of the system might not agree with the developers' implicit values about what kinds of errors are most important to avoid and what frequency of errors is acceptable.

A fifth form of content about PLF systems is even more fraught with difficulties. This content concerns the operation of the machine learning algorithms associated with some PLF systems. Because the working of machine learning algorithms is typically not comprehensible to human beings, it raises particularly important issues related to transparency [33–35]. Some users of PLF systems might want to know, for example, how confident they should be that the algorithms will generate reliable conclusions in the context of their farms. They might also want to know whether the concerns about implicit values discussed in the previous paragraph might apply to the machine learning algorithms. For example, although most farmers probably would not be interested in the details about how these algorithms were developed and how they operate, they might have the general worry that the algorithms might not be prioritizing exactly the same features of animal welfare as the farmers. For instance, the algorithms might be designed to promote the animals' growth or freedom from disease, whereas some farmers might be more concerned about the animals' activities or subjective experiences. Users might also be concerned with the likelihood that the algorithms are finding specious connections between inputs and outcomes that a human would correctly judge as an artifact of limited or incorrect data. Without receiving more information about the nature of these machine learning algorithms, many users might feel uncomfortable making use of them in their farming operations [33].

3.3. *Challenges*

The next component of our proposed framework consists of the challenges associated with pursuing transparency. Many of these challenges have already emerged from our discussion of different audiences and content. For example, one challenge is that those developing PLF systems might not have a clear understanding of the different audiences they need to consider and the kinds of content they want to know. Even if they could identify the relevant audiences, they might have difficulty, in some cases, explaining the technical details of how their systems work. Especially in the case of closed-loop systems that gather information about the animals and make automatic corrections in response to the available information, the systems might be too complex to explain easily to those who might want to know how they operate.

But even if the developers could find a way to disclose all the detailed information that some audiences might want, another challenge is that the developers themselves might be ignorant of some relevant information. As discussed above, some audiences might want to know how PLF systems implicitly promote some values (e.g., particular conceptions of animal welfare) over others. However, when developing new technological innovations, it is often unclear—even to the developers—how they promote particular values or interests over others. One might think that this is not specifically a problem related to *transparency*; rather, it might seem to be merely a function of the complexity of PLF systems and the limited perspective associated with particular disciplinary approaches. However, it is important to keep in mind the full breadth of the conception of transparency associated with Elliott's taxonomy [25] and the framework that we are developing here. In Elliott's view, the audiences for transparency can include not only external stakeholders but also the scientists and engineers who are working on a project. Important values and assumptions associated with the project may not be clear to them, and one of the goals of a transparency initiative could be to help them develop a richer understanding of these project features so that they can in turn be more transparent with others outside the project.

Another challenge to successful communication is trust between developers and potential users. If those trying to communicate the information do not trust the receivers, it is possible that they will try to manipulate them to obtain a desired outcome or to protect themselves. The communicators might also simplify the information they are communicating to the point that it becomes unhelpful, misleading, or even incorrect. If the receivers do not trust the communicators, they may not be able to take up any of the information they are being provided, even if it is in their own interest to do so. While some trust issues have to do with presentation style and other tools of rhetoric, and some trust issues have to do with the creation and maintenance of relationships between the various groups, it is also the case that previous harmful incidents can lead to justifiably low trust in ways that are quite difficult to overcome [36].

A separate but related challenge to achieving transparency involves the motivations and interests of those offering or receiving the information. For example, those promoting PLF systems might prefer not to acknowledge some of the systems' weaknesses or the ways the systems prioritize some values over others. In addition, those using the systems might not want to disclose certain kinds of data generated by the systems (e.g., about rates of disease or injuries or other animal welfare concerns). Although this reticence to share information might sometimes be narrow-minded and self-serving, it could also reflect the legitimate concern that those receiving the information could misinterpret it and draw illegitimate conclusions. It could also reflect companies' concerns about protecting their intellectual property and safeguarding confidential business information. These IP and related concerns might differ between different countries' IP regimes, making communication across national boundaries more difficult. For farmers, the fear of providing detailed information about their operations might even motivate some of them not to adopt PLF systems at all.

Finally, the use of machine learning algorithms in PLF systems raises special challenges. In some cases, the challenge might just be that the recipients of information about the algorithms do not have the background knowledge to understand them. Although this might not initially seem to be a failure of transparency (because information about the algorithms is available), we would classify it as a failure of transparency because the available information is not understandable or usable for the intended audience. An additional challenge is that in the case of machine learning algorithms, even the developers might not know what factors are responsible for the algorithms' outputs. As a result, the developers might be unable to provide a number of other relevant pieces of information. For example, they might not be able to identify the precise conditions under which the systems could become unreliable. They also might be unable to identify important biases, limitations, or "blind spots" that affect the systems. These limitations could be caused by biases in the training data used to develop the system, or they could be a function of the particular phenomena that the algorithms focus on. Without understanding how the algorithms work, it could be very difficult to provide detailed information about their strengths and weaknesses and the implicit values associated with them. This is particularly the case for the subset of machine learning commonly referred to as "deep learning" algorithms, in which programmers do not set which aspects of the environment the system is tracking.

3.4. Strategies

The final component of our transparency framework is to explore strategies for providing relevant content to audiences in a manner that overcomes major challenges. Although different situations and challenges are likely to call for different strategies, some general ones could prove helpful under a variety of circumstances. For example, one important kind of strategy is for the developers of PLF systems to collaborate with end users during the development process. One benefit of this co-creation process is that it helps the end users understand the major features of PLF systems, and thus it provides a form of transparency that would be difficult to provide otherwise. In addition, when users and developers collaborate, they are more likely to identify ways in which implicit values could be embedded in the operation of the systems. Thus, this process of collaboration "early and often" [37] can be especially helpful for uncovering features of PLF systems that could be important to disclose but that would not have even been recognized otherwise. (See Thompson et al.'s discussion of an "ethical matrix" for a useful framework for these stakeholders to express their ethical concerns [6]).

Another, and somewhat less ambitious, kind of strategy to address transparency challenges is to design PLF systems in ways that require less transparency. One way to perform this is to make systems less complex. For example, whereas a closed-loop system might leave farmers "in the dark" about why particular changes are being made by the system, an open-loop system might give the farmers more control and understanding of what is happening. For instance, in an open-loop system, farmers might be notified that something is wrong or sub-optimal. The farmers could then investigate and decide whether there is indeed a problem and how they want to address it. Because the PLF system would not be making as many decisions on their behalf, farmers would not need to demand as much transparency from it. Similarly, if a PLF system were designed so that the farmers could control more features of how it operated, that could also obviate some of the need for transparency. For example, if farmers could choose what temperatures they wanted the system to maintain, what amount of food to provide, what levels of activity they expected from the animals, and so on, they would be in control over more variables and thus less dependent on receiving assurances that the system would handle these variables in accordance with their preferences. This is a tradeoff with the kinds of efficiencies and emergent animal management practices sometimes promised by PLF but might be worth that loss for those with less trust in the technologies, at least initially.

Another general strategy for addressing transparency challenges is to provide acknowledgements of the major values that guided the development of particular PLF systems. This strategy is an alternative to providing numerous details that could become overwhelming and impractical to disclose. For example, rather than providing extensive details about all the input and output variables associated with a PLF system and the ways those values were analyzed, the developers of a system could focus on the main features of the system (e.g., animal welfare or economic efficiency) that the system was designed to maximize. In order to avoid being misleading, it would also typically be important to clarify how those features were defined or conceptualized (e.g., how animal welfare was measured) and how trade-offs are handled (e.g., what the system does when it would cost more to maximize particular features of animal welfare). Along these lines, the developers could also provide general information about how carefully their systems had been developed and under what range of conditions they had been implemented, which could serve as a proxy for more detailed information about the reliability of the systems. As mentioned above, it is sometimes the case that people might not be aware of the implicit values they hold or are expressing in their work. In such cases, going through guided processes of dialogue to draw these out could be an important step before acknowledging them [6,38].

To address issues of trust and motivated communication, several strategies can be employed. For example, third-party experts who have higher trust from multiple parties and who have different interests can verify claims. This could include regulators, activist watchdogs, the media, or engineers and designers working for a non-profit institution such as a university. Alternately, increasing the community epistemic capacities of both potential users and developers of a PLF system so that they can verify claims, formulate better questions, and communicate their values more clearly can mitigate a lack of trust [39].

Finally, although machine learning algorithms create significant complications for achieving transparency, there are some steps that can be taken to address these challenges. For example, the field of "explainable AI" (XAI) explores ways to clarify some of the features of machine learning algorithms even if they cannot be fully understood [34], such as by running "sensitivity analyses" to determine which input variables make the most difference to the predictions that a system provides. Sometimes uncertainty estimates can also be provided so that the users of a machine learning system have a sense of its reliability. Thus, the use of machine learning algorithms in PLF systems is not a barrier to providing at least some forms of transparency. Additionally, in keeping with the strategy discussed previously of making PLF less complex, machine learning could be added only in a gradual fashion so that its predictions could initially be compared to human judgment in open-loop processes. This comparison could provide its own form of transparency, as users could see the ways in which machine learning complemented, recapitulated, or differed from their own judgment.

4. Conclusions

Transparency is an important tool for promoting public engagement with emerging technologies and navigating ethical disagreements or concerns between various stakeholders. However, the subject matter of PLF technologies—animals and food—makes transparency even more important in this context than in many other scientific and technological contexts. People have a (literally) visceral reaction to the food they eat, which makes them wary of novelty. Moreover, PLF is wading into the complex relationship our food system has with non-human animals raised to be food; we acknowledge that they have welfare concerns that we must respect, but we eventually take their lives for the benefit of humans. In addition, producers, distributors, regulators, and all the other actors in the food system must balance desires for adaptation and innovation with a need to preserve the current system's ability to function. The fraught discourse around genetically modified organisms and other forms of agricultural biotechnology provides a good example of how a lack of meaningful transparency early in research and development can contribute to catastrophic loss of trust; it is an example that PLF developers would do well to learn from and avoid.

Author Contributions: Conceptualization, K.C.E. and I.W.; research—K.C.E. and I.W.; writing—original draft, K.C.E. and I.W.; writing—review and editing, K.C.E. and I.W. Authors are ordered alphabetically. All authors have read and agreed to the published version of the manuscript.

Funding: This research received no external funding.

Institutional Review Board Statement: Not applicable.

Informed Consent Statement: Not applicable.

Data Availability Statement: Not applicable.

Conflicts of Interest: The authors declare no conflict of interest.

References

1. Benjamin, M.; Yik, S. Precision livestock farming in swine welfare: A review for swine practitioners. *Animals* **2019**, *9*, 133. [CrossRef]
2. Berckmans, D. General introduction to precision livestock farming. *Anim. Front.* **2017**, *7*, 6–11. [CrossRef]
3. Neethirajan, S.; Kemp, B. Digital livestock farming. *Sens. Bio-Sens. Res.* **2021**, *32*, 100408. [CrossRef]
4. Norton, T.; Chen, C.; Larsen, M.L.V.; Berckmans, D. Precision livestock farming: Building 'digital representations' to bring the animals closer to the farmer. *Animal* **2019**, *13*, 3009–3017. [CrossRef]
5. Brier, D.; Eastwood, C.R.; Dela Rue, B.T.; Viehland, D.W. Foresighting for responsible innovation using a delphi approach: A case study of virtual fencing innovation in cattle farming. *J. Agric. Environ. Ethics* **2020**, *33*, 549–569. [CrossRef]
6. Thompson, P.B.; Thorp, L.; Ginsburg, B.L.; Zivku, T.M.; Benjamin, M. Early Ethical Assessment: An Application to the Sustainability of Swine Body Scanners. *Sustainability* **2021**, *13*, 14003. [CrossRef]
7. Vaintrub, M.O.; Levit, H.; Chincarini, M.; Fusaro, I.; Giammarco, M.; Vignola, G. Precision livestock farming, automats and new technologies: Possible applications in extensive dairy sheep farming. *Animal* **2021**, *15*, 100143. [CrossRef]

8. Frost, A.R.; Parsons, D.J.; Stacey, K.F.; Robertson, A.P.; Welch, S.K.; Filmer, D.; Fothergill, A. Progress towards the development of an integrated management system for broiler chicken production. *Comput. Electron. Agric.* **2003**, *39*, 227–240. [CrossRef]
9. Pomar, C.; Remus, A. Fundamentals, limitations and pitfalls on the development and application of precision nutrition techniques for precision livestock farming. *Animal* **2023**, *17*, 100763. [CrossRef]
10. Upinder, K.; Malacco, V.M.R.; Bai, H.; Price TPDatta, A.; Xin, L.; Sen, S.; Nawrocki, R.A.; Chiu, G.; Sundaram, S.; Min, B.C.; et al. Invited review: Integration of technologies and systems for precision animal agriculture—A case study on precision dairy farming. *J. Anim. Sci.* **2023**, *101*, skad206. [CrossRef]
11. Fernandes, A.F.A.; Dórea, J.R.R.; Rosa, G.J.D.M. Image analysis and computer vision applications in animal sciences: An overview. *Front. Vet. Sci.* **2020**, *7*, 551269. [CrossRef]
12. Gauthier, R.; Largouët, C.; Dourmad, J.Y. Prediction of litter performance in lactating sows using machine learning, for precision livestock farming. *Comput. Electron. Agric.* **2022**, *196*, 106876. [CrossRef]
13. Tuyttens, F.A.; Molento, C.F.; Benaissa, S. Twelve threats of precision livestock farming (PLF) for animal welfare. *Front. Vet. Sci.* **2022**, *9*, 889623. [CrossRef]
14. Werkheiser, I. Precision livestock farming and farmers' duties to livestock. *J. Agric. Environ. Ethics* **2018**, *31*, 181–195. [CrossRef]
15. Lovarelli, D.; Bacenetti, J.; Guarino, M. A review on dairy cattle farming: Is precision livestock farming the compromise for an environmental, economic and social sustainable production? *J. Clean. Prod.* **2020**, *262*, 121409. [CrossRef]
16. Tullo, E.; Finzi, A.; Guarino, M. Environmental impact of livestock farming and Precision Livestock Farming as a mitigation strategy. *Sci. Total Environ.* **2019**, *650*, 2751–2760. [CrossRef]
17. Pomar, C.; Hauschild, L.; Zhang, G.H.; Pomar, J.; Lovatto, P.A. Applying precision feeding techniques in growing-finishing pig operations. *Rev. Bras. De Zootec.* **2009**, *38*, 226–237. [CrossRef]
18. Pomar, C.; Hauschild, L.; Zhang, G.H.; Pomar, J.; Lovatto, P.A. Precision feeding can significantly reduce feeding cost and nutrient excretion in growing animals. In *Modelling Nutrient Digestion and Utilisation in Farm Animals*; Sauvant, D., van Milgen, J., Faverdin, P., Friggens, N., Eds.; Wageningen Academic Publishers: Wageningen, The Netherlands, 2011; pp. 327–334.
19. Bovenkerk, B.; Brom, F.W.; Van Den Bergh, B.J. Brave new birds: The use of 'animal integrity'in animal ethics. *Hastings Cent. Rep.* **2002**, *32*, 16–22. [CrossRef] [PubMed]
20. Yli-Huumo, J.; Ko, D.; Choi, S.; Park, S.; Smolander, K. Where is current research on blockchain technology?—A systematic review. *PLoS ONE* **2016**, *11*, e0163477. [CrossRef]
21. Elliott, K.C.; Resnik, D.B. Making open science work for science and society. *Environ. Health Perspect.* **2019**, *127*, 75002. [CrossRef]
22. National Academies of Sciences, Engineering, and Medicine. *Open Science by Design: Realizing a Vision for 21st Century Research*; The National Academies Press: Washington, DC, USA, 2018. [CrossRef]
23. Royal Society of London. *Science as an Open Enterprise*; The Royal Society: London, UK, 2012.
24. Elliott, K.C. The value-ladenness of transparency in science: Lessons from Lyme disease. *Stud. Hist. Philos. Sci.* **2021**, *88*, 1–9. [CrossRef]
25. Elliott, K.C. A taxonomy of transparency in science. *Can. J. Philos.* **2022**, *52*, 342–355. [CrossRef]
26. Bright, L.K.; Heesen, R. To be scientific is to be communist. *Soc. Epistemol.* **2023**, *37*, 249–258. [CrossRef]
27. Leonelli, S. Open science and epistemic diversity: Friends or foes? *Philos. Sci.* **2022**, *89*, 991–1001. [CrossRef]
28. Muñoz-Tamayo, R.; Nielsen, B.L.; Gagaoua, M.; Gondret, F.; Krause, E.T.; Morgavi, D.P.; Olsson, I.A.S.; Pastell, M.; Taghipoor, M.; Tedeschi, L.; et al. Seven steps to enhance open science practices in animal science. *PNAS Nexus* **2022**, *1*, pgac106. [CrossRef] [PubMed]
29. Sassenrath, G.F.; Halloran, J.M.; Archer, D.; Raper, R.L.; Hendrickson, J.; Vadas, P.; Hanson, J. Drivers impacting the adoption of sustainable agricultural management practices and production systems of the Northeast and Southeast United States. *J. Sustain. Agric.* **2010**, *34*, 680–702. [CrossRef]
30. Thompson, P.B.; Appleby, M.; Busch, L.; Kalof, L.; Miele, M.; Norwood, B.F.; Pajor, E. Values and public acceptability dimensions of sustainable egg production. *Poult. Sci.* **2011**, *90*, 2097–2109. [CrossRef] [PubMed]
31. Piso, Z.; Werkheiser, I.; Noll, S.; Leshko, C. Sustainability of what? Recognising the diverse values that sustainable agriculture works to sustain. *Environ. Values* **2016**, *25*, 195–214. [CrossRef]
32. Longino, H.E. *Science as Social Knowledge: Values and Objectivity in Scientific Inquiry*; Princeton University Press: Princeton, NJ, USA, 1990.
33. Creel, K.A. Transparency in complex computational systems. *Philos. Sci.* **2020**, *87*, 568–589. [CrossRef]
34. Nyrup, R. The Limits of Value Transparency in Machine Learning. *Philos. Sci.* **2022**, *89*, 1054–1064. [CrossRef]
35. Sullivan, E. Inductive Risk, Understanding, and Opaque Machine Learning Models. *Philos. Sci.* **2022**, *89*, 1065–1074. [CrossRef]
36. Whyte, K.P.; Crease, R.P. Trust, expertise, and the philosophy of science. *Synthese* **2010**, *177*, 411–425. [CrossRef]
37. Yosie, T.F.; Herbst, T.D. Using Stakeholder Processes in Environmental Decision making. In *An Evaluation of Lessons Learned, Key Issues, and Future Challenges*; Ruder Finn: Washington, DC, USA, 1998.

38. O'Rourke, M.; Crowley, S.J. Philosophical intervention and cross-disciplinary science: The story of the Toolbox Project. *Synthese* **2013**, *190*, 1937–1954. [CrossRef]
39. Werkheiser, I. Community epistemic capacity. *Soc. Epistemol.* **2016**, *30*, 25–44. [CrossRef]

animals

MDPI

Article

Analysis of the Drinking Behavior of Beef Cattle Using Computer Vision

Md Nafiul Islam, Jonathan Yoder, Amin Nasiri, Robert T. Burns and Hao Gan *

Department of Biosystems Engineering and Soil Science, University of Tennessee, Knoxville, TN 37996, USA; mislam50@vols.utk.edu (M.N.I.); jyoder5@vols.utk.edu (J.Y.); anasiri@utk.edu (A.N.); rburns@utk.edu (R.T.B.)
* Correspondence: hgan1@utk.edu

Simple Summary: Monitoring drinking behavior allows producers to assess the health and well-being of their beef cattle. Changes in regular drinking behavior can serve as an indicator of potential health issues. Detecting these issues early on enables the application of timely interventions, mitigating the likelihood of severe complications and enhancing the prospects for prompt and efficient treatments. In the current study, we used computer vision techniques to study and analyze the drinking behavior of beef cattle. Two different camera positions were used to identify the drinking behavior. Our proposed method was able to successfully identify both the drinking behavior and drinking time of beef cattle.

Abstract: Monitoring the drinking behavior of animals can provide important information for livestock farming, including the health and well-being of the animals. Measuring drinking time is labor-demanding and, thus, it is still a challenge in most livestock production systems. Computer vision technology using a low-cost camera system can be useful in overcoming this issue. The aim of this research was to develop a computer vision system for monitoring beef cattle drinking behavior. A data acquisition system, including an RGB camera and an ultrasonic sensor, was developed to record beef cattle drinking actions. We developed an algorithm for tracking the beef cattle's key body parts, such as head–ear–neck position, using a state-of-the-art deep learning architecture DeepLabCut. The extracted key points were analyzed using a long short-term memory (LSTM) model to classify drinking and non-drinking periods. A total of 70 videos were used to train and test the model and 8 videos were used for validation purposes. During the testing, the model achieved 97.35% accuracy. The results of this study will guide us to meet immediate needs and expand farmers' capability in monitoring animal health and well-being by identifying drinking behavior.

Keywords: animal behavior; beef cattle; drinking time; computer vision; precision livestock farming

Citation: Islam, M.N.; Yoder, J.; Nasiri, A.; Burns, R.T.; Gan, H. Analysis of the Drinking Behavior of Beef Cattle Using Computer Vision. *Animals* **2023**, *13*, 2984. https://doi.org/10.3390/ani13182984

Academic Editor: Cédric Sueur

Received: 25 August 2023
Revised: 16 September 2023
Accepted: 19 September 2023
Published: 21 September 2023

1. Introduction

The US is the world leader in annual beef production value, and maintains the third-largest cattle herd and the largest cattle industry globally; it produced USD 86.1 billion in gross income in 2022 [1,2]. In the US, heat stress is one of the major challenges in beef cattle production and management. Due to heat stress, cattle experience health problems as well as decreases in feed intake and animal growth, which result in great economic losses to cattle farmers [3]. Based on previous studies, heat stress may also affect daily activity, including drinking and feeding [4]. Early detection of a disease is crucial because it allows for timely interventions and care based on the clinical signs and symptoms associated with that specific disease. Automatic disease detection, using advanced technologies such as artificial intelligence and machine learning, holds promise in facilitating early detection. Conventionally, direct contact (i.e., attaching sensors to animals' bodies) has been the primary method of monitoring animal behavior [5–7]. However, this method faces problems such as a limited sampling frequency, low accuracy, and inconsistent readings from multiple units [8]. To help with those issues, computer vision techniques can be used to identify and classify specific animal behaviors [9,10]. Therefore, in recent years, vision-based analysis has been examined to monitor the animal behavior and health of beef cattle [8,11].

Different deep learning techniques have been developed to automate animal behavior detection methods using computer vision and video analysis technologies. Tsai et al. [12] investigated dairy cow heat stress by monitoring drinking behavior using a convolutional neural network (CNN) with an imaging system and found that drinking behavior reflects the effects of heat stress on dairy cows. Wu et al. [13] investigated a method that was proposed to detect the breathing frequency of standing resting dairy cows by using computer vision and video analysis. A Deeplab V3+ semantic segmentation model was developed using the framework of ResNet-101. Li et al. [10] studied basic motion behaviors based on cow skeletons and a hybrid convolution algorithm. The multi-resolution module was used to extract cow skeletons. The skeleton is a visual representation of a pose formed by connecting key points with lines or curves. In previous attempts, various hybrid deep learning tools were used to investigate animal behaviors. Optimizing hybrid approaches may involve fine-tuning hyperparameters to strike the ideal equilibrium between computational efficiency and model accuracy. Determining the right combination of convolution strategies can be time-consuming and resource-intensive. To simplify the tuning of hyperparameters, the pose estimation technique using different pre-trained models might be a solution. Pose estimation is a basic computer vision technique that identifies the location of a series of key body parts. In 2014, Toshev and Szegedy [14] first applied a 2D human pose estimation technique using a deep learning method. Over time, researchers developed many updated pose estimation methods such as YOLO, DeepLabCut, LEAP, and DeepPoseKit [15]. Among them, DeepLabCut is the first tool for animal pose estimation, which uses ResNet's transfer learning to reduce the training times [16,17]. In this study, the DeepLabCut pose estimation technique was used to train and validate the model for tracking cattle body parts.

Recently, in identifying and classifying livestock activity, different artificial neural networks have been used to improve the performance of recognition tasks. Chen et al. [18] investigated pig drinking and drinker-playing behavior recognition based on ResNet50 and long short-term memory (LSTM), and the classification accuracy for the body and head regions was found to be 87% and 93%, respectively. Wu et al. [19] used a fusion of convolutional neural networks and LSTM to recognize the basic behaviors (drinking, ruminating, walking, standing, and lying) of a single cow. Nasiri et al. [20] proposed a technique to identify pose-estimation-based lameness recognition for broilers using the CNN-LSTM model. Du et al. [21] used the Resnet50-LSTM model to investigate broodstock breeding behavior, and the investigated method achieved an average accuracy of 97.32% for five types of breeding behavior recognition.

Deep learning and vision analysis play significant roles in the automated recognition of livestock behavior, as highlighted in the previously mentioned works [13]. Although researchers have identified different behaviors for various purposes, no effective method has been reported for recognizing beef cattle and/or cow drinking behavior. Identifying this behavior is crucial for detecting heat stress [22]. Consequently, as an alternative to previous behavior recognition methods, this project intended to evaluate beef cattle drinking behavior by considering skeleton-based body parts using a CNN-based model (DeepLabCut) and a time-series network (LSTM). Therefore, the specific objective of this research was to develop a computer vision system for monitoring beef cattle water drinking behavior.

2. Materials and Methods

2.1. Ethical Considerations

The University of Tennessee (UT) Animal Care and Use Committee (IACUC) approved the experiment on the use of cattle subjects under the following protocol title: Computer vision characterization of respiration as an indicator of cattle health; Protocol #: 2932-0922.

2.2. Experimental Site

The experiment was conducted at the University of Tennessee, Middle Tennessee AgResearch and Education Center (MTREC), between 5 October 2022 and 12 October 2022. The experiment involved four purebred black Angus cattle, all of which were 1 year old at the time. The cattle were kept in a barn equipped with a roof and automatic fans that activated when the temperature reached predefined conditions, ensuring they were protected from direct sunlight, rain, wind, and muddy conditions when needed. An open space was connected to the barn, offering plenty of room for the animals to move about without restraint. Surrounding both the open space and the barn were wooden fences, ensuring that the animals remained within their designated boundaries. MTREC staff diligently supervised the well-being and health of all animals during the experiment, adhering to the farm's standard operating procedures and following veterinary recommendations.

The waterer was placed in the open space to provide free access to drinking water. The vision system was installed over the top of the waterer. The vision system consisted of a mainframe, a ball waterer, and a sensing camera unit (Figure 1). The sensing camera unit included a Raspberry Pi 4 microprocessor, a Raspberry Pi camera (8MP IMX219, Arducam Technology Co. Limited, Kowloon, Hong Kong, China), and two ultrasonic sensors. System power was provided using Power over Ethernet (PoE). To optimize power usage, an ultrasonic sensor (Figure 1C) was used to trigger camera recording when an animal was in the field of view of the vision system. The camera was positioned at a height of 1.8 m above the ground. Another ultrasonic sensor was used to monitor the drinking behavior of the animal. This ultrasonic sensor was mounted on the side of the waterer at a height that was just above the ball of the waterer. The operating principle was that when the animal was pushing the ball to drink water, the ultrasonic sensor would detect the animal's head, thus confirming drinking behavior.

Figure 1. Structure of the vision system: (**A**) mainframe, (**B**) camera unit, (**C**) first ultrasonic sensor (trigger the camera), (**D**) second ultrasonic sensor (for ground truth), and (**E**) animal entering the waterer.

2.3. Data Collection and Annotation

In this study, animal drinking behavior was recorded from two camera positions. This was to ensure that our algorithm was not sensitive to a specific camera position and could potentially handle flexible camera mounting options. A total of 78 videos, including 39 videos of each position, were recorded. The videos were recorded as H264 files. After that, the MP4 (RGB) and Hierarchical Data Format (HDF) files were extracted from each H264 file. The MP4 (RBG) files were used to track the key body points. A total of five key points were used to observe the pose skeleton (Figure 2). Each skeleton line that connects two key body points represents the relationship between the two points. The skeleton lines were defined in the training dataset to help with model accuracy. An open-source 'VGG Image Annotator' online annotator tool was used to label the drinking time for the recorded video as ground-truth data [23]. The second ultrasonic sensor (Figure 1D) was also used to collect the ground-truth drinking time for each animal. The two types of ground-truth data were compared to ensure the accuracy of the labeled data.

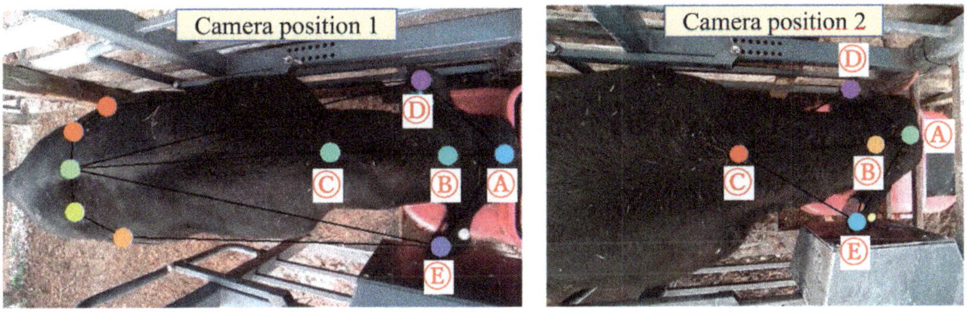

Figure 2. The annotated key points and beef cattle pose skeleton: (**A**) head, (**B**) upper neck, (**C**) lower neck, (**D**) left ear, and (**E**) right ear.

2.4. CNN-Based Pose Recognition

Based on the literature review, key body points and pose were identified using the DeepLabCut CNN-based pose estimation tool [16]. DeepLabCut has two ResNet (50 and 101) architectures to choose from, and ResNets facilitate the substitution of deconvolutional layers with dense layers to enhance the process of feature extraction [24]. The network has the capability to learn labeled key body points, which allows greater probabilities of recognition and reduced likelihoods of misidentifying other points. In the last stage, the trained model analyzed the videos and estimated the pose for the whole dataset. In this study, the pre-trained ResNet50 architecture was utilized as the transfer learning approach for pose estimation. Figure 3 demonstrates the diagram that outlines the pose estimation workflow.

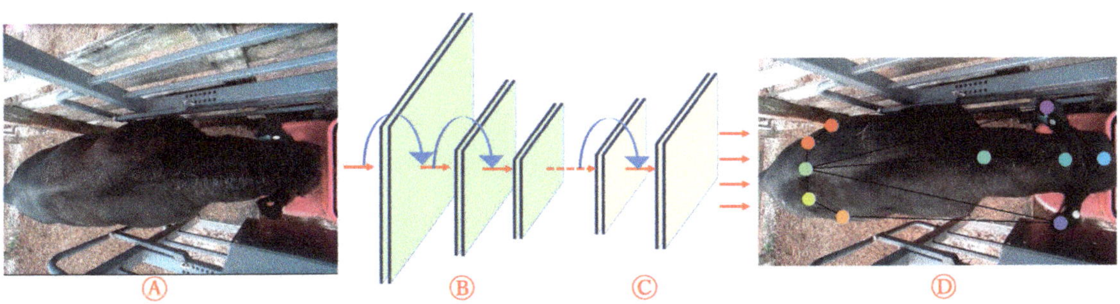

Figure 3. The diagram outlines the pose estimation workflow: (**A**) input, (**B**) pre-trained model (ResNet50), (**C**) deconvolutional layers, and (**D**) output with pose skeleton.

2.5. LSTM-Based Drinking Behavior Estimation

The LSTM-based model was used to classify drinking and non-drinking behavior. This model used the output from the previous DeepLabCut model, which were the coordinates of the key body parts, as the input, and produced either drinking or non-drinking classes as the output. The input included coordinates from 30 consecutive frames (1 s of video). In this study, 30-frame sequences were constantly sampled and predicted, and a 30-frame step size was set to make sure that the video sequence was repetitive each time. Therefore, after setting the fixed window size to 30 and sliding the window with a step size of 30, the proposed LSTM algorithm was used to detect the drinking behavior from small video segments obtained from the sequence. Figure 4 shows the schematic diagram of the sliding-window video sequence sampling technique in this study.

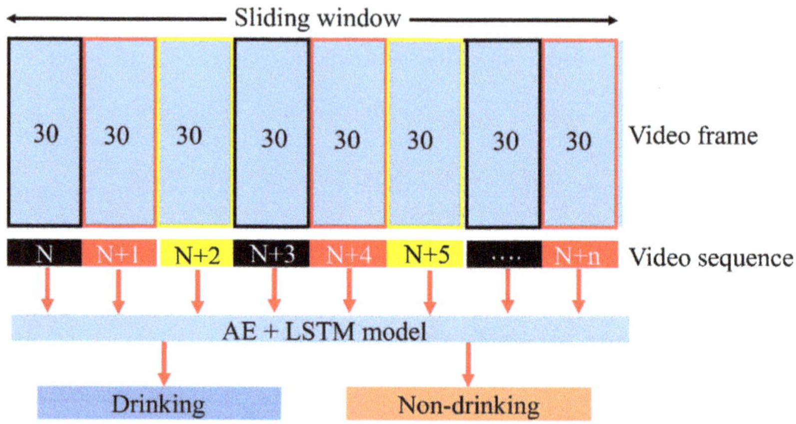

Figure 4. The schematic diagram of the sliding-window sampling technique.

In deep learning studies, the augmentation of data is a very useful method to enhance the efficiency of the model training process by reducing over-sampling and augmenting random transformations [20]. Therefore, before starting the LSTM training, a convolutional autoencoder (AE) was applied to augment the training dataset. Figure 5 shows the architecture of the proposed AE and LSTM models. For the AE, the entire training dataset was randomly split into two sets (train and test) at a proportion of 8:2. The selected AE model was trained for 300 epochs. The mean absolute error loss function and Adam optimizer were used for the AE model. After applying the augmentation process, the proposed LSTM model was trained for 1000 epochs, including a cross-entropy loss function and Adam optimizer. The learning rate and learning rate decay were used to train the LSTM model, set at 1×10^{-5} and 1×10^{-7}, respectively. The input dimension of the LSTM was 30×10, which represented 30 frames $\times$ 5 key points in each frame $\times$ 2 coordinates (x and y) per point.

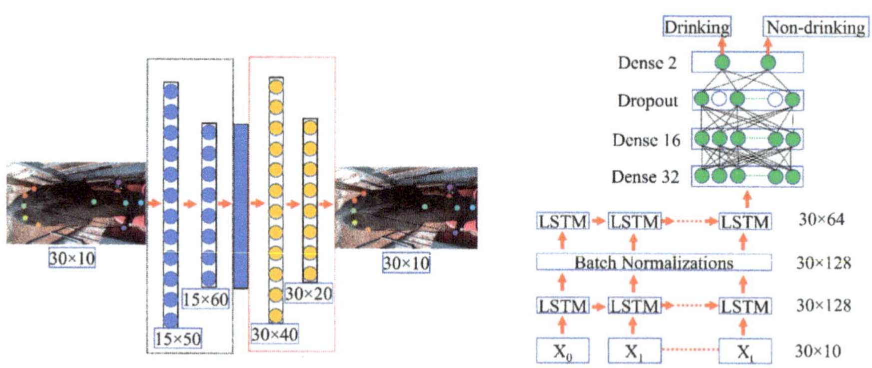

Figure 5. The architecture of the proposed AE (**left**) and LSTM models (**right**).

Evaluation indices including accuracy, precision, recall, specificity, F1 score, and AUC were used to validate the proposed algorithm. The evaluation indices were calculated as shown in Equations (1)–(5) [20]:

$$\text{Accuracy} = \frac{\text{TP} + \text{TN}}{\text{FP} + \text{FN} + \text{TP} + \text{TN}} \quad (1)$$

$$\text{Precision} = \frac{\text{TP}}{\text{TP} + \text{FP}} \quad (2)$$

$$\text{Recall} = \frac{\text{TP}}{\text{TP} + \text{FN}} \quad (3)$$

$$\text{Specificity} = \frac{\text{TN}}{\text{TN} + \text{FP}} \quad (4)$$

$$\text{F1 score} = 2 \times \frac{\text{Precision} \times \text{Recall}}{\text{Precision} + \text{Recall}} \quad (5)$$

where TP = true positive, TN = true negative, FP = false positive, and FN = false negative.

3. Results and Discussion

3.1. Pose Estimation

Figure 6 shows the loss values of the DeepLabCut-based model for identifying key points. The training procedure took nearly five days for each model, and the selected weights of the DeepLabCut were attained at 1,030,000 iterations along with loss values of 0.0033 and 0.0023 at the learning rate of 0.001 for camera positions 1 and 2, respectively.

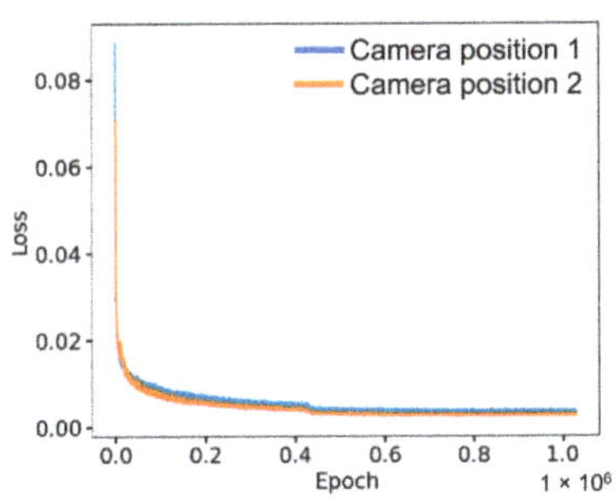

Figure 6. The loss values throughout the DeepLabCut model training for two camera positions.

3.2. Evaluating the AE and LSTM Model Performance

Figure 7 (left) indicates the AE model loss values. The loss values for the best model were 17.90 and 11.77 for training and testing, respectively. There was a decreasing trend in the loss values with increasing epoch numbers. After epoch 150, the training and testing loss curves were stabilized, which indicates that the proposed AE model has gained sufficient convergence. Figure 7 (right) shows the performance of the proposed LSTM model. After 200 epochs, the accuracy increased, and the loss decreased simultaneously. The accuracy values for the best LSTM model were achieved at 97.35% and 97.37% for training and testing, respectively.

Eight 1-minute-long individual videos were used for validation of the proposed LSTM model. The videos were collected from both camera positions. Table 1 shows the validation results of the proposed LSTM algorithm. Our evaluation includes a range of metrics, including efficiency, precision, recall, specificity, F1 score, AUC, and confusion matrices (as shown in Supplementary Materials, Figures S1–S8), to comprehensively assess the model's performance. The highest accuracy was obtained at 98% for video numbers 1, 4, 5, 6, and 7. The lowest accuracy was obtained at 95% for video number 2. In addition, this proposed algorithm was able to calculate the drinking and non-drinking times.

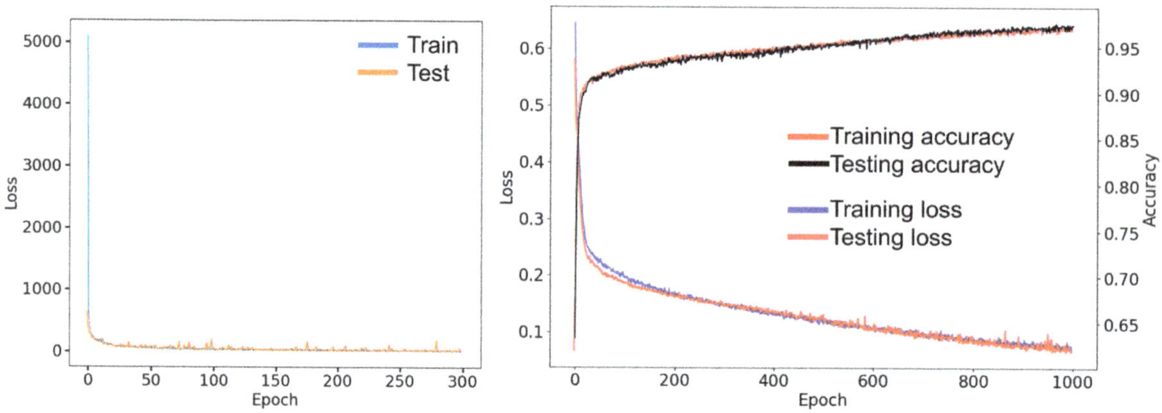

Figure 7. The changes in loss values during the AE model (**left**), and loss and accuracy values during the LSTM model (**right**).

Table 1. Validation results of the proposed LSTM algorithm.

Video Number	Accuracy (%)	Precision (%)	Recall (%)	Specificity (%)	F1 Score (%)	AUC (%)	Drinking Time (s)	Non-Drinking Time (s)
1	98.25	100.00	95.24	100.00	97.56	97.62	20	37
2	96.49	97.50	97.50	94.12	97.50	95.81	40	17
3	94.74	100.00	90.32	100.00	94.92	95.16	28	29
4	98.25	100.00	97.62	100.00	98.80	98.81	42	15
5	98.28	100.00	97.30	100.00	98.63	98.65	36	22
6	98.28	100.00	94.44	100.00	97.14	97.22	17	41
7	98.28	100.00	92.86	100.00	96.30	96.43	13	45
8	96.56	95.12	100.00	89.48	97.50	94.74	41	17

Video numbers 1 to 4 were recorded at camera position 1, and 5 to 8 were recorded at camera position 2.

3.3. Comparison of Different Related Studies

Efforts to develop automated dairy cattle drinking behavior models have been underway by various research groups. Most researchers were mainly focused on automatic basic behavior recognition methods including for drinking, ruminating, walking, standing, and lying. Among them, Tsai et al. [12] developed an imaging system for monitoring and analyzing drinking behavior. When a dairy cow head was detected in the drinking area, the drinking status and the duration of drinking were recorded until the dairy cow left the drinking area. A YOLOv3 model was used to detect the head movements and a recognition accuracy of 90% was achieved. Wu et al.'s [19] CNN-LSTM network was proposed to recognize the basic behaviors of a single cow. The developed algorithm mainly included two parts. The VGG16 framework was first used as the network skeleton to extract the feature sequence corresponding to each video. The second part was basic behavior recognition using the designed LSTM model. The combination of the two parts formed the final output to realize the recognition of basic behaviors. Shu et al. [25] developed a video system for monitoring different behaviors (drinking, eating, lying, and standing). YOLOv5 architectures were trained using the transfer learning method. The results showed that the recognition accuracy of drinking behavior was 97.50%. Zhang et al. [26] proposed a SlowFast-based cow behavior recognition algorithm to identify cow behaviors such as standing, lying down, walking, drinking, and eating. The SlowFast algorithm was designed to address the challenges of recognizing actions and events in videos. The key idea behind the SlowFast architecture was to tackle the temporal resolution trade-off problem in video action recognition. In videos, actions can occur at different speeds, and capturing both fast and slow temporal dynamics is essential for accurate recognition. The accuracy of drinking estimation was found to be 92.60%.

Table 2 summarizes the outcomes of the computer vision models for cattle drinking behavior recognition from previous research. Currently, there has been limited research on cattle drinking behavior recognition. Most studies to date have applied computer vision models to recognize some basic behaviors by training the image for a particular action. Detecting drinking needs recognition of different actions including the motion of the head and the initial movement of drinking. Some of the previous research did not focus on specifically drinking actions. Due to those limitations, the previous methods of drinking recognition accuracy were lower than for the other behaviors. In the current study, we used a combination of a video annotator (head movement) and an ultrasonic sensor (drinking status) to produce the ground-truth model.

Table 2. Comparison of computer vision model outcomes for cattle drinking behavior from previous research and the current study.

Dataset	Model	Recognition Accuracy (%)	Reference
Dairy cow image	YOLOv3	90.00	[12]
Dairy cow video	VGG16 and LSTM	95.00	[19]
Dairy cow video	YOLOv5	97.50	[25]
Dairy cow video	Slowfast	92.60	[26]
Beef cattle video	DeepLabCut and LSTM	98.25	Current study

In contrast to the previous research, the proposed method in this study was a rapid and non-invasive technique. The DeepLabCut pose-estimation-based model accurately categorized the drinking behavior and could analyze drinking time tracked through videos with different lengths. Addressing the continuous detection and tracking of key points on constantly moving cattle, without the need for expert pre-marking, is an essential challenge. In this study, the focus was on tracking the key points of interest in the head area. DeepLabCut has the capability to effectively acquire body part information, even when faced with challenges such as a complex and changing background, uneven lighting conditions, or distortions caused by the camera. DeepLabCut also offers several key benefits, including cost reduction in manual behavior analysis, achieving high accuracy with a minimal number of training images, and eliminating the necessity of placing visible markers on specific locations of interest [16]. Creating a DeepLabCut model entails a potentially intricate and resource-intensive procedure. It necessitates a robust GPU, and the training phase may consume a substantial amount of time, particularly when dealing with large datasets. This can create challenges for individuals who have restricted computational capabilities.

4. Conclusions

This study proposed a skeleton-based computer vision method for beef cattle drinking behavior recognition. It used cameras from different positions and orientations for beef cattle pose estimation using DeepLabCut with a ResNet50 backbone. A dataset containing 70 videos was evaluated using the model to create the sequential key body points data. After that, an LSTM model was used to classify the drinking and non-drinking behaviors. The accuracy of the model was 98.25%, which was higher than the previous studies using other computer vision methods. We are currently conducting a preliminary study focused on addressing a significant and practical challenge within the field of livestock farming. This research has the potential to be applied to various livestock species, including but not limited to beef cattle, sheep, horses, and more. By employing the same vision system and pose estimation techniques, we can additionally quantify the respiration rates of animals.

Currently, our group is collecting more video data from different farms using different camera positions. More collected data will further increase the accuracy of the LSTM model as well as different classification models. They will also help prepare the vision system as a practical tool for beef farms worldwide.

Supplementary Materials: The following supporting information can be downloaded at: https://www.mdpi.com/article/10.3390/ani13182984/s1, Figure S1: Confusion matrix for video number 2; Figure S2: Confusion matrix for video number 2; Figure S3: Confusion matrix for video number 3; Figure S4: Confusion matrix for video number 4; Figure S5: Confusion matrix for video number 5; Figure S6: Confusion matrix for video number 6; Figure S7: Confusion matrix for video number 7; Figure S8: Confusion matrix for video number 8.

Author Contributions: Conceptualization, H.G., R.T.B. and M.N.I.; methodology, H.G. and M.N.I.; software, M.N.I.; validation, H.G. and A.N.; formal analysis, M.N.I. and A.N.; investigation, H.G.; resources, H.G.; data curation, M.N.I. and J.Y.; writing—original draft preparation, M.N.I.; writing—review and editing, M.N.I. and H.G.; visualization M.N.I. and J.Y.; supervision, H.G.; project administration, H.G.; funding acquisition, R.T.B. and H.G. All authors have read and agreed to the published version of the manuscript.

Funding: This research was funded by USDA NIFA (Award No. 2022-67021-37863) and the University of Tennessee AgResearch. Project title: PLF SPRINT: Precision Livestock Farming: Investing for Future Success.

Institutional Review Board Statement: Not applicable.

Informed Consent Statement: Informed consent was obtained from all subjects involved in the study.

Data Availability Statement: Data are presented in this article in the form of figures and tables.

Acknowledgments: Thanks to the Middle Tennessee AgResearch and Education Center (MTREC), Spring Hill, TN, for allowing us to conduct our research.

Conflicts of Interest: The authors declare no conflict of interest.

References

1. Martinez, C.C.; Maples, J.G.; Benavidez, J. Beef Cattle Markets and COVID-19. *Appl. Econ. Perspect. Policy* **2021**, *43*, 304–314. [CrossRef]
2. USDA. Cattle/Calf Receipts Comprised the Largest Portion of U.S. Animal/Animal Product Receipts in 2022. Available online: https://www.ers.usda.gov/data-products/chart-gallery/gallery/chart-detail/?chartId=76949 (accessed on 11 September 2023).
3. Brown-Brandl, T.M. Understanding heat stress in beef cattle. *Rev. Bras. Zootec.* **2018**, *47*, e20160414. [CrossRef]
4. Tuan, S.-A.; Rustia, D.J.A.; Hsu, J.-T.; Lin, T.-T. Frequency modulated continuous wave radar-based system for monitoring dairy cow respiration rate. *Comput. Electron. Agric.* **2022**, *196*, 106913. [CrossRef]
5. González, L.; Bishop-Hurley, G.; Handcock, R.N.; Crossman, C. Behavioral classification of data from collars containing motion sensors in grazing cattle. *Comput. Electron. Agric.* **2015**, *110*, 91–102. [CrossRef]
6. Smith, D.; Rahman, A.; Bishop-Hurley, G.J.; Hills, J.; Shahriar, S.; Henry, D.; Rawnsley, R. Behavior classification of cows fitted with motion collars: Decomposing multi-class classification into a set of binary problems. *Comput. Electron. Agric.* **2016**, *131*, 40–50. [CrossRef]
7. Wang, S.; Li, Q.; Peng, J.; Niu, H. Effects of Long-Term Cold Stress on Growth Performance, Behavior, Physiological Parameters, and Energy Metabolism in Growing Beef Cattle. *Animals* **2023**, *13*, 1619. [CrossRef]
8. Guo, Y.; He, D.; Chai, L. A Machine Vision-Based Method for Monitoring Scene-Interactive Behaviors of Dairy Calf. *Animals* **2020**, *10*, 190. [CrossRef]
9. Jorquera-Chavez, M.; Fuentes, S.; Dunshea, F.R.; Warner, R.D.; Poblete, T.; Jongman, E.C. Modelling and Validation of Computer Vision Techniques to Assess Heart Rate, Eye Temperature, Ear-Base Temperature and Respiration Rate in Cattle. *Animals* **2019**, *9*, 1089. [CrossRef]
10. Li, Z.; Song, L.; Duan, Y.; Wang, Y.; Song, H. Basic motion behaviour recognition of dairy cows based on skeleton and hybrid convolution algorithms. *Comput. Electron. Agric.* **2022**, *196*, 106889. [CrossRef]
11. Gonzalez, L.F.; Montes, G.A.; Puig, E.; Johnson, S.; Mengersen, K.; Gaston, K.J. Unmanned Aerial Vehicles (UAVs) and Artificial Intelligence Revolutionizing Wildlife Monitoring and Conservation. *Sensors* **2016**, *16*, 97. [CrossRef]
12. Tsai, Y.-C.; Hsu, J.-T.; Ding, S.-T.; Rustia, D.J.A.; Lin, T.-T. Assessment of dairy cow heat stress by monitoring drinking behaviour using an embedded imaging system. *Biosyst. Eng.* **2020**, *199*, 97–108. [CrossRef]
13. Wu, D.; Yin, X.; Jiang, B.; Jiang, M.; Li, Z.; Song, H. Detection of the respiratory rate of standing cows by combining the Deeplab V3+ semantic segmentation model with the phase-based video magnification algorithm. *Biosyst. Eng.* **2020**, *192*, 72–89. [CrossRef]

14. Toshev, A.; Szegedy, C. DeepPose: Human Pose Estimation via Deep Neural Networks. In Proceedings of the IEEE Conference on Computer Vision and Pattern Recognition (CVPR), Columbus, OH, USA, 23–28 June 2014; pp. 1653–1660.
15. Mathis, M.W.; Mathis, A. Deep learning tools for the measurement of animal behavior in neuroscience. *Curr. Opin. Neurobiol.* **2020**, *60*, 1–11. [CrossRef] [PubMed]
16. Mathis, A.; Mamidanna, P.; Cury, K.M.; Abe, T.; Murthy, V.N.; Mathis, M.W.; Bethge, M. DeepLabCut: Markerless pose estimation of user-defined body parts with deep learning. *Nat. Neurosci.* **2018**, *21*, 1281–1289. [CrossRef]
17. Mathis, A.; Biasi, T.; Schneider, S.; Yuksekgonul, M.; Rogers, B.; Bethge, M.; Mathis, M.W. Pretraining Boosts Out-of-Domain Robustness for Pose Estimation. In Proceedings of the IEEE/CVF Winter Conference on Applications of Computer Vision (WACV), Waikoloa, HI, USA, 3–8 January 2021; pp. 1859–1868.
18. Chen, C.; Zhu, W.; Steibel, J.; Siegford, J.; Han, J.; Norton, T. Classification of drinking and drinker-playing in pigs by a video-based deep learning method. *Biosyst. Eng.* **2020**, *196*, 1–14. [CrossRef]
19. Wu, D.; Wang, Y.; Han, M.; Song, L.; Shang, Y.; Zhang, X.; Song, H. Using a CNN-LSTM for basic behaviors detection of a single dairy cow in a complex environment. *Comput. Electron. Agric.* **2021**, *182*, 106016. [CrossRef]
20. Nasiri, A.; Yoder, J.; Zhao, Y.; Hawkins, S.; Prado, M.; Gan, H. Pose estimation-based lameness recognition in broiler using CNN-LSTM network. *Comput. Electron. Agric.* **2022**, *197*, 106931. [CrossRef]
21. Du, L.; Lu, Z.; Li, D. Broodstock breeding behaviour recognition based on Resnet50-LSTM with CBAM attention mechanism. *Comput. Electron. Agric.* **2022**, *202*, 107404. [CrossRef]
22. Sejian, V.; Shashank, C.G.; Silpa, M.V.; Madhusoodan, A.P.; Devaraj, C.; Koenig, S. Non-Invasive Methods of Quantifying Heat Stress Response in Farm Animals with Special Reference to Dairy Cattle. *Atmosphere* **2022**, *13*, 1642. [CrossRef]
23. Dutta, A.; Gupta, A.; Zissermann, A. VGG Image Annotator (VIA). Available online: https://www.robots.ox.ac.uk/~vgg/software/via/ (accessed on 20 September 2023).
24. Russakovsky, O.; Deng, J.; Su, H.; Krause, J.; Satheesh, S.; Ma, S.; Huang, Z.; Karpathy, A.; Khosla, A.; Bernstein, M.; et al. ImageNet Large Scale Visual Recognition Challenge. *Int. J. Comput. Vis.* **2015**, *115*, 211–252. [CrossRef]
25. Shu, H.; Bindelle, J.; Guo, L.; Gu, X. Determining the onset of heat stress in a dairy herd based on automated behaviour recognition. *Biosyst. Eng.* **2023**, *226*, 238–251. [CrossRef]
26. Zhang, Y.; Ibrayim, M.; Hamdulla, A. Research on Cow Behavior Recognition Based on Improved SlowFast with 3DCBAM. In Proceedings of the 2023 5th International Conference on Communications, Information System and Computer Engineering (CISCE), Guangzhou, China, 14–16 April 2023; pp. 470–475. [CrossRef]

animals

MDPI

Article

Effects of a Sprinkler and Cool Cell Combined System on Cooling Water Usage, Litter Moisture, and Indoor Environment of Broiler Houses

Jonathan Moon [1], Jan DuBien [2], Reshma Ramachandran [1], Yi Liang [3], Sami Dridi [4] and Tom Tabler [5,*]

1 Department of Poultry Science, Mississippi State University, Mississippi State, MS 39762, USA
2 Department of Mathematics and Statistics, Mississippi State University, Mississippi State, MS 39762, USA
3 Center of Excellence for Poultry Science, Department of Biological and Agricultural Engineering, University of Arkansas, Fayetteville, AR 72701, USA
4 Department of Poultry Science, University of Arkansas, Fayetteville, AR 72701, USA
5 Department of Animal Science, University of Tennessee, Knoxville, TN 37996, USA
* Correspondence: gtabler@utk.edu; Tel.: +1-931-486-2129

Simple Summary: Water scarcity is an increasing problem facing the global agricultural industry, particularly livestock production. The poultry industry is actively seeking opportunities to reduce the use of water for cooling broiler houses during hot summer months. We investigated a sprinkler system combined with a cool cell system and a cool-cell only system for cooling heavy broiler chickens for two summer flocks. We found the sprinkler/cool cell combination exhibited a higher house temperature, lower relative humidity, and a 64% reduction in average cooling water use. Litter moisture also tended to be lower in the combination system. Findings are similar to previous reported research and offer additional confirmation that sprinklers in conjunction with cool cells maintain litter conditions while reducing cooling water use, thus lessening the threat to the economic and environmental sustainability of the poultry industry and improving its water efficiency efforts.

Abstract: Climate change is a serious challenge to food production around the world. Sustainability and water efficiency are critical to a poultry industry faced with global production concerns including increased demands for high-quality, affordable animal protein and greater environmental pressures resulting from rising global temperatures, flock heat stress, and limits on water availability. To address these concerns, a commercial sprinkler system used in combination with a cool cell system was evaluated against a cool cell system alone for two summer flocks of heavy broilers at Mississippi State University to determine effects of sprinkler technology on cooling water usage, litter moisture, and in-house environments. Environmental data were calculated and recorded throughout the flocks. The combination house exhibited a 2.2 °C (4 °F) increase in daily maximum temperature, lower coincident relative humidity, and a 64% (62,039 L/flock) reduction in average cooling water usage over the cool cell-only house. Litter moisture for the combination house tended to be numerically lower but showed no significant difference at several time points between and across flocks. A combined sprinkler/cool cell system reduced cooling water use by 64% over two flocks compared to a cool cell alone system and decreased in-house relative humidity levels.

Keywords: broiler cooling; climate change; sprinkler; sustainability; water efficiency

Citation: Moon, J.; DuBien, J.; Ramachandran, R.; Liang, Y.; Dridi, S.; Tabler, T. Effects of a Sprinkler and Cool Cell Combined System on Cooling Water Usage, Litter Moisture, and Indoor Environment of Broiler Houses. *Animals* **2023**, *13*, 2939. https://doi.org/10.3390/ani13182939

Academic Editor: Alessandro Dal Bosco

Received: 4 August 2023
Revised: 10 September 2023
Accepted: 13 September 2023
Published: 16 September 2023

1. Introduction

Poultry meat represented almost 40 percent of global meat production in 2020 [1]. Poultry is one of the main sources of animal protein due to its universal acceptability, high nutritional value, and health benefits [2]. The growth in global population has pressured the poultry industry to increase its production capacity [3–5]. Global meat production doubled from 1980 to 2004 and is projected to double again from 2000 to 2050 [6]. This rapid growth in global meat production puts pressure on water resources because livestock production is a very water-intensive agricultural activity, i.e., about one-third of the total water that is utilized in global agricultural production is assigned to animal production [2]. Broiler chickens are one of the most efficient animals in terms of growth rate and feed conversion ratio [7]. Broiler chicken meat is considered to have a relatively low carbon footprint among all farmed meat products [8]. According to previous research, feed is the largest contributor to climate change impact associated with broiler production [9–16]. However, water scarcity, always a concern in developing countries, is rapidly emerging as a global concern [17,18]. Estimates are that by 2025, half the world's population will be living in countries facing considerable water stress or scarcity issues [19–21]. A changing environment threatens water security and can affect the future of poultry production, making water conservation efforts critical to the sustainability of the poultry industry. Climate change is creating new challenges such as increasing the earth's temperature by 0.2 °C per decade with significant fluctuation in the amount and distribution of rainfall, resulting in longer and more intense heat waves and increasing global water scarcity concerns [2].

Commercial broiler chickens are raised in specially designed houses capable of maintaining an environment that allows for optimum performance even during long periods of high environmental challenges. These houses are the result of decades worth of research to determine the correct combination of cooling and ventilation [22–24]. In recent years, consumers and the poultry industry have placed increasing emphasis on raising chickens in a more sustainable manner. Water conservation is a major emphasis for the poultry industry today, as sustainability and global food security issues challenge the industry to meet consumers' demands for safe high quality affordable meat protein and lessen the industry's overall environmental footprint [16].

Climate change and heat stress are additional challenges to sustainable poultry production. Water is crucial in poultry production not only in bird consumption but also in alleviating heat stress during cooling periods [25,26]. Evaporative cooling pad systems, while effective at reducing the temperature of the air entering the poultry house, often result in excessive relative humidity levels of 80% or higher in the house, require large volumes of water [27], and negatively affect the ability for broilers to dissipate heat through evaporative respiration during periods of high environmental temperatures [28–32]. Under normal conditions, broilers typically achieve heat dissipation primarily through respiratory evaporation [33,34], which is severely hindered by high in-house humidity levels. High in-house humidity level is a known factor that negatively affects litter quality and thereby, animal welfare [35–39]. High environmental temperatures combined with high in-house humidity levels can create life-threatening conditions to broiler chickens as they near targeted market weight and age.

Animals dissipate body heat through two mechanisms, i.e., sensible heat dissipation and evaporation (latent heat dissipation). The temperature gradient between the animal and its surroundings governs the sensible heat dissipation. Evaporative cooling pad systems employed in modern tunnel-ventilation poultry houses aim to lower temperature of the surroundings, hence increase the sensible portion of the heat dissipation. In comparison, evaporative heat dissipation from birds is governed by the vapor pressure gradient rather than by temperature differences. As the ambient temperature comes close to the body temperature, evaporative heat flux becomes the only pathway for an animal to dissipate heat to maintain a constant body temperature. Chickens do not have sweat glands but lose heat mainly through panting, with some heat also lost by skin surface evaporation [40]. Directly applying controlled amounts of water onto poultry using sprinkler systems partially wets the birds, allowing direct evaporation of water from the birds' surface. Water evaporation absorbs heat directly from the body of the birds and the surrounding air, similar to water evaporation from evaporative cooling pads at the tunnel-ventilation air inlet. For water to evaporate from the chicken surface efficiently, it is crucial to maintain a large vapor pressure gradient by keeping the humidity of the surrounding air low. In contrast, evaporative cooling pads, along with high-pressure or low-pressure fogging systems would increase the humidity inside the poultry house and be less able to reduce the effective environmental temperature.

Research has reported efficacy of surface wetting and its effect on body temperature of laying hens [41,42] and broilers [30] under controlled thermal conditions, and commercial production conditions [43–46]. Sprinkler systems offer water conservation advantages without sacrificing flock performance [31,45–47]. However, it is a challenge to incorporate sprinkler systems into the existing evaporative cooling systems which have been in use for more than two decades in the southern United States. As mentioned earlier, surface wetting by sprinklers requires low in-house humidity to function properly. It is necessary to develop a management strategy such as a set point and application schedule in a sprinkler and cool cell combined system to achieve not only cooling water savings, but also satisfactory air and litter quality.

The objective of this study was to develop and evaluate a management strategy under a combined sprinkler and cool cell technology on cooling water conservation, in-house environments, and litter moisture conditions when raising heavy broilers on a commercial broiler farm.

2. Materials and Methods

2.1. Broiler Houses

The study was conducted at two commercial broiler houses on the Mississippi State University poultry research farm for two summer flocks from May through October of 2020. For the May-placed flock, each house received 13,700 straight run Ross × Ross 708 broiler chicks on day of hatch. For the August-placed flock, each house received 14,700 straight run Ross × Ross 708 broiler chicks on day of hatch. Both flocks were heavy broilers with a final target body weight of 4.42 kg. The two houses were each 13 m × 122 m (42 ft × 400 ft) and equipped with three lines of pan type feeders and four lines of nipple-type drinkers. Both houses were drop ceiling houses and neither house contained ceiling baffles.

Each house contained 15 m (50 ft) of 1.5 m × 15 cm × 0.30 m (5 ft × 6 in × 1 ft) cool cell on each side of the house. Ten 122 cm (48-in) diameter tunnel ventilation fans (Acme Engineering and Manufacturing Corp., Muskogee, OK, USA) were at the opposite end from the cool cells in each house. Each house was also equipped with two lines of commercial sprinklers (Weeden Environments, Woodstock, ON, Canada) mounted to the ceiling and located 3 m (10 ft) from each sidewall above the two outside feed lines. The sprinkler lines consisted of 19 mm (3/4 in) PVC pipe running the length of the house with 275 kPa pressure regulators. Sprinkler spinner heads with flexible droppers of nominal flow rate of 1.3 L/min were located every 6 m (20 ft) down each line and were directly across from one another (e.g., not staggered). There were 20 spinner heads on each line; a total of 40 per house located 2.1 m (7 ft) above the litter.

The evaporative cooling system remained intact in each house. The cool cell system was controlled by a Chore-Tronics 3 (Chore-Time, Milford, IN, USA) poultry house controller. An integrator-developed cooling program using their set point temperatures and run times was followed in the cool cell only house. For the two summer flocks, one house was cooled by the evaporative cooling system only. For this house, the set point temperature on the cool cell pads was 28 °C (82 °F). The tunnel set point temperature was always 3.3 °C (6 °F) above the house set point temperature for any given day.

The sprinkler system was operated in combination with the cool cell system and not as a stand-alone cooling system, although previous research has demonstrated successful sprinkler use in a stand-alone setting [46]. The combination-cooling house was cooled by the sprinkler system as the first stage of cooling with assistance from the cool cell system once house temperature reached 32 °C (90 °F). This was accomplished by modifying the operational settings on the cool cell set point in the poultry house controller program. The cool cell set point temperature was raised to run water over the pads for 15–20 s, but only when the house temperature reached 32 °C. The goal of the cool cell in the combination-cooling house was not to cool the house temperature and increase the humidity as in the cool cell only house, but to prevent the house temperature from going any higher than 32 °C. The two houses were switched between the two flocks to remove any house effect (e.g., the cool cell house on the first flock became the sprinkler/cool cell combination house on the second flock and vice versa). The sprinkler system and the evaporative cool cells were allowed to operate from 9:00 am to 9:00 pm.

2.2. Sprinkler System

Both houses were equipped with a low-pressure commercial sprinkler system (Weeden Environments, Woodstock, ON, Canada) capable of three levels of cooling intensity that is explained below. The sprinkler controller was mounted in the control room of each house where the main house controller was located. However, there was no communication between the two controllers. The sprinkler system consisted of two zones, with 20 spinner heads in each zone and one temperature sensor (at bird height) located approximately in the center of each zone near the north side feed line. The brood half of the house containing the cool cells was one zone, and the non-brood half containing the tunnel fans was the second zone. Each zone was operated independently by activating an electronic solenoid valve assigned to that zone depending on the temperature in that zone. As a result, the two zones might run on different schedules and be in different intensity levels at a given time.

Sprinklers in the combination-cooling house and cool cells in the cool cell only house were allowed to operate from d 37 until harvest (d 61) for the first summer flock (Table 1). Cool cells in the combination house were allowed to operate from d 53 until harvest (d 61). During the second summer flock, sprinklers and cool cells in the combination house and cool cells in the cool cell only house were all allowed to operate from d 27 until harvest (d 60).

Table 1. Summary of days during grow-outs when the cooling systems were used.

Flock #	Harvest Date	Cooling Allowed in Combination House		Cooling Allowed in Cool Cell Only House
		Sprinkler	Cool Cell	
1	20 July 2020	37–61 d	53–61 d	37–61 d
2	5 October 2020	27–61 d	27–61 d	27–61 d

The three levels of cooling programmed into the sprinkler controller served different functions. The levels recommended by the manufacturer were as follows: Level 1 begins at 1.1 °C above the tunnel set point temperature (TT) and operates for 10 sec every 30 min (Table 2). When sprinkled, birds stand up and release trapped heat between and under the birds. Additionally, upon standing, numerous birds were observed to move to the feeder and drinker lines to eat and drink. Level 2 activates at 2.8 °C above the TT and operates for 20 s every 15 min. It combines getting the birds up to release trapped heat with increased wind chill on the birds from additional tunnel fans operating and increased amount of sprinkler droplets on the heads and feathers of the birds. Level 3 activates at 4.4 °C above TT and operates for 20 s every 5–7 min, depending on conditions, and creates bird surface wetting that allows maximum wind chill because of the nearly constant evaporative cooling of water droplets off the birds and a steady wind speed above 2.5 m/s (500 ft/min). For this study, 8 tunnel fans were running during Level 1, 9 fans during Level 2, and 10 fans during Level 3. Even though the tunnel fan set points were staged 0.56 °C apart, fans 9 and 10 were withheld until the sprinkler system reached levels 2 and 3, respectively. Table 2 lists temperature and run time settings suggested by the manufacturer. Table 3 lists sprinkler and cool cell settings used in the study.

Table 2. Temperature and run time settings suggested by sprinkler manufacturer.

Level	Degrees Offset above TT [+]	Run Time (s)	Idle Time (min)
1	1.1 °C (2 °F)	10	30
2	2.8 °C (5 °F)	20	15
3	4.4 °C (8 °F)	20	5–7

[+] TT = tunnel temperature set point.

Table 3. Set point temperatures (T) of the sprinkler system (SS)/cool cell (CC) used in this study.

Flock Day	Set-Point T (ST)	Tunnel T (TT)	SS Level 1	SS Level 2	SS Level 3	CC-ON T [†]
	Temperature, °C					
27–61	23.7–15.6	ST + 3.3	TT + 2.2	TT + 3.9	TT + 5.6	TT + 12.2 (ST + 15.6)
	Example					
56	16.7	20	25.6	27.2	28.9	32.2

[†] Cool cell is allowed to be wetted for a short duration of 20 s at a time when CC setpoint temperature is reached in order to prevent substantial decrease of indoor temperature.

We operated the sprinklers under similar idle times of each cooling level but at temperature settings slightly higher than recommended by the manufacturer (Table 3). In the combination house, the cool cell (CC) was allowed to be wetted for a short duration of 15–20 s at a time when CC setpoint temperature of 32 °C (90 °F) was reached in order to prevent substantial decrease of indoor temperature. The goal is to allow the sprinklers to do as much cooling as possible and use the cool cells only in extreme situations. It is critical to not run the cool cell system too soon, too often, or too long when sprinklers are used. Doing so tends to keep the house too cool and too humid for the sprinklers to be most effective. Ideally, cool cells should not operate before a house temperature of 31 °C is reached for the sprinklers to be most efficient. While this higher temperature may seem frightening to some, it comes with a lower humidity level that is beneficial to house environment, allowing birds to utilize evaporative respiration more efficiently.

2.3. Measurements

In-house temperature and relative humidity data were monitored and recorded by an Intelia data collection system (Intelia Technologies Inc., Quebec, QC, Canada) in each house with collection interval of 15 min. Cooling water use by sprinkler and evaporative cooling systems were monitored using water meters containing an electrical pulse output (1 pulse = 3.8 L (1 gal)).

Initial bedding material was kiln-dried pine shavings. However, at the time of the study, the two broiler houses had not received a total litter clean out in over 10 years. Every two years, approximately one half of the litter material was removed from each house to re-establish a litter depth of four to six inches and the remaining litter was evenly spread back out and was used to grow additional flocks. No topdressing of litter was performed between flocks. Chicks were placed directly on old litter material after caked litter was removed and the litter re-leveled. Down time between flocks was 22 d for the first flock and 17 d for the second flock. Litter moisture content was measured by sampling litter at the end of weeks 7 and 9 during both flocks. Litter was collected separately from 16 locations in the cool cell and fan ends of the houses. The eight subsamples from each end (16 total subsamples) were collected from the top 1–2 cm of the litter surface using a round point shovel and thoroughly mixed in a 19 L (5-gal) bucket. From this mixed sample, a 946 mL composite subsample was placed in a plastic bag and transported to the Mississippi State University Chemical Laboratory for moisture content analysis [48]. Production data such as feed conversion ratio (FCR), body weight, mortality, and paw quality data were collected by processor at harvest. However, there were no significant differences in production data; therefore, production data are not discussed in the current paper.

2.4. Statistical Analysis

Maximum indoor temperatures of each day (collected every 15-min) for the time period 9:00 am to 9:00 pm (when water cooling systems were operational) were collected and averaged as daily averaged maximum temperature from each house. Coincident relative humidity (RH) values when the maximum indoor temperatures occurred were collected and averaged for this duration. The temperature and RH values from two cooling regimes were analyzed using one-way ANOVA using SAS 9.4 with significance indicated by $p \leq 0.05$. Litter moisture contents of the two flocks from either cool cell half or the tunnel fan half for weeks 7 and 9 were analyzed separately. Although low replication was an issue in this study, as is always the case with whole-house treatments on commercial poultry farms, results are similar to those reported previously [31,47,49].

3. Results

3.1. Relative Humidity and Temperature

Maintaining cool cell set point temperature in the combination-cooling house at 32 °C (90 °F) resulted in limited use of cool cells in the combination house. As a result, the daily averaged maximum temperature in the combination house was 2.2 °C higher ($p < 0.0001$) in the combination house than in the cool cell only house (Figure 1a). However, the higher temperature was offset by a 12.2% lower coincident humidity ($p < 0.0001$) in the combination house (Figure 1b). The effect of sprinklers on the in-house environment showed the trend of lower relative humidity and higher temperature in the combination house than those in the cool cell only house during day-time cooling period, consistent with previous studies [31,50].

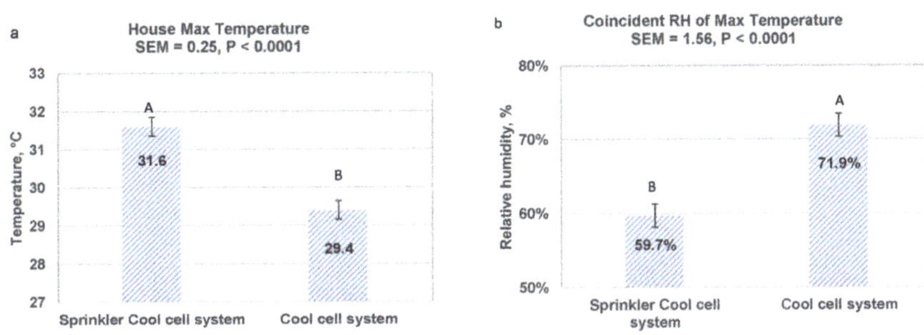

Figure 1. Thermal environment when cooling was in effect to raise heavy broilers during May–Oct 2020. (**a**) Average daily maximum house temperature, and (**b**) average of coincident relative humidity as maximum temperature occurred. SEM: Pooled standard error of mean. Letters A or B showing significant difference.

Even though maximum house temperatures in the combination house were higher (Figure 1a), this should not be equated with actual bird comfort temperature. The direct cooling effect of the sprinklers on the birds, indicated by the lower surface temperature (Figure 2a,b), was effective in facilitating heat release from birds and compensating for higher environmental ambient temperature [30–34,45,51]. High ambient temperatures resulted in use of evaporative cooling pads in the cool cell house and sprinklers in the sprinkler/cool cell combination house.

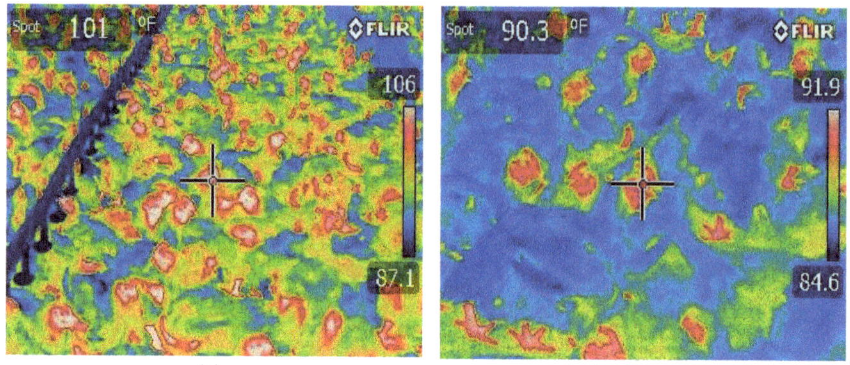

(a) (b)

Figure 2. Thermal images showing surface temperature of birds in a combination-cooling house at 58 d before sprinkling (**a**) and after sprinkling (**b**). The wetted surface due to sprinkler operation lowered body surface temperature overall (blue color (**b**)) with most of chickens' feet and head/neck shown as warm red color.

3.2. Water Used for Cooling

Water is essential for a variety of physiological functions and the productive performance of chickens. However, with recent uncertainties exacerbated by climate change, plentiful water availability is no longer a certainty in many locations and water scarcity is becoming a major global concern. Several factors affect the daily water requirement for poultry: e.g., housing conditions (temperature, lighting program and intensity, etc.), performance level, and feeding-related factors (type and ingredients) [2]. A major benefit associated with the sprinkler system is the potential cooling water savings compared to a cool cell only system. Daily cooling water use for each house is reported per flock in Figure 3. Cumulative cooling water for each flock is reported in Table 4. Water usage in the sprinkler/cool cell combination house demonstrated water savings that averaged 64% over the two summer flocks in comparison to the cool cell only house. These savings are in close agreement with [31] where savings of 67% were reported and [50] where savings of 58% were reported. The greatest water savings were observed on days when only sprinklers were in use in the combination house while evaporative cooling pads were being used in the cool cell house.

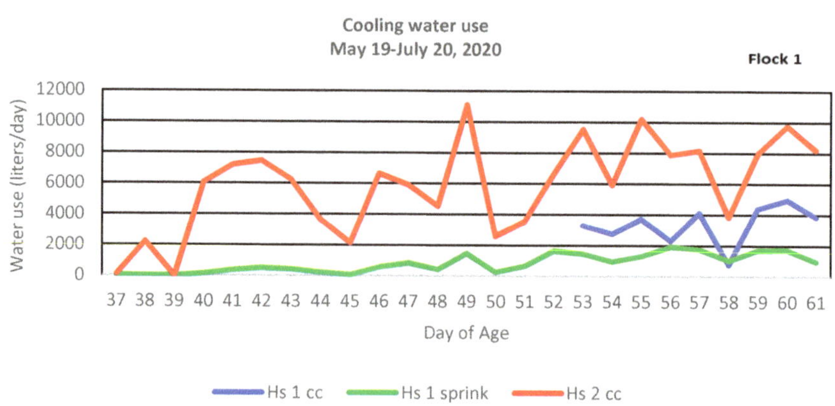

Figure 3. *Cont.*

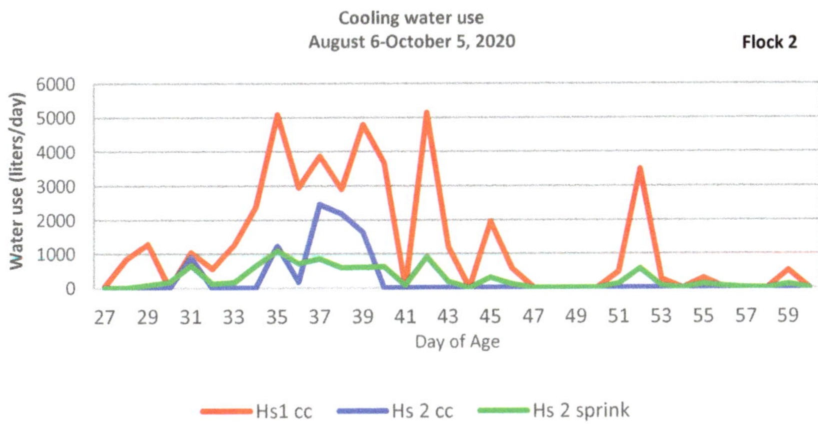

Figure 3. Daily cooling water use of sprinklers and cool cells for two flocks in two houses from May to October 2020. House 1 cool cell (Hs 1 cc), House 1 sprinkler (Hs 1 sprink), House 2 cool cell (Hs 2 cc), and House 2 sprinkler (Hs 2 sprink).

Table 4. Cumulative cooling water used in the cool cell (CC) only house and the combination house when raising heavy broilers during May–October 2020.

Flock #	Water Used in CC-Only House (L)	Water Used in Combination House (L)	Portion of Water Used by Sprinklers (L)	Water Saved in Combination House
1 †	147,465	50,266	20,138	66%
2 ‡	44,342	17,459	8877	61%

† 30 d > 30 °C, including 6 d > 35 °C (daily maximum ambient temperatures). ‡ 13 d > 30 °C (daily maximum ambient temperatures).

3.3. Litter Moisture

We saw no significant effect of sprinklers on litter moisture contents in either the fan half (Figure 4a) or cool cell half (Figure 4b) of the house at either week 7 or week 9 of the flocks. This is in agreement with findings from [31] who reported no significant effect by sprinklers on litter moisture content. However, it does not agree with research by [50] who found moisture content differed with a two-way interaction between growout × sprinklers ($p = 0.002$), with slightly drier litter in the sprinkler house. A weaker relationship was also found by [50] when sprinklers were considered as a main effect ($p = 0.046$), where litter moisture was slightly lower in the sprinkler houses. In the current study, we did see slightly drier litter in the sprinkler/cool cell combination house. Litter moisture content is a result of excreted moisture, normal drinker spillage, leaking drinkers, and water evaporation into the air [39,51]. The potential lower relative humidity and higher temperature in the sprinkler/cool cell combination house, hence higher water vapor deficit [51], may have encouraged higher evaporation from the litter.

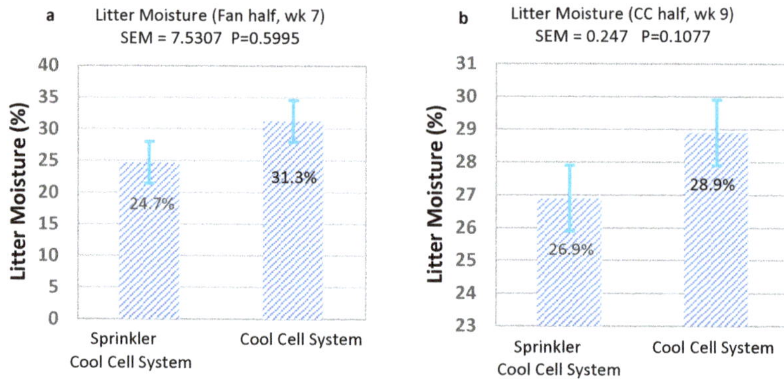

Figure 4. Average litter moisture contents over two flocks. (**a**) fan half of houses in week 7; (**b**) cool cell half of houses in week 9. SEM: Pooled standard error of mean.

4. Discussion

4.1. Understanding Sprinkler Cooling

Understanding a sprinkler-type surface-wetting system requires a complete rethinking of how to cool chickens and overcoming the initial fear of a slightly elevated house temperature. Sprinkler cooling attempts to cool individual chickens, unlike cool cell systems that attempt to cool the environment where the chickens live. Sprinkler cooling requires a slightly warmer house temperature and an associated lower house humidity other than a standard cool cell system usually provides. This warmer, drier environment does two things: (1) it creates a situation where the bird can more effectively and efficiently cool itself through evaporative respiration, and (2) it creates an auxiliary heat release mechanism, replacing the loss of sensible heat dissipation due to hot weather [31,50,52].

It may surprise some that the sprinkler system did not have a more obvious effect on litter moisture, as one might think applying water directly to the birds would increase litter moisture content. However, the quantity of water reaching the litter from the sprinklers is generally much less than the amount of water that the birds add to the litter in the manure [53]. The quantity of water added to the floor (including onto the birds) by the sprinklers (median 0.07 L/m^2/day, maximum 1.04 L/m^2/day) is less than the amount of water the birds add to the litter in their excreta (estimated to be 1.6 L/m^2/day to 3.3 L/m^2/day) [51].

In addition, even though the sprinkler system applied greater amounts of water on hotter days, the house was operating in tunnel ventilation mode on these hotter days with maximum air speed from all tunnel fans in operation. These conditions were favorable for rapidly evaporating the water applied to the birds by the sprinklers before it had the opportunity to reach the litter (while also ensuring a maximum evaporative cooling effect). A key feature is to maintain an interval between sprinkler operations that is sufficient to allow the water applied during one sprinkler application to be evaporated just before the next sprinkler application. It is important to allow the birds to dry off between sprinkler applications.

The phase change from liquid to water vapor taking place at the birds' surfaces is much more efficient at dissipating heat than convective heat transfer between chicken bodies and the warm surroundings [48]. Furthermore, as birds age and more completely fill the house, only a small portion of the water applied by the sprinklers actually reaches the litter due to water droplets landing only on the birds that now cover all the floor area. We did observe a tendency for litter moisture to be slightly drier in the sprinkler/cool cell combination house compared to the cool cell only house. In general, litter moisture content is influenced by multiple factors [39], such as those associated with drinkers, ventilation rate, bird health, and litter properties which were beyond control within the scope of this study.

4.2. Sprinkler Advantages

The reason the combination cooling achieved a large amount of cooling water saving lies in the way water is employed. The quantity of water used by a cool cell system is largely determined by the psychrometric process and is a function of the outside temperature and humidity conditions, the amount of heat released by birds, and the amount of ventilation air exchange through the house. High cooling water usage from a cool cell system is a consequence of the high ventilation rate at approximately one air exchange per minute to exhaust humidified air. The quantity of water required to cool birds through surface wetting followed by evaporation is governed by the heat production demand of the birds and the temperature surrounding the birds. Birds' surface acted as a local evaporative device, hence high efficiency with relatively small and controlled amount of water in achieving birds' cooling. Even though houses experience the same amount of air exchange under tunnel ventilation, the exhaust air in the combination house was minimally altered and remained similar to the psychrometric condition as the outside air.

Sustainability has become a key focus for the poultry industry in recent years. In addition, water scarcity is a serious concern that can no longer be ignored. Sprinkler cooling of broilers offers considerable water savings without sacrificing flock performance. As the poultry industry seeks opportunities to lessen its environmental footprint and become more sustainable, sprinkler cooling should be considered an important tool to address water scarcity, improve water conservation efforts, and help the industry reach its sustainability goals.

5. Conclusions

The combination cooling of heavy broilers used the sprinklers as the first line of cooling defense, with cool cells used only to prevent extreme warm conditions, to best achieve the full potential of sprinkler cooling. The management employed in this study allowed 64% of cooling water saving compared to cool cells alone. The houses with combination cooling were warmer during the daytime period when supplemental cooling by water is necessary, with an average daily maximum temperature of 2.2 °C above that in a house typically operated under a cool cell system. The relative humidity remained lower in the combination houses than in typical cool cell houses. This study indicated that an intermittently operated surface wetting method in tunnel ventilated houses can be effective to compensate for indoor house temperature up to 31.6 °C (89 °F) in cooling floor-raised broiler chickens up to 4.42 kg. Operating the house slightly warmer and drier by delaying the cool cell system is critical in maintaining similar litter quality under the two cooling regimes.

Author Contributions: All authors were involved in the conceptualization of this study; methodology, J.M. and T.T.; formal analysis, J.D., J.M. and R.R.; investigation, J.M., Y.L. and S.D.; resources, J.M. and T.T.; data curation, J.M. and Y.L.; writing—original draft preparation, J.M., Y.L. and T.T.; writing—review and editing, J.D., R.R. and S.D.; visualization, J.M. and T.T.; supervision, J.M. and T.T.; project administration, J.M. and T.T.; funding acquisition, T.T. All authors have read and agreed to the published version of the manuscript.

Funding: This research was funded by USDA-NIFA and USDA-AFRI Sustainable Agriculture Systems (Grant number 2019-69012-29905).

Institutional Review Board Statement: The animal study protocol was approved by the IACUC Committee of Mississippi State University (protocol code IACUC-19-063 approved on 8 February 2019).

Informed Consent Statement: Not applicable.

Data Availability Statement: Restrictions apply to the availability of these data. Data were obtained in collaboration with a major poultry integrator and are available from the authors only upon permission from the integrator.

Acknowledgments: The authors gratefully acknowledge the support of USDA-NIFA and Weeden Environments, Inc. (Woodstock, ON, Canada) for their assistance in making this research possible. Thanks are also extended to the Mississippi State University Poultry Science Department farm crew for flock care and management throughout the study.

Conflicts of Interest: The authors declare no conflict of interest.

References

1. FAO. Production: Gateway to Poultry Production and Products. 2023. Food and Agriculture Organization of the United Nations. Available online: https://www.fao.org/poultry-production-products/en/ (accessed on 5 June 2023).
2. El Sabry, M.I.; Romeih, Z.U.; Stino, F.K.R.; Khosht, A.R.; Aggrey, S.E. Water scarcity can be critical limitation for the poultry industry. *Trop. Anim. Health Prod.* **2023**, *55*, 215. [CrossRef]
3. Marangoni, F.; Corsello, G.; Cricelli, C.; Ferrara, N.; Ghiselli, A.; Lucchin, L.; Poli, A. Role of poultry meat in a balanced diet aimed at maintaining health and wellbeing: An Italian consensus document. *Food Nutr. Res.* **2015**, *59*, 27606. [CrossRef]
4. El Sabry, M.I.; Charal, J.W.; McMillin, K.W.; Lavergne, T.A. Nanotechnology considerations for poultry and livestock production systems—A review. *Annals Anim. Sci.* **2018**, *18*, 319–334. [CrossRef]
5. El Sabry, M.I.; Hassan, S.S.; Zaki, M.M.; Stino, F.K. Stocking density: A clue for improving social behavior, welfare, health indices along with productivity performances of quail (*Coturnix coturnix*)—A review. *Trop Anim. Health Prod.* **2022**, *54*, 83. [CrossRef]
6. Steinfeld, H.; Gerber, P.; Wassenaar, T.D.; Castel, V.; Rosales, M.; Rosales, M.; de Haan, C. Livestock's Long Shadow: Environmental Issues and Options. 2006. Food and Agriculture Organization of the United Nations. 2006. Available online: http://www.virtualcentre.org/enlibrary/key_pub/longshad/A0701E00 (accessed on 22 June 2023).
7. Peters, C.J.; Picardy, J.A.; Darrouzet-Nardi, A.; Griffin, T.S. Feed conversions, ration compositions, and land use efficiencies of major livestock products in U.S. agricultural systems. *Agric. Syst.* **2014**, *130*, 35–43. [CrossRef]
8. Poore, J.; Nemecek, T. Reducing food's environmental impacts through producers and consumers. *Science* **2018**, *6392*, 987–992. [CrossRef]
9. Martinelli, G.; Vogel, E.; Decian, M.; Farinha, M.J.U.S.; Bernardo, L.V.M.; Borges, J.A.R.; Gimenes, R.M.T.; Garcia, R.G.; Ruviaro, C.F. Assessing the eco-efficiency of different poultry production systems: An approach using life cycle assessment and economic value added. *Sustain. Prod. Consum.* **2020**, *24*, 181–193. [CrossRef]
10. Skunca, D.; Tomasevic, I.; Nastasijevic, I.; Tomovic, V.; Djekic, I. Life cycle assessment of the chicken meat chain. *J. Clean. Prod.* **2018**, *184*, 440–450. [CrossRef]
11. Cesari, V.; Zucali, M.; Sandrucci, A.; Tamburini, A.; Bava, L.; Toschi, I. Environmental impact assessment of an Italian vertically integrated broiler system through a Life Cycle approach. *J. Clean. Prod.* **2017**, *143*, 904–911. [CrossRef]
12. Kalhor, T.; Rajabipour, A.; Akram, A.; Sharifi, M. Environmental impact assessment of chicken meat production using life cycle assessment. *Inf. Process Agric.* **2016**, *4*, 262–271. [CrossRef]
13. Prudêncio da Silva, V.; van der Werf Hayo, M.G.; Soares, S.R.; Corson, M.S. Environmental impacts of French and Brazilian broiler chicken production scenarios: An LCA approach. *J. Environ. Manag.* **2014**, *133*, 222–231. [CrossRef]
14. González-García, S.; Gomez-Fernández, Z.; Diad, A.C.; Feijoo, G.; Moreira, M.T.; Arroja, L. Life cycle assessment of broiler chicken production: A Portuguese case study. *J. Clean. Prod.* **2014**, *74*, 125–134. [CrossRef]
15. Leinonen, I.; Williams, A.G.; Wiseman, J.; Guy, J.; Kyriazakis, I. Predicting the environmental impacts of chicken systems in the United Kingdom through a life cycle assessment: Broiler production systems. *Poult Sci.* **2012**, *91*, 8–25. [CrossRef]
16. Pelletier, N. Environmental performance in the US broiler poultry sector: Life cycle energy use and greenhouse gas, ozone depleting, acidifying and eutrophying emissions. *Agric. Syst.* **2008**, *98*, 67–73. [CrossRef]
17. Beekman, G.B. Water conservation, recycling and reuse. *Water Res. Dev.* **1998**, *14*, 353–364. [CrossRef]
18. Casani, S.; Rouhany, M.; Knøchel, S. A discussion paper on challenges and limitations to water reuse and hygiene in the food industry. *Water Res.* **2005**, *39*, 1134–1146. [CrossRef] [PubMed]
19. Alcamo, J.; Döll, P.; Kaspar, F.; Siebert, S. *Global Change and Global Scenarios of Water Use and Availability: An Application of Water GAP 1.0*; Center for Environmental Systems Research (CESR), University of Kassel: Kassel, Germany, 1997.
20. Alcamo, J.; Henrichs, T.; Rösch, T. *World Water in 2025: Global Modeling and Scenario Analysis for the World Commission on Water for the 21st Century*; Center for Environmental Systems Research (CESR), University of Kassel: Kassel, Germany, 2000.
21. Rijsberman, F.R. Water scarcity: Fact or fiction? *Agri. Water Manag.* **2006**, *80*, 5–22. [CrossRef]
22. Dozier, W.A.; Lott, B.D.; Branton, S.L. Growth responses of male broilers subjected to increasing air velocities at high ambient temperatures and a high dew point. *Poult. Sci.* **2005**, *84*, 962–966. [CrossRef] [PubMed]
23. Yahav, S.; Straschnow, A.; Lugar, D.; Shinder, D.; Tanny, I.; Cohen, S. Ventilation, sensible heat loss, broiler energy, and water balance under harsh environmental conditions. *Poult. Sci.* **2004**, *83*, 253–258. [CrossRef]
24. Dozier, W.A., III; Purswell, J.L.; Branton, S.L. Growth responses of male broilers subjected to high air velocity for either twelve or twenty-four hours from thirty-seven to fifty-one days of age. *J. Appl. Poult. Res.* **2006**, *15*, 362–366. [CrossRef]
25. Simmons, J.D.; Lott, B.D. Evaporative cooling performance resulting from changes in water temperature. *Appl. Eng. Agric.* **1996**, *12*, 497–500. [CrossRef]

26. Ryder, A.A.; Feddes, J.J.R.; Zuidhof, M.J. Field study to relate heat stress index to broiler performance. *J. Appl. Poult. Res.* **2004**, *13*, 493–499. [CrossRef]
27. Purswell, J.L.; Linhoss, J.E.; Edge, C.M.; Davis, J.D.; Campbell, J.C. Water supply rates for recirculating evaporative cooling systems. *Appl. Eng. Agric.* **2018**, *34*, 581–590. [CrossRef]
28. Berry, I.L.; Costello, T.A.; Benz, R.C. *Cooling Broiler Chickens by Surface Wetting. American Society of Agricultural Engineers Meeting Presentation*; Paper No. 90-4024; ASAE: St. Joseph, MI, USA, 1990.
29. Xin, H.; Berry, I.L.; Tabler, G.T.; Barton, T.L. Temperature and humidity profiles of broiler houses with experimental conventional and tunnel ventilation systems. *Appl. Eng. Agric.* **1994**, *10*, 535–542. [CrossRef]
30. Tao, X.; Xin, H. Acute synergistic effects of air temperature, humidity, and velocity on homeostasis of market-size broilers. *Trans. ASAE* **2003**, *46*, 491–497.
31. Liang, Y.; Tabler, G.T.; Costello, T.A.; Berry, I.L.; Watkins, S.E.; Thaxton, Y.V. Cooling broiler chickens by surface wetting: Indoor thermal environment, water usage, and bird performance. *Appl. Eng. Agric.* **2014**, *30*, 249–258.
32. Dunlop, M.W. *Effect of An In-Shed Sprinkler Cooling System on Temperature, Relative Humidity, Water Usage, Litter Conditions, Live Weight and Mortality*; AgriFutures Australia: Wagga, NSW, Australia, 2018.
33. Lin, H.; Zhang, H.F.; Du, R.; Gu, X.H.; Zhang, Z.Y.; Buyse, J.; Decuypere, E. Thermoregulation responses of broiler chickens to humidity at different ambient temperatures. II. Four weeks of age. *Poult. Sci.* **2005**, *84*, 1173–1178. [CrossRef]
34. Hillman, P.E. *Thermoregulatory Physiology in Livestock Energetics and Thermal Environmental Management*; De Shazer, J.A., Ed.; American Society of Agricultural and Biological Engineers: St. Joseph, MI, USA, 2009; Chapter 2.
35. Payne, C.G. Factors influencing environmental temperature and humidity in intensive broiler houses during the post-brooding period. *Brit. Poult. Sci.* **1967**, *8*, 101–118. [CrossRef]
36. Weaver, W.D.; Meijerhof, R. The effect of different levels of relative humidity and air movement on litter conditions, ammonia levels, growth, and carcass quality for broiler chickens. *Poult. Sci.* **1991**, *70*, 746–755. [CrossRef]
37. Jones, T.A.; Donnelly, C.A.; Stamp Dawkins, M. Environmental and management factors affecting the welfare of chickens on commercial farms in the United Kingdom and Denmark stocked at five densities. *Poult. Sci.* **2005**, *84*, 1155–1165. [CrossRef]
38. Shepherd, E.M.; Fairchild, B.D. Footpad dermatitis in poultry. *Poult. Sci.* **2010**, *89*, 2043–2051. [CrossRef] [PubMed]
39. Dunlop, M.W.; Moss, A.F.; Groves, P.J.; Wilkinson, S.J.; Stuetz, R.M.; Selle, P.H. The multidimensional causal factors of 'wet litter' in chicken-meat production. *Sci. Total Environ.* **2016**, *562*, 766–776. [CrossRef] [PubMed]
40. Richards, S.A. Evaporative water loss in domestic fowls and its partition in relation to ambient temperature. *J. Agric. Sci. Camb.* **1976**, *87*, 527–532. [CrossRef]
41. Mutaf, S.; Seber Kahraman, N.; First, M.Z. Surface wetting and its effect on body and surface temperatures of domestic laying hens at different thermal conditions. *Poult. Sci.* **2008**, *87*, 2441–2450. [CrossRef]
42. Wolfenson, D.; Bachrach, D.; Maman, M.; Graber, Y.; Rozenboim, I. Evaporative cooling of ventral regions of the skin in heat-stressed laying hens. *Poult. Sci.* **2001**, *80*, 958–964. [CrossRef]
43. Chepete, H.J.; Xin, H. Cooling laying hens by intermittent partial surface sprinkling. *Trans. ASAE* **2000**, *43*, 965–971. [CrossRef]
44. Ikeguchi, A.; Xin, H. Field evaluation of a sprinkling system for cooling commercial laying hens in Iowa. *Appl. Eng. Agric.* **2001**, *17*, 217. [CrossRef]
45. Tao, X.; Xin, H. Surface wetting and its optimization to cool broiler chickens. *Trans. ASAE* **2003**, *46*, 483–490.
46. Tabler, G.T.; Berry, I.L.; Liang, Y.; Costello, T.A.; Xin, H. Cooling broiler chickens by direct sprinkling. *Avian Advice* **2008**, *10*, 10–15.
47. Moon, J.W.; DuBien, J.; Brown, A.T.; Liang, Y.; Tabler, T. Water conservation and production benefits of sprinkling broilers. In Proceedings of the 2020 International Poultry Scientific Forum Georgia World Congress Center, Atlanta, Georgia, 27–28 January 2020. Available online: https://poultryscience.org/files/galleries/2020_IPSF_Abstracts.pdf (accessed on 3 July 2023).
48. ASTM International. Standard Test Method for Determination of Water Content of Soil by Direct Heating. D4959-16. Available online: https://compass.astm.org/document/?contentCode=ASTM%7CD4959-16%7Cen-US&proxycl=https%3A%2F%2Fsecure.astm.org&fromLogin=true (accessed on 6 September 2023).
49. Liang, Y.; Tabler, G.T.; Dridi, S. Sprinkler technology improves broiler sustainability: From heat stress alleviation to water conservation: A mini review. *Front. Vet. Sci.* **2020**, *77*, 544814. [CrossRef] [PubMed]
50. Dunlop, M.W.; McAuley, J. Direct surface wetting sprinkler system to reduce the use of evaporative cooling pads in meat chicken production: Indoor thermal environment, water usage, litter moisture content, live market weights, and mortalities. *Poult. Sci.* **2021**, *100*, 101078. [CrossRef] [PubMed]
51. Dunlop, M.W.; Blackwell, P.J.; Stuetz, R.M. Water addition, evaporation and water holding capacity of poultry litter. *Sci. Total Environ.* **2015**, *538*, 979–985. [CrossRef] [PubMed]
52. Liang, Y. *Water Use Patterns Differ between Pad and Sprinkler Cooling. Factsheet*; FSA 1068; University of Arkansas Division of Agriculture: Little Rock, AR, USA, 2017.
53. Moon, J.W. The Effect of Sprinkler Cooling on Water Conservation, House Environment, and Broiler Performance. Master's Thesis, Mississippi State University, Mississippi State, MS, USA, 2022.

animals

MDPI

Article

US Swine Industry Stakeholder Perceptions of Precision Livestock Farming Technology: A Q-Methodology Study

Babatope E. Akinyemi [1,*], Faical Akaichi [2], Janice M. Siegford [1] and Simon P. Turner [3]

[1] Department of Animal Science, Michigan State University, East Lansing, MI 48824, USA; siegford@msu.edu
[2] Rural Economy, Environment and Society Department, Scotland's Rural College (SRUC), Edinburgh EH9 3JG, UK; faical.akaichi@sruc.ac.uk
[3] Animal and Veterinary Sciences Department, Scotland's Rural College, Easter Bush, Edinburgh EH25 9RG, UK; simon.turner@sruc.ac.uk
* Correspondence: akinyem3@msu.edu

Simple Summary: Understanding the views and concerns of precision livestock farming (PLF) held by the key players in the US swine industry is particularly important for its widespread acceptability. Using a Q-methodology approach following initial individual in-depth interviews 6 months earlier, we found that while most stakeholders view PLF as good and beneficial to pigs and humans, some expressed concerns about its limitations and pitfalls. Specifically, stakeholders who are optimistic about PLF think it should augment but not replace effective management. They also think it will improve pig health, welfare, and make pig keepers' jobs simpler, safer, and better. However, stakeholders who are concerned about PLF view it as a poor proxy for farmers to keep an eye on the herd, see issues, and resolve them. They also believe PLF does little to lessen the negative effects of swine farming on the environment. We contend that competing perceived benefits and concerns about PLF may hinder widespread adoption.

Abstract: This study used the Q-methodology approach to analyze perceptions of precision livestock farming (PLF) technology held by stakeholders directly or indirectly involved in the US swine industry. To see if stakeholders' perceptions of PLF changed over time as PLF is a rapidly evolving field, we deliberately followed up with stakeholders we had interviewed 6 months earlier. We identified three distinct points of view: PLF improves farm management, animal welfare, and laborer work conditions; PLF does not solve swine industry problems; PLF has limitations and could lead to data ownership conflict. Stakeholders with in-depth knowledge of PLF technology demonstrated elevated levels of optimism about it, whereas those with a basic understanding were skeptical of PLF claims. Despite holding different PLF views, all stakeholders agreed on the significance of training to enhance PLF usefulness and its eventual adoption. In conclusion, we believe this study's results hold promise for helping US swine industry stakeholders make better-informed decisions about PLF technology implementation.

Keywords: precision livestock farming; Q-methodology; stakeholder; pork industry

Citation: Akinyemi, B.E.; Akaichi, F.; Siegford, J.M.; Turner, S.P. US Swine Industry Stakeholder Perceptions of Precision Livestock Farming Technology: A Q-Methodology Study. *Animals* **2023**, *13*, 2930. https://doi.org/10.3390/ani13182930

Academic Editors: Roy Neville Kirkwood and Cesare Castellini

Received: 9 August 2023
Revised: 12 September 2023
Accepted: 13 September 2023
Published: 15 September 2023

1. Introduction

Global meat consumption has increased significantly over the past few decades, with pork and poultry showing the most substantial growth in consumption rates [1]. In fact, the number of animals kept on the farm has grown substantially, while the number of farmers has dwindled [2]. As a result, each farmer has considerably larger herds, which makes it challenging for them to consistently monitor every animal in such large groupings in a way that assures each animal's welfare while minimizing environmental impact. Specifically in the swine industry, the low-profit, high-risk [3] nature of pork production further exacerbates this situation.

Precision Livestock Farming (PLF) is a management system that involves using technology to continuously collect and process data from individual animals [2]. This is significantly different from other approaches that monitor animal welfare by human experts scoring animal-based indicators on an infrequent basis. The application of PLF technology has recently been proposed as one of the ways to help address public concerns transparently and objectively [4]. According to the proponents of PLF, these technologies can assist with many issues, including farmer profitability, environmental concerns, and animal welfare, and might even be able to improve all these issues at once [5].

Although PLF holds many promises, it also raises many concerns. As PLF evolves, further farm consolidation is possible since only those with the financial wherewithal to invest in PLF may benefit from it [6,7]. The loss and "de-skilling" of agricultural employment is another issue associated with the use of PLF in animal agriculture [8]. As livestock monitoring and care become more automated, PLF will need less labor and those jobs will not require many of the skills that are currently needed [7].

According to Werkheiser [5], there are other broader societal implications of PLF. Firstly, there is a worry that resources are being allocated to develop technology that could be used for monitoring humans as well [5]. Secondly, there is a risk that this technology becomes normalized, which could decrease resistance to its use, both in terms of legal regulations and public perception [5].

Hitherto, very few studies have investigated stakeholders' perceptions of PLF—exceptions are Giersberg and Meijboom [9], Klerkx and colleagues [10], and Pfeiffer and colleagues [11]. Specifically, there is no empirical study on the suitability of PLF technologies for different stakeholders within the US swine industry, and the perspectives of various stakeholders on the implementation of these technologies and the utilization of PLF data are not yet known [10]. Moreover, there is a paucity of information on the benefits PLF offers, as well as the concerns and limitations associated with its implementation [9]. Consequently, there is a significant knowledge gap that must be addressed to fully comprehend the broader implications and viability of PLF technologies within the US swine industry.

This study examined US swine industry stakeholders' perceptions of and needs for PLF across the pig production system ranging from areas of genetic improvement to welfare certification. We adopted Q-methodology to quantify swine stakeholders' perceptions of PLF technology. The growing recognition that the opinions of stakeholders involved in the development and usage of technology must be considered for decisions to be recognized as legitimate is a major factor in Q-methodology's rising popularity in technology-related studies [12–14].

2. Materials and Methods

The Q-methodology approach combines quantitative and qualitative techniques to examine people's subjective beliefs scientifically and enables researchers to statistically identify and categorize opinions on a given topic [15]. In a Q-study, an individual is presented with a set of statements about a given topic and then asked to rank-order the statements. In our study, participants ranked statements from "Most Like What I Think" to "Least Like What I Think " using the Q-grid in Figure 1. This operation is referred to as Q-sorting.

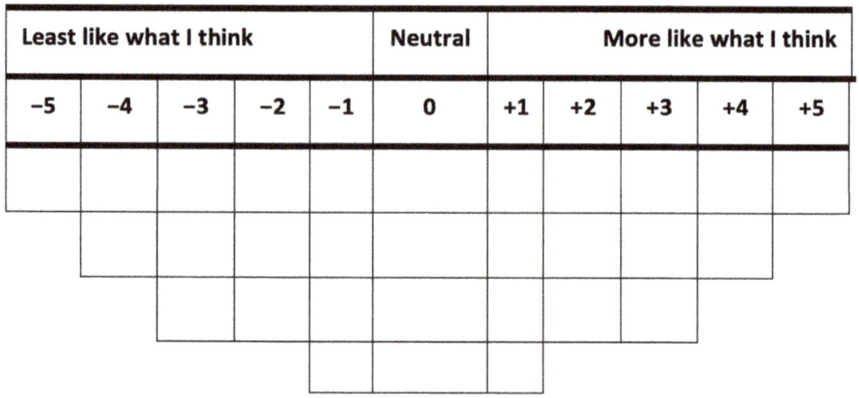

Figure 1. Q-sort grid.

The Q-methodology procedure shown in Figure 2 is briefly described here. The first step in the Q-method entailed defining and building a concourse. The concourse refers to the collection of subjective statements (from the Latin concursus, meaning "a running together" as when ideas run together in thought) [16]. Concourses are statements describing the subjective views, opinions, and arguments put forward by individuals, stakeholders, professionals, and scientists about a particular topic (in this study, PLF). Meanings attributed to the concourse are inherently social and contextual; hence, it is assumed that stakeholders would inevitably draw from the ongoing PLF social discourse in constructing and articulating their experiences and perceptions of PLF [16].

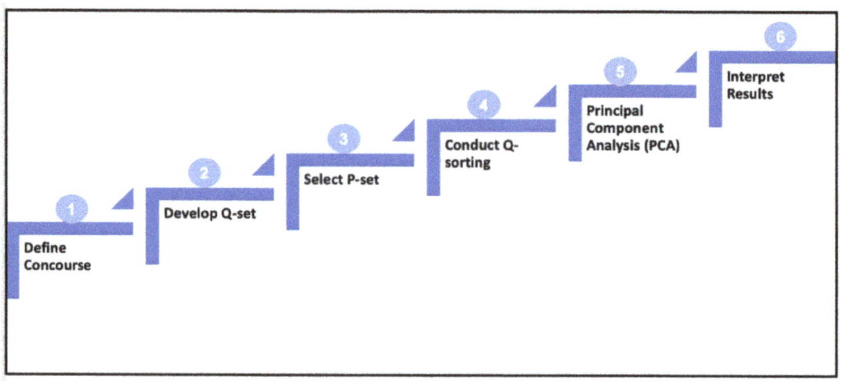

Figure 2. Q-methodology procedure.

The second step was the development of the Q-set, which was a subset of statements drawn from the larger concourse, and it was this set of statements that was presented to participants in the form of a Q-sort. The Q-set for this study was developed from in-depth interviews reported by Akinyemi and colleagues [17]. The Q-set comprised subjective statements on PLF made by 12 individuals. These individuals were purposefully chosen to reflect the diversity of PLF perspectives across the swine industry ranging from pig producers to pork consumers. The stakeholders were domain experts in swine welfare, swine veterinary medicine, animal welfare auditing, animal health regulation, technology development, agricultural engineering, animal care, and compliance, animal breeding and genetics, and consumer advocacy. The demographic characteristics of the participants are shown in Table 1.

Table 1. Participants' demographics.

	Experience (Years)	Organizational Role	Position	Relationship with PLF
P1	26–30	Veterinary services Technology development Farming Academic/research Welfare certification/auditing	Owner	PLF technology developer
P2	1–10	Welfare certification and regulation	Director animal care compliance	Potential off-farm PLF user
P3	31 or more	Regulation Food retailing Welfare certification/auditing Regulation Education	Executive	Potential off-farm PLF user
P4	1–10	Welfare certification and compliance	Consumer services manager	Potential off-farm PLF user
P5	11–15	Large-scale swine farming	Chief technology officer	PLF technology developer and farmer
P6	16–20	Technology development Veterinary services Technology development Pharmaceutical	Director	PLF technology developer
P7	31 or more	Technology development	Director	PLF technology developer and farmer
P8	16–20	Swine veterinarians association	Director	Potential off-farm PLF user
P9	31 or more	Government regulation	Leader	Potential off-farm PLF user
P10	11–15	Academic/research institution	Senior researcher	PLF technology developer
P11	31 or more	Technology development Academic/research institution Technology development	Professor	PLF technology developer

The 30 Q-set statements that were chosen for the Q-sorting exercise were derived from 55 subjective statements on PLF that were obtained from the 12 interviews. Ten main themes emerged from the 55 statements after careful analysis and categorization. These themes were not a strict categorization; rather, they represent coverage of the most relevant topics related to stakeholders' views on PLF. After several rounds of discussion among authors, redundant statements were either rephrased clearly, closely similar themes were merged, and vague statements were deleted. At the end of the process, the initial 55 statements were refined into a more comprehensive Q-sample composed of 30 statements. We also felt these 30 statements would reduce the cognitive burden of ranking 55 statements in 30 min on the participants. We ensured that the Q-samples were composed of statements that were "natural" in the language of stakeholders and "comprehensive" in their PLF views [18] to provide them with the opportunity to best express their personal views of PLF [19].

Gender and level of education were omitted to preserve participants' anonymity. The third step in this process entailed choosing the Q-participants which are referred to as the 'P-set'. In our case, the selection of the P-set was not performed at random; rather, these participants were 11 of the 12 original stakeholders who informed the Q-set and had been chosen to reflect diverse PLF perspectives across the US swine industry. This sample size is considered appropriate for this type of study [20–23] since P-set is expected to be composed of individuals who are theoretically relevant to the research question [24,25]. We deliberately made the decision to use the same group of stakeholders to develop the concourse underlying the Q-set and to participate in ranking the Q-set statements for two reasons. First, we wished to see how US swine industry stakeholders would view their own statements as well as statements from others in the group to examine how the stakeholders as a whole were similar to or different from one another. Furthermore, we wanted to examine whether their individual perspectives about PLF, which is a rapidly changing landscape, had changed in the 6 months between their original interviews and the Q-sort exercise.

The actual Q-sorting was conducted in step four. During Q-sorting, each participant ranked the 30 Q-statements into a Q-grid (see Figure 1). The Q-grid created a forced quasi-normal distribution using a pre-determined pattern grid with a scale labeled "More Like What I Think", "Neutral", and "Least Like What I Think". The Q-sort was concurrently administered to a group of 9 persons in a conference room while 2 joined online via Zoom. The instructions (see Appendix B) for the Q-sort exercise were read and explained to all the participants who were also given the chance to ask clarifying questions. The Q-sorting lasted for approximately 30 min. During the Q-sorting, participants were given a set of 30 index cards (9 persons in the room were given printed index cards while the 2 participants on Zoom were given electronic index cards). Each index card had one of the 30 Q-set statements written on it and randomly numbered. They were given the Q-sort grid with the instructions, and they were first asked to sort the cards into three piles of "More Like What I Think", "Least Like What I Think", and "Neutral (statements for which participants had no opinion)", based on their opinion of PLF. Next, they ranked the 30 statements on the 30-item forced normal distribution ranging from +5 (More Like What I Think) to −5 (Least Like What I Think). Participants began by taking the statements they grouped as "More Like What I Think" and chose the one statement that was "most like what I think" to place on the worksheet in the "+5" column on the far right. They were then asked to place the next two statements "More Like What I Think" into the "+4" column (in no particular order) until all statements from their "More Like What I Think" pile were placed on the Q-sort grid. The process was repeated with "Least Like What I Think" statements, starting with a statement in the "−5" column. When the participants were satisfied with how they had arranged the concourse statements on the Q-sort grid, they wrote the number on the statement card inside the corresponding squares in the Q-sort grid. A completed Q-sort grid is shown in Appendices E.1–E.3. Finally, participants provided basic demographic information (Table 1). The entire process lasted for about 30 min.

In step five, the 11 completed Q-sorts were analyzed using a by-person correlation and factor analysis approach using PQMethod free online software [26]. Factor analysis is the practice of condensing many variables into just a few so that research data are easier to work with. The idea is to identify and work with deeper factors driving the underlying concepts within data and then uncover and work with these rather than the lower-level variables that flow from them. Factor analysis is also sometimes called "dimension reduction analysis". With factor analysis, data "dimensions" can be reduced into one or more "super-variables", also known as unobserved variables or latent variables. For more information on factor analysis, please see Mulaik [27].

In this study, factor analysis was used to identify correlations (see Table 2) between the 11 Q-sorts and to reduce the many 'viewpoints' of PLF expressed through the sorting pattern down to 3 factors. These 3 factors represent unique ways of thinking about PLF among the 11 participants who shared the same factor. These factors are then interpreted as perspectives. In Q-methodology, 'viewpoints' are defined as a common set of perceptions profiles from participants that form a cluster of correlations [24]. By default, the analysis produced eight unrotated factors, which accounted for 95% of the total variance in the Q-sorts. Second, a varimax rotation was used to identify a small number of factors with significant factor loadings.

Table 2. By-person correlation matrix.

	P6	P4	P8	P3	P7	P9	P1	P10	P11	P2	P5
P6	1	−0.29	0.75	−0.38	0.73	0.28	0.58	0.64	0.24	0.50	0.60
P4		1	−0.21	0.59	−0.54	0.5	−0.46	−0.7	−0.3	−0.4	−0.50
P8			1	−0.25	0.46	0.43	0.58	0.64	0.34	0.50	0.52
P3				1	−0.62	0.17	−0.46	−0.18	−0.17	−0.17	−0.54
P7					1	0.17	0.55	0.62	0.14	0.24	0.68
P9						1	0.25	0.48	0.43	0.32	0.9
P1							1	0.49	0.25	0.50	0.56
P10								1	0.31	0.62	0.58
P11									1	0.28	0.4
P2										1	0.38
P5											1

Before factors were extracted, we had to decide the number of factors to be extracted and retained as being significant via the PQMethod software. Following Brown's [25] guidelines on factor stability, distinctness, clarity, and simplicity, we determined the number of factors to rotate based on (1) the eigenvalue being equal to or greater than one and (2) each factor should have at least two significant factor loadings in the unrotated factor matrix. Two of the eight factors met both criteria (see Table 2). The third factor was included in the rotation because its eigenvalue was close to one (0.94) and satisfied Humphrey's Rule, which states that a factor is significant if the absolute value of "the cross-product of the two highest loadings exceeds the standard error" [19]. After rotating the three factors (see Table 2), we found that at least two participants were loaded on each factor, the eigenvalue of the three factors was higher than one, and the three factors accounted for 73% of the total variance.

In the last step, we focused the interpretations of the factor analysis on the rotated solution factor Q-sort values. The unrotated factor solution is presented in Appendix F. For each factor, we considered the statements with the highest and the lowest Q-sort values (see Table 3). The highest-ranking statements (i.e., Q-sort values that are between +3 and +5) indicate what participants who loaded on the factor of interest think of PLF and least like what participants think of PLF for the lowest-ranking statements. We also paid close attention to the distinguishing statements (see Appendix D) which are the statements that were found to be significantly different among factors. The consensus statements (Appendix D) were also found to be helpful in identifying the PLF-related issues that all participants could agree on even though they loaded on varied factors.

Table 3. More and least like what participants think of PLF.

Scores	Perspective 1	Perspective 2	Perspective 3
5	PLF should be a supplement to good management and not replace good management	PLF technology can disconnect caretakers from pigs and make them less caring about pigs	PLF usage is limited by poor internet connection on most swine farms
4	PLF will improve pig health	PLF technology is a poor proxy of a farmer looking over the herd, detecting problems, and addressing them	PLF usage can lead to data privacy and data ownership conflicts
4	PLF makes the pig caretaker's job easier, safer, and better	PLF technology does not minimize the environmental impact of swine farming	PLF's data often cannot be used within existing management practices

Table 3. *Cont.*

Scores	Perspective 1	Perspective 2	Perspective 3
3	PLF will improve pig welfare	PLF technology may digitize swine production and make it look less natural and more artificial	PLF is cost-prohibitive to use across the entire livestock system
3	PLF will help in controlling disease outbreaks through traceability	PLF technology should be a supplement to good management and not replace good management	PLF requires a significant effort to change processes within an existing production system
−3	PLF is displacing labor and reducing job opportunities in the swine industry	PLF technology will make it possible for consumers to verify welfare certification claims	PLF is displacing labor and reducing job opportunities in the swine industry
−3	PLF often generates data that conflict with farmer expert opinions thus making PLF data less useful	PLF technology will address public concerns and increase consumer trust in pork production	PLF will address labor shortages in the swine industry
−4	PLF can be used without any form of training	PLF can be used without any form of training	PLF is a poor proxy of a farmer looking over the herd, detecting problems, and addressing them
−4	PLF can disconnect caretakers from pigs and make them less caring about pigs	PLF will improve pig health	PLF technology will address public concerns and increase consumer trust in pork production
−5	PLF's data often cannot be used within existing management practices	PLF will improve pig welfare	PLF technology can be used without any form of training

3. Results and Discussions

The by-person correlation matrix in Table 2 shows the correlation between the 11 participants who participated in this study. A perfect positive and negative correlation (r) corresponds to +1, and −1, respectively. As a rule of thumb, correlations are generally considered to be statistically significant if they are 2 to 2.5 times the standard error (i.e., between 0.44 and 0.56 irrespective of sign but the correlation coefficients considered significant in this study are r = 56 and above) [16]. Hence, participant P6 in the first row/column was statistically significantly and positively correlated with P8 (r = 75), P7 (r = 73), P10 (r = 64), and P5 (r = 60), but weakly and negatively correlated with P3 (r = −38) and P4 (r = −29). In addition, P6 moderately positively correlated with P1 (r = 58) and P2 (r = 50); however, its positive correlation with P11 (r = 24) was not substantial. Moreover, participants P5 and P7, P8 and P10, and P2 and P10 were strongly positively correlated with each other with the correlation coefficients of r = 68, r = 64, and r = 62, respectively. It is worth noting that the statistics associated with Q are not intended as a substitute for the apparent fact that the correlation is suffused with subjectivity, each Q-sort being a transformation of a person's opinion, and the coefficients merely indicate the degree of similarity or dissimilarity in perspective.

Presented in Table 4 are the factor loadings from the PCA. In essence, PCA uses the correlation matrix shown in Table 2 to determine how many fundamentally distinct Q-sorts are present. Q-sorts that are highly correlated with one another may be considered to have a family resemblance, with members within one family being strongly correlated with one another but not with those of other families. The number of various families (factors) is revealed via PCA. Therefore, the number of factors is entirely reliant on the actual performance of the Q-sorters and is simply empirical. In this study, the factors represent different perspectives of PLF, with those individuals sharing a common perspective defining the same factor. As previously noted, each factor represents different perspectives or conceptualizations of PLF. Factor 1 perspective of PLF is shared by seven participants, namely P1, P2, P5, P6, P7, P8, and P10. Factor 2, however, is a manifestation of a strong bipolarity between P3 and P4 on the positive side of the pole and P1, P5, and P7 on the negative side of the pole. Factor 3 represents a single perspective of PLF shared by participants P9 and P11. A sort is significantly associated with a factor at $p < 0.01$ statistical significance if the absolute value of the factor loading is greater than 2.58 ÷ √ (the number of statements used in the Q-sort)), which is equal to 0.47 in this study [25]. Based on the Q-sort association with each factor, P1, P2, P5, P6, P7, P8, and P10 are correlated with Factor 1. However, P3 and P4 are correlated with Factor 2 whereas P9 and P11 are correlated with Factor 3.

Table 4. Rotated solutions of participants' Q-sorts.

Participants/Q-Sorts	Factor 1	Factor 2	Factor 3
P1	0.57 *	−0.51	0.2
P2	0.73 *	0.02	0.19
P3	−0.1	0.87 *	−0.03
P4	−0.01	0.86 *	−0.01
P5	0.62 *	−0.59	−0.17
P6	0.79 *	−0.37	0.08
P7	0.55 *	−0.68	−0.02
P8	0.76 *	−0.21	0.28
P9	0.46	0.22	0.63 *
P10	0.88 *	−0.08	0.17
P11	0.11	−0.13	0.91 *
% Eigenvalue	3.75	2.82	1.45
% Explained Variance	34	26	13

Factor loadings with an asterisk (*) are significantly associated with each factor. A sort was significantly associated with a factor at $p < 0.01$ statistical significance if the absolute value of the factor loading was greater than 0.47 [25].

Based on the results in Table 3 and Appendices C–E, three perspectives were identified. Factor 1 represents Perspective 1: "PLF improves management, animal welfare, and laborer work conditions"; Factor 2 represents Perspective 2: "PLF does not solve the problems"; and Factor 3 represents Perspective 3: "PLF has limitations and could lead to data ownership conflict". Table 3 shows the bipolar ranked factor score of five statements of "more like what I think" of PLF (corresponding to the positive factor scores) and "least like what I think" of PLF (corresponding to the negative factor scores). The higher positive factor scores imply strong agreement with the statement whereas higher negative factor scores mean strong rejection of the statement. For example, the statement "PLF should be a supplement to good management and not replace good management" with the highest factor score of +5 indicates what participants sharing Perspective 1 strongly think of PLF.

The two statements "PLF will improve pig health" and "PLF makes the pig caretaker's job easier, safer, and better" with factor score +4 also reflect what Perspective 1 participants think of PLF. This group strongly rejects the following statements: "PLF's data often cannot be used within existing management practices" (−5), "PLF can disconnect caretakers from pigs and make them less caring about pigs" (−4), and "PLF can be used without any form of training" (−4).

Conversely, Perspective 2 participants strongly think "PLF can disconnect caretakers from pigs and make them less caring about pigs" (+5), "PLF is a poor proxy of a farmer looking over the herd, detecting problems and addressing them" (+4), and "PLF technology does not minimize the environmental impact of swine farming" (+4). They rejected the notion that "PLF will improve pig welfare "(−5), "PLF will improve pig health (−4), and "PLF can be used without any form of training" (−4).

Participants sharing Perspective 3 strongly think "PLF usage is limited by poor internet connection on most swine farms" (+5). They share the concern that" PLF usage can lead to data privacy and data ownership conflicts" (+4), and that "PLF often generates data that cannot be used within existing management practices" (+4). They strongly rejected the idea that "PLF technology can be used without any form of training" (−5). They also disagreed with the notion that "PLF technology will address public concerns and increase consumer trust in pork production" (−4), and that "PLF is a poor proxy of a farmer looking over the herd, detecting problems, and addressing them" (−4).

The 30 Q-set statements and the sources of each statement are shown in Appendix G. As expected, all stakeholders tend to agree with their own statements as well as statements from other stakeholders sharing similar roles or relationships with PLF. The PLF technology developers and large-scale farmers generally agreed with each other but disagreed sharply with statements from potential off-farm PLF users. Their Q-sorting was consistent with the views of PLF they expressed earlier during the interview conducted six months before the Q-sort exercise.

For example, the statements "PLF will improve pig health" and "PLF will improve pig welfare" were made by technology developers and large-scale farmers during their individual interviews six months earlier and were later ranked highly with factor scores of (+4) and (+3) (Table 4), respectively, indicating these statements were more like what this group thinks of PLF. However, they rejected the statements "PLF can be used without any form of training" (−4) and "PLF can disconnect caretakers from pigs and make them less caring about pigs" (−4), which were statements made by potential off-farm users during individual interviews six months earlier.

Likewise, the potential off-farm PLF user group held on to the views of PLF previously shared during the individual interview six months earlier and ranked the statements "PLF technology can disconnect caretakers from pigs and make them less caring about pigs" (+5) and "PLF technology is a poor proxy of a farmer looking over the herd, detecting problems, and addressing them" (+4) highly. They rejected the two statements from the PLF technology developer and farmer's group that "PLF will improve pig health" (−4) and "PLF will improve pig welfare" (−5) and ranked them very low.

Perspective 1: PLF improves management, animal welfare, and labor work conditions.

Seven stakeholders (P1, P2, P5, P6, P7, P8, and P10) shared Perspective 1. They included three technology developers, two swine veterinarians, a swine farmer, and an animal care and compliance specialist. This group is distinctly different from those with Perspectives 2 and 3 in that participants who shared this view were optimistic and enthusiastic about PLF and were less worried about its limitations and concerns. This group emphasized PLF benefits such as improvements to swine farm management, pig health, and animal welfare.

For this group, it was important that PLF should supplement good animal husbandry and not replace caregivers. According to [28,29], existing swine production management practices require skilled workers to work long hours in a challenging environment in ways that can both harm the workers' mental and physical health as well as increase the animals' biosecurity risks. Therefore, PLF can improve laborers' work conditions through automated non-invasive monitoring and management [30,31] that allows targeted, problem-solving work rather than repetitive manual labor. In addition, in view of the anticipated labor shortage in the swine industry in the near future [32], PLF technology that improves laborer work conditions might serve as an incentive for retaining and attracting farm workers.

Perspective 1 shares the findings of Benjamin and Yik [33] that PLF will help in controlling diseases by closely monitoring animals which allows farmers, veterinarians, and others to ensure pigs' welfare. Further potential to track animals may assist with the traceability of pork products through the production and distribution value chain. This is particularly important in view of the low-profit margins in swine production and the increasing pressures related to the responsible sourcing of pork. These pressures may create a compelling need and impetus for swine farmers to adopt cost-efficient technologies that can monitor the actual state of pigs to provide transparency, traceability, and evidence of improved welfare needed to meet consumer demands.

Most of the Perspective 1 stakeholders had practical knowledge of PLF which may have informed their beliefs about its value and potential and explained their high optimism about PLF's benefits to the swine industry. It is noteworthy that this group shared the views of Perspective 2 because they do not see PLF as disconnecting caretakers from pigs and making them less caring, or that PLF data are not usable within existing farm management practices, and as displacing labor and reducing job opportunities in the swine industry.

Perspective 2: PLF does not solve the problems.

Two stakeholders (P3 and P4) share Perspective 2. These stakeholders had no prior direct knowledge of PLF. This perspective is dissimilar to the first perspective in that it is less optimistic about PLF technology. These respondents thought PLF would not fix the problems facing the swine industry. This group was worried about the ways PLF could change the relationships between farmers and their livestock by eliminating the bond between farmers and their animals that develops as they spend time directly assessing their well-being [34]. They were particularly concerned that animal caretakers may become under- or over-reliant on PLF technology and spend less (quality) time with their pigs which could cause a loss of animal-oriented husbandry skills.

Moreover, Perspective 2 disagreed with Tullo and colleagues [35] and Lovarelli and colleagues' [36] claims that PLF can promote sustainability by enabling farmers to feed their animals with more precision, thereby avoiding waste [37], thus increasing both environmental and economic sustainability for the farm. On the contrary, they thought PLF technology does not reduce the negative externalities of swine farming but digitizes the swine industry, giving it a more artificial and unnatural outlook. Stakeholders in the Perspective 2 group felt that PLF technology should be used with effective management, not as a replacement. Finally, respondents favoring Perspective 2 were pessimistic about the claims that PLF could improve pig welfare and pig health. There was also skepticism among this group that PLF will allay public concerns, boost consumer confidence in pork production, and enable consumers to independently verify welfare certification claims. However, they agreed with Perspectives 1 and 3 that PLF technology cannot be used without training.

Perspective 3: PLF limitation and data ownership conflict.

Two stakeholders (P9 and P11) shared Perspective 3. This group was concerned about limitations, data ownership, and conflict issues associated with PLF usage. Echoing previous studies [38–41], this group considered poor internet connection a major problem on most swine farms in Australia, Europe, and the USA. Although internet connectivity is becoming widespread on farms across the world, the connection is sometimes still unreliable and too slow to carry large amounts of data. Hence, implementing PLF technologies on commercial farms poses challenges because transferring data to the cloud might not be optimal or even feasible in all circumstances. Edge computing, which enables data to be processed at the farm level rather than being sent over the internet, is one potential solution to these challenges [42].

Another issue associated with PLF emphasized by the Perspective 1 and 3 groups was data privacy and data ownership conflicts. A previous study by Neethirajan and Kemp [43] emphasized the need to ensure data privacy and security in precision livestock farming. According to Wolfert and colleagues [44], data collection on or from farms is currently limited because farmers prioritize privacy. To address this, new advances in machine learning are being developed that utilize privacy-preserving data exchange systems. Successful commercialization of PLF technologies has been hampered by a lack of open access to and proprietary control in data ownership held by a few commercial companies [30].

In contrast to the Perspective 1 group, the Perspective 3 group believed PLF data cannot be utilized within current management practices and that it would take substantial modification to adapt current production system processes to fully utilize PLF data. Stakeholders with Perspective 3 concurred with Morrone and colleagues [4] that PLF technologies are too expensive and may reduce the profitability of farming operations. In line with Perspectives 1 and 2, Perspective 3 stakeholders saw the need for training to effectively utilize PLF technology, echoing [38]. They rejected the claim that PLF technology would allay public worries and boost consumer confidence in pork production. They neither saw PLF technology as a solution to labor problems nor as displacing labor and reducing job availabilities in the swine industry. Likewise, they did not see PLF technology as a poor proxy for a farmer detecting problems and addressing them.

4. Conclusions

This study investigated how persons with varied involvement in the swine industry saw precision livestock farming technology. Three distinct viewpoints were identified by this study using Q-methodology: PLF improves farm management, animal welfare, and labor work conditions; PLF does not solve swine industry problems; and PLF has limitations and could lead to data ownership conflicts. The results uncovered the diversity of perspectives held by swine stakeholders about PLF technology and provided common descriptions of viewpoints.

We noticed dissimilarities in the opinions of stakeholders with direct and indirect PLF knowledge. Stakeholders with direct and advanced knowledge of PLF (i.e., technology developers and large-scale farmers) consistently displayed high optimism about PLF technology over a six-month period whereas stakeholders with indirect and limited knowledge (i.e., potential off-farm PLF users) tended to be skeptical of PLF claims over the same period. Nevertheless, these perspectives were also influenced by their employment role, as those most familiar with PLF were more actively involved in its development. This study emphasized the significance of training to enhance perceptions of PLF's usefulness and its eventual implementation.

5. Limitations of This Study

Although the 11 individuals that constituted the P-set for this study meet Brouwer's [45] recommendation of 10 to 40 people and these individuals perform different roles across the swine industry, there was some imbalance in the perspectives represented. Veterinarians, technology developers, and large-scale farmers dominated the study population, while government officials and consumers were least represented, whereas smallholder farmers, nutritionists, NGOs, and extension agents were not represented at all. These limitations should be taken into consideration when interpreting findings from this study. Future studies confirming the three dominant perspectives in underrepresented populations are needed. In addition, given that the current study focused on the US swine industry, generalizing findings from this study to other regions of the world should be carried out with caution. Nevertheless, future research may benefit from implementing similar studies in other regions of the world.

Author Contributions: Conceptualization, J.M.S., S.P.T., F.A. and B.E.A.; methodology, J.M.S., S.P.T., F.A. and B.E.A.; formal analysis, F.A.; investigation, B.E.A.; resources, J.M.S.; data curation, S.P.T., F.A. and B.E.A.; writing—original draft preparation, B.E.A.; writing—review and editing, J.M.S., S.P.T., F.A. and B.E.A.; supervision, J.M.S.; project administration, J.M.S.; funding acquisition, J.M.S., S.P.T. and F.A. All authors have read and agreed to the published version of the manuscript.

Funding: This research was funded by the USDA National Institute of Food and Agriculture, IDEAS award, grant number 2021-68014-34140. SRUC receives support from the Scottish Government Strategic Research Programme.

Institutional Review Board Statement: The study protocol was approved by the Institutional Review Board of Michigan State University under 45 CFR 46.104(d) 2i. with the MSU Study ID: STUDY00005432.

Informed Consent Statement: Informed consent was obtained from all human subjects involved in this study.

Data Availability Statement: The data presented in this study are available on request from the corresponding author. The data are not publicly available to preserve participant identities.

Conflicts of Interest: The authors declare no conflict of interest.

Appendix A. 30 Statements for Q-Set

No	Statements
1	PLF technology is a poor proxy of a farmer looking over the herd and detecting problems and addressing them
2	PLF technologies cannot manage individual pigs on commercial farms
3	PLF technologies can be used without any form of training
4	PLF technology will improve pig health
5	PLF technology will improve pig welfare
6	PLF technology will help in controlling disease outbreaks through traceability
7	PLF technologies can lower the cost of pork production
8	PLF technologies are cost-prohibitive to use across the entire livestock system
9	PLF technology will increase swine producers' profit margin
10	PLF technology may digitize swine production and make it look less natural and more artificial
11	PLF technology usage can lead to data privacy and data ownership conflicts
12	PLF technology can disconnect caretakers from pigs and make them less caring about pigs
13	PLF technology will make it possible for consumers to verify welfare certification claims
14	PLF technology may discourage pork consumption because it is probably profitable on large farms which some
15	PLF technology will ensure producers are transparently accountable to consumers
16	PLF technology does not minimize the environmental impact of swine farming
17	PLF technologies are not as precise as technology development companies claim them to be
18	PLF technology usage is limited by poor internet connection on most swine farms
19	PLF technologies' data often cannot be used within existing management practices
20	PLF technologies require a significant effort to change processes within an existing production system
21	PLF technologies make the pig caretaker's job easier, safer, and better
22	PLF technologies are displacing labor and reducing job opportunities in the swine industry
23	PLF technology will address labor shortages in the swine industry
24	PLF technology should be a supplement to good management and not replace good management
25	PLF technology is an environmentally friendly production system
26	PLF technology will address public concerns and increase consumer trust in pork production
27	PLF technology will enable producers to accurately plan their feed conversion ratio (FCR) upfront
28	PLF technologies on a farm are made by different vendors and often don't speak to one another thus making integration and management difficult
29	PLF technology is not necessarily useful to small-scale farmers operating outdoor mixed housing system
30	PLF technology often generates data that conflict with farmer expert opinions thus making PLF data less useful

Appendix B. Q-Sort Condition of Instruction

Box A1. Conditions of Instructions for Plf 30 Statements Q Sorting.

A Study on Precision Livestock Farming
In this study, we are interested in your opinion about Precision Livestock Farming (PLF). You will be asked to order 30 cards that contain statements based on what PLF means to you. Rank and order the statements according to least or most like what you think of PLF. We are only interested in your opinions, so there are no right or wrong answers.

Instructions

1. There are 30 cards numbered from 1 to 30. As you read the statement on each card, sort them into three piles:
    - most like what I think;
    - least like what I think; and
    - neutral (statements of which you have no opinion)
2. From the "most like what I think" pile, select one card which you think is the most important and place the card in the box under column +5.
3. From the remaining cards in the "most like what I think" pile, select two cards that you think are the next most important and arrange them in the boxes (in no particular order) under column +4.
4. From the remaining cards in the "most like what I think" pile, select three cards that you think are the next most important and arrange them in the boxes (in no particular order) under column +3.
5. From the remaining cards in the "most like what I think" pile, select three cards that you think are the next most important and arrange them in the boxes (in no particular order) under column +2.
6. From the remaining cards in the "most like what I think" pile, select four cards that you think are the next most important and arrange them in the boxes (in no particular order) under column +1.

    You may find that you do not have enough cards to completely fill these columns. In that case, pick cards from the neutral pile to fill in the columns. In the event that you have too many cards, place the extras in the neutral pile.
7. Now from the "least like what I think" pile, select one card which you think is the most important and place the card in the box under column −5.
8. From the remaining cards in the "least like what I think" pile, select two cards that you think are the next most important and arrange them in the boxes (in no particular order) under column −4.
9. From the remaining cards in the "least like what I think" pile, select three cards that you think are the next most important and arrange them in the boxes (in no particular order) under column −3.
10. From the remaining cards in the "least like what I think" pile, select three cards that you think are the next most important and arrange them in the boxes (in no particular order) under column −2.
11. From the remaining cards in the "least like what I think" pile, select four cards that you think are the next most important and arrange them in the boxes (in no particular order) under column −1.

    You may find that you do not have enough cards to completely fill these columns. In that case, pick cards from the neutral pile to fill in the columns. In the event that you have too many cards, place the extras in the neutral pile.

Finally, arrange the cards in the neutral pile in the boxes (in no particular order) under column 0. When you are finished, you should have no cards left and no blank spaces on the grid. If you wish to change the position of certain cards, you may do this at any time

Appendix C. Factor Arrays

		Factors		
No.	**Statement**	**1**	**2**	**3**
1	PLF technology is a poor proxy of a farmer looking over the herd and detecting problems and addressing them	−3	4	−4
2	PLF technologies cannot manage individual pigs on commercial farms	−2	1	0
3	PLF technologies can be used without any form of training	−4	−4	−5
4	PLF technology will improve pig health	4	−4	0
5	PLF technology will improve pig welfare	3	−5	2
6	PLF technology will help in controlling disease outbreaks through traceability	3	−2	−1
7	PLF technologies can lower the cost of pork production	2	−3	−1
8	PLF technologies are cost-prohibitive to use across the entire livestock system	−1	1	3
9	PLF technology will increase swine producers' profit margin	1	0	0
10	PLF technology may digitize swine production and make it look less natural and more artificial	−2	3	−2
11	PLF technology usage can lead to data privacy and data ownership conflicts	−1	0	4
12	PLF technology can disconnect caretakers from pigs and make them less caring about pigs	−4	5	−2
13	PLF technology will make it possible for consumers to verify welfare certification claims	2	−3	2
14	PLF technology may discourage pork consumption because it is probably profitable on large farms which some	0	2	−1
15	PLF technology will ensure producers are transparently accountable to consumers	1	−2	0
16	PLF technology does not minimize the environmental impact of swine farming	−2	4	−3
17	PLF technologies are not as precise as technology development companies claim them to be	0	0	2
18	PLF technology usage is limited by poor internet connection on most swine farms	2	2	5
19	PLF technologies' data often cannot be used within existing management practices	−5	0	4
20	PLF technologies require a significant effort to change processes within an existing production system	−1	1	3
21	PLF technologies make the pig caretaker's job easier, safer, and better	4	−1	3
22	PLF technologies are displacing labor and reducing job opportunities in the swine industry	−3	1	−3
23	PLF technology will address labor shortages in the swine industry	1	−1	−3
24	PLF technology should be a supplement to good management and not replace good management	5	3	1
25	PLF technology is an environmentally friendly production system	1	−2	−2
26	PLF technology will address public concerns and increase consumer trust in pork production	−1	−3	−4
27	PLF technology will enable producers to accurately plan their feed conversion ratio (FCR) upfront	0	−1	1
28	PLF technologies on a farm are made by different vendors and often don't speak to one another thus making	3	−1	1
29	PLF technology is not necessarily useful to small-scale farmers operating outdoor mixed housing system	0	3	−1
30	PLF technology often generates data that conflict with farmer expert opinions thus making PLF data less useful	−3	2	1

Appendix D. Distinguishing and Consensus Statements

No.	Statements	Factor 1	Factor 2	Factor 3
	Distinguishing statements			
1	PLF technology is a poor proxy of a farmer looking over the herd and detecting problems and addressing them		✓(4)	
4	PLF technology will improve pig health	✓(4)	✓(−4)	✓(0)
5	PLF technology will improve pig welfare		✓(−5)	
6	PLF technology will help in controlling disease outbreaks through traceability	✓(3)		
7	PLF technologies can lower the cost of pork production	✓(2)	✓(−3)	✓(−1)
8	PLF technologies are cost-prohibitive to use across the entire livestock system	✓(−1)		
10	PLF technology may digitize swine production and make it look less natural and more artificial		✓(3)	
11	PLF technology usage can lead to data privacy and data ownership conflicts			✓(4)
12	PLF technology can disconnect caretakers from pigs and make them less caring about pigs		✓(5)	
13	PLF technology will make it possible for consumers to verify welfare certification claims		✓(−3)	
14	PLF technology may discourage pork consumption because it is probably profitable on large farms which some		✓(2)	
16	PLF technology does not minimize the environmental impact of swine farming		✓(4)	
18	PLF technology usage is limited by poor internet connection on most swine farms			✓(5)
19	PLF technologies' data often cannot be used within existing management practices	✓(−5)	✓(0)	✓(4)
21	PLF technologies make the pig caretaker's job easier, safer, and better		✓(−1)	
22	PLF technologies are displacing labor and reducing job opportunities in the swine industry		✓(1)	
24	PLF technology should be a supplement to good management and not replace good management	✓(5)		
25	PLF technology is an environmentally friendly production system	✓(1)		
29	PLF technology is not necessarily useful to small-scale farmers operating outdoor mixed housing system		✓(3)	
30	PLF technology often generates data that conflict with farmer expert opinions thus making PLF data less useful	✓(−3)		
	Consensus statements			
3	PLF technologies can be used without any form of training	(−4)	(−4)	(−5)
9	PLF technology will increase swine producers' profit margin	(1)	(0)	(0)
17	PLF technologies are not as precise as technology development companies claim them to be	(0)	(0)	(2)
27	PLF technology will enable producers to accurately plan their feed conversion ratio (FCR) upfront	(0)	(−1)	(1)

Note that only statements whose Z-score value is statistically significant at 5%. ✓indicates that the statement distinguishes the factor. The value in parentheses () is the Q-sort value.

Appendix E.

Appendix E.1. Perspective 1: PLF Improves Management, Animal Welfare, and Labor Conditions

Least like what I think **Neutral** **More like what I think.**

−5	−4	−3	−2	−1	0	1	2	3	4	5
19	3	1	2	8	14	9	7	5	4	24
	12	22	10	11	17	15	13	6	21	
		30	16	20	27	23	18	28		
				26	29	25				

Note: Numbers 1–30 are the numbers randomly assigned to the Q-set statements in Appendix A.

The Factor Arrays values in Appendix D were used to sort the 30 random numbers into the grid.

Appendix E.2. Perspective 2: PLF Does Not Solve Swine Industry Problems

Least like what I think **Neutral** **More like what I think.**

−5	−4	−3	−2	−1	0	1	2	3	4	5
5	3	7	6	21	9	2	14	20	1	12
	4	13	15	23	11	8	18	24	16	
		26	25	27	17	20	30	29		
				28	19	22				

Note: Numbers 1–30 are the numbers randomly assigned to the Q-set statements in Appendix A

The Factor Arrays values in Appendix D were used to sort the 30 random numbers into the grid.

Appendix E.3. Perspective 3: PLF Has Limitations and Could Lead to Data Ownership Conflict

Least like what I think **Neutral** **More like what I think.**

−5	−4	−3	−2	−1	0	1	2	3	4	5
3	1	16	10	6	2	24	5	8	11	18
	26	22	12	7	4	27	13	20	19	
		23	25	14	9	28	17	21		
				29	15	30				

Note: Numbers 1–30 are the numbers randomly assigned to the Q-set statements in Appendix A.

The Factor Arrays values in Appendix D were used to sort the 30 random numbers into the grid.

Appendix F. Unrotated Factor Solutions

Participants/Q-Sorts	Factor 1	Factor 2	Factor 3
P1	0.85	0.06	−0.17
P2	−0.49	0.64	−0.30
P3	0.79	0.27	−0.06
P4	−0.57	0.61	−0.28
P5	0.81	−0.32	−0.04
P6	0.39	0.68	0.22
P7	0.78	−0.07	0.08
P8	0.79	0.36	−0.24
P9	0.37	0.43	0.73
P10	0.62	0.39	−0.20
P11	0.78	−0.30	−0.23
% Eigenvalue	5.10	1.97	0.95
% Explained Variance	46	18	9

Appendix G. Sources of Q-Sort Statements

No	Statements	Statement from:
1	PLF technology is a poor proxy of a farmer looking over the herd and detecting problems and addressing them	Potential off-farm PLF user
2	PLF technologies cannot manage individual pigs on commercial farms	PLF technology developer
3	PLF technologies can be used without any form of training	Potential off-farm PLF user
4	PLF technology will improve pig health	PLF technology developer and farmer
5	PLF technology will improve pig welfare	PLF technology developer and farmer
6	PLF technology will help in controlling disease outbreaks through traceability	PLF technology developer
7	PLF technologies can lower the cost of pork production	PLF technology developer and farmer
8	PLF technologies are cost-prohibitive to use across the entire livestock system	PLF technology developer and farmer
9	PLF technology will increase swine producers' profit margin	PLF technology developer
10	PLF technology may digitize swine production and make it look less natural and more artificial	Potential off-farm PLF user
11	PLF technology usage can lead to data privacy and data ownership conflicts	Potential off-farm PLF user
12	PLF technology can disconnect caretakers from pigs and make them less caring about pigs	Potential off-farm PLF user
13	PLF technology will make it possible for consumers to verify welfare certification claims	PLF technology developer
14	PLF technology may discourage pork consumption because it is probably profitable on large farms which some	Potential off-farm PLF user
15	PLF technology will ensure producers are transparently accountable to consumers	PLF technology developer and farmer
16	PLF technology does not minimize the environmental impact of swine farming	All groups
17	PLF technologies are not as precise as technology development companies claim them to be	PLF technology developer
18	PLF technology usage is limited by poor internet connection on most swine farms	All groups
19	PLF technologies' data often cannot be used within existing management practices	PLF technology developer
20	PLF technologies require a significant effort to change processes within an existing production system	PLF technology developer
21	PLF technologies make the pig caretaker's job easier, safer, and better	All groups
22	PLF technologies are displacing labor and reducing job opportunities in the swine industry	Potential off-farm PLF user
23	PLF technology will address labor shortages in the swine industry	
24	PLF technology should be a supplement to good management and not replace good management	All groups
25	PLF technology is an environmentally friendly production system	PLF technology developer and farmer
26	PLF technology will address public concerns and increase consumer trust in pork production	PLF technology developer
27	PLF technology will enable producers to accurately plan their feed conversion ratio (FCR) upfront	PLF technology developer and farmer
28	PLF technologies on a farm are made by different vendors and often don't speak to one another thus making integration and management difficult	PLF technology developer and farmer
29	PLF technology is not necessarily useful to small-scale farmers operating outdoor mixed housing system	Potential off-farm PLF user
30	PLF technology often generates data that conflict with farmer expert opinions thus making PLF data less useful	PLF technology developer

References

1. Parlasca, M.; Qaim, M. Meat Consumption and Sustainability. *Annu. Rev. Resour. Econ.* **2022**, *14*, 17–41. [CrossRef]
2. Berckmans, D. General Introduction to Precision Livestock Farming. *Anim. Front.* **2017**, *7*, 6–11. [CrossRef]
3. Plain, R.; Lawrence, J. Swine Production. *Vet. Clin. Food Anim.* **2003**, *19*, 319–337. [CrossRef]
4. Morrone, S.; Dimauro, C.; Gambella, F.; Capper, M. Industry 4.0 and Precision Livestock Farming (PLF): An up to Date Overview across Animal Productions. *Sensors* **2022**, *22*, 4319. [CrossRef]
5. Werkheiser, I. Precision Livestock Farming and Farmers' Duties to Livestock. *J. Agric. Environ. Ethics* **2018**, *31*, 181–195. [CrossRef]
6. Röling, N. Pathways for Impact: Scientists' Different Perspectives on Agricultural Innovation. *Int. J. Agr. Sustain.* **2009**, *7*, 83–94. [CrossRef]
7. Werkheiser, I. Technology and Responsibility: A Discussion of Underexamined Risks and Concerns in Precision Livestock Farming. *Anim. Front.* **2020**, *10*, 51–57. [CrossRef]
8. Heffernan, W. Sociological Dimensions of Agricultural Structures in the United States. *Sociol. Rural.* **1972**, *12*, 481–499. [CrossRef]
9. Giersberg, M.F.; Meijboom, F.L. Smart Technologies Lead to Smart Answers? On the Claim of Smart Sensing Technologies to Tackle Animal Related Societal Concerns in Europe Over Current Pig Husbandry Systems. *Front. Vet. Sci.* **2021**, *7*, 588214. [CrossRef]
10. Klerkx, L.; Jakku, E.; Labarthe, P. A Review of Social Science on Digital Agriculture, Smart Farming and Agriculture 4.0: New Contributions and a Future Research Agenda. *NJAS-Wagening. J. Life Sci.* **2019**, *90*, 100315. [CrossRef]
11. Pfeiffer, J.; Gabriel, A.; Gandorf, M. Understanding the Public Attitudinal Acceptance of Digital Farming Technologies: A Nationwide Survey in Germany. *Agric. Human. Values* **2021**, *38*, 107–128. [CrossRef]
12. Choi, S.; Moon, J.M. Disruptive Technologies, and Future Societies: Perspectives and Forecasts Based on Q-Methodology. *Futures* **2023**, *145*, 103059. [CrossRef]
13. Gauttier, M.F. A Primer on Q-Method and the Study of Technology. In *Research Anthology on Innovative Research Methodologies and Utilization across Multiple Disciplines*; Khosrow-Pour, M., Clarke, S., Jennex, M.E., Anttiroiko, A.V., Eds.; IGI Global: Hershey, PA, USA, 2022; Volume 17033, pp. 498–509.
14. Yenilmez, T.; Aydin, F.; Es, H. Science Teacher's Perceptions of the Nature of Technology: A Q-Methodology Study. *Int. J. Technol. Des. Educ.* **2022**, *32*, 2671–2696. [CrossRef] [PubMed]
15. Brown, S. Q-Methodology and Qualitative Research. *Qual. Health Res.* **1996**, *6*, 561–567. [CrossRef]
16. Brown, S.R. A Primer on Q Methodology. *Operant. Subj.* **1993**, *16*, 91–138. [CrossRef]
17. Akinyemi, B.E.; Vigors, B.; Turner, S.P.; Akaichi, F.; Benjamin, M.; Johnson, A.K.; Paris-Garcia, M.D.; Rozeboom, D.W.; Steibel, J.P.; Thompson, D.P.; et al. Precision Livestock Farming: A Qualitative Exploration of Swine Industry Stakeholders. *Front. Anim. Sci.* **2023**, *4*, e1150528. [CrossRef]
18. McKeown, B.; Thomas, D.B. *Q Methodology*, 2nd ed.; Sage: Newbury Park, CA, USA, 2013.
19. Watts, S.; Stenner, P. *Doing Q Methodological Research: Theory, Method, and Interpretation*; SAGE Publications Ltd.: Thousand Oaks, CA, USA, 2012.
20. Pilcher, E. "Moving in More Closely": Using Q Methodology to Explore Preservice Teacher Identity Formation as a Complex Dynamic System. *Int. J. Educ. Res. Open* **2023**, *5*, 100265. [CrossRef]
21. Hensel, D.; Cifrino, S. Using Q Methodology to Understand Faculty Development Needs to Prepare for Next Generation NCLEX. *Nurse Educ.* **2023**, *48*, 225–226. [CrossRef]
22. Kenward, L.; Whiffin, C.; Townend, M. The Needs of Clients Coming to Counselling Following Second Harm: A Q Methodology Study. *Couns. Psychother. Res.* **2023**, *23*, 404–416. [CrossRef]
23. Nezami, M.R.; de Bruijne, M.L.C.; Hertogh, M.J.C.M.; Bakker, H.L.M. Inter-Organizational Collaboration in Interconnected Infrastructure Projects. *Sustainability* **2023**, *15*, 6721. [CrossRef]
24. van Exel, J.; de Graaf, G. Q Methodology: A Sneak Preview. 2005. Available online: www.jobvanexel.nl (accessed on 28 August 2023).
25. Brown, S.R. *Political Subjectivity: Applications of Q Methodology in Political Science*; Yale University Press: New Haven, CO, USA, 1980.
26. Schmolck, P. PQ Method Manual. 2014. Available online: http://schmolck.org/qmethod/ (accessed on 16 December 2022).
27. Mulaik, S. *Foundations of Factor Analysis*, 2nd ed.; Routledge: Milton, UK, 2009.
28. Rima, D.; Macan, J.; Varnai, V.; Vučemilo, M.; Matković, K.; Prester, L.; Orct, T.; Trosić, I.; Pavicić, I. Exposure to Poultry Dust and Health Effects in Poultry Workers: Impact of Mould and Mite Allergens. *Int. Arch. Occup. Environ. Health* **2010**, *83*, 9–19. [CrossRef]
29. Viegas, S.; Faísca, V.; Dias, H.; Clérigo, A.; Carolino, E.; Viegas, C. Occupational Exposure to Poultry Dust and Effects on the Respiratory System in Workers. *J. Toxicol. Environ. Health* **2013**, *76*, 230–239. [CrossRef]
30. Banhazi, T. Environmental and Management Effects Associated with Improved Production Efficiency in a Respiratory Disease Free Pig Herd in Australia. In *Livestock Housing: Modern Management to Ensure Optimal Health and Welfare of Farm Animals*; Aland, A., Banhazi, T., Eds.; Wagenigen Academic Publishers: Wagenigen, The Netherlands, 2013; pp. 297–317.
31. Cardoso, C.S.; Hötzel, M.J.; Weary, D.M.; Robbins, J.A.; Von Keyserlingk, M.A. Imagining the Ideal Dairy Farm. *J. Dairy. Sci.* **2016**, *99*, 1663–1671. [CrossRef] [PubMed]

32. Boessen, C.; Artz, G.; Schulz, L. *A Baseline Study of Labor Issues and Trends in Us Pork Production*; Ames Publishing: Ames, IA, USA, 2018.
33. Benjamin, M.; Yik, S. Precision Livestock Farming in Swine Welfare: A Review for Swine Practitioners. *Animals* **2019**, *9*, 133. [CrossRef] [PubMed]
34. Tuyttens, F.; Molento, C.; Benaissa, S. Twelve Threats of Precision Livestock Farming (PLF) for Animal Welfare. *Front. Vet. Sci.* **2022**, *9*, 889623. [CrossRef] [PubMed]
35. Tullo, E.; Finzi, A.; Guarino, M. Review: Environmental Impact of Livestock Farming and Precision Livestock Farming as a Mitigation Strategy. *Sci. Total Environ.* **2019**, *650*, 2751–2760. [CrossRef] [PubMed]
36. Lovarelli, D.; Bacenetti, J.; Gu, M. A Review on Dairy Cattle Farming: Is Precision Livestock Farming the Compromise for an Environmental, Economic, and Social Sustainable Production? *J. Clean. Prod.* **2020**, *262*, 121409. [CrossRef]
37. Pomar, C.; Hauschild, L.; Zhang, G.; Pomar, J.; Lovatto, P. Applying Precision Feeding Techniques in Growing-Finishing Pig Operations. *Rev. Bras. Zootec.* **2009**, *38*, 226–237. [CrossRef]
38. Banhazi, T.; Dunn, M.; Banhazi, A. Weight-Detect^{TM}: On-Farm Evaluation of the Precision of Image Analysis Based Weight Prediction System. In *Practical Precision Livestock Farming Hands-On Experiences with PLF Technologies in Commercial and R&D Settings*; Banhazi, T., Halas, V., Maroto-Molina, F., Eds.; Wageningen University Press: Wageningen, The Netherlands, 2022; pp. 29–39.
39. Makinde, A.; Islam, M.; Wood, K.; Conlin, E.; Williams, M.; Scott, S. Investigating Perceptions, Adoption, and Use of Digital Technologies in the Canadian Beef Industry. *Comput. Electron. Agric.* **2022**, *198*, 107095. [CrossRef]
40. O'Grady, M.; Langton, D.; O'Hare, G. Edge Computing: A Tractable Model for Smart Agriculture? *Artif. Intell. Agric.* **2019**, *3*, 42–51. [CrossRef]
41. Rosa, G. Grand Challenge in Precision Livestock Farming. *Front. Anim. Sci.* **2021**, *2*, e650324. [CrossRef]
42. Shi, W.; Cao, J.; Zhang, Q.; Li, Y.; Xu, L. Edge Computing: Vision and Challenges. *IEEE Internet Things J.* **2016**, *3*, 637–646. [CrossRef]
43. Neethirajan, S.; Kemp, B. Digital Livestock Farming. *Sens. Biosensing Res.* **2021**, *32*, 100408. [CrossRef]
44. Wolfert, S.; Ge, L.; Verdouw, C.; Bogaardt, M. Big Data in Smart Farming—A Review. *Agric. Syst.* **2017**, *153*, 69–80. [CrossRef]
45. Brouwer, M. Q Is Accounting for Tastes. *J. Advert. Res.* **1999**, *39*, 35–39.

animals MDPI

Communication

Training and Adaptation of Beef Calves to Precision Supplementation Technology for Individual Supplementation in Grazing Systems

Joshua L. Jacobs [1,†], Matt J. Hersom [1,†], John G. Andrae [2,†] and Susan K. Duckett [1,*,†]

1 Department of Animal and Veterinary Sciences, Clemson University, Clemson, SC 29631, USA; jjacobs9@g.clemson.edu (J.L.J.); mhersom@clemson.edu (M.J.H.)
2 Department of Plant and Environmental Sciences, Clemson University, Clemson, SC 29631, USA; jandrae@clemson.edu
* Correspondence: sducket@clemson.edu
† These authors contributed equally to this work.

Simple Summary: Understanding the relationship between supplementation and animal production requires a systems approach to understand all variables affecting this relationship. Many publications regarding the utilization of consumed supplements have been published, as have studies focused on cattle production response. Supplementation costs strongly influence farm profitability, yet many supplementation strategies aim to supply nutrients to the average of the group, though inter-animal variation within groups can differ vastly meaning both over- and under-supplementation is occurring, potentially decreasing production. Measuring real-time cattle intake is historically labor intensive and typically focuses on total intake instead of individual supplement intake. New precision feeding technologies such as the C-Lock SuperSmart Feeder (SSF, C-Lock Inc., Rapid City, SD, USA) allow for real-time supplement intake data collection in field settings but require adaptation periods to facilitate the usage of these technologies by cattle. New technologies, such as the SSF, offer insight into behavioral information relating to cattle adaptation rates, feeding patterns, and the relationship between supplementation and animal production. The objective of this research was to assess the training and adoption rates of three different groups of cattle (suckling calves, weaned steers, replacement heifers) to the SSF.

Abstract: Supplementation of beef cattle can be used to meet both nutrient requirements and production goals; however, supplementation costs influence farm profitability. Common supplementation delivery strategies are generally designed to provide nutrients to the mean of the group instead of an individual. Precision individual supplementation technologies, such as the Super SmartFeed (SSF, C-Lock Inc., Rapid City, SD, USA), are available but are generally cost prohibitive to producers. These systems require adaptation or training periods for cattle to utilize this technology. The objective of this research was to assess the training and adoption rates of three different groups of cattle (suckling calves, weaned steers, replacement heifers) to the SSF. Successful adaptation was determined if an individual's supplement intake was above the group average of total allotted feed consumed throughout the training period. Suckling calves (n = 31) underwent a 12 d training period on pasture; 45% of suckling calves adapted to the SSF and average daily intake differed ($p < 0.0001$) by day of training. Weaned steers (n = 79) were trained in drylot for 13 d. Of the weaned steers, 62% were trained to the SSF, and average daily intake differed ($p < 0.0001$) by day of training. Replacement heifers (n = 63) grazed tall fescue pastures and had access to SSF for 22 d of training. The success rate of replacement heifers was 73%. For replacement heifers, the daily intake did not differ ($p < 0.0001$) by day of training. Results indicate production stage may influence cattle adaptation to precision technologies.

Keywords: precision supplementation; technology adaptation; beef cattle

Citation: Jacobs, J.L.; Hersom, M.J.; Andrae, J.G.; Duckett, S.K. Training and Adaptation of Beef Calves to Precision Supplementation Technology for Individual Supplementation in Grazing Systems. *Animals* **2023**, *13*, 2872. https://doi.org/10.3390/ani13182872

Academic Editors: Brett Ramirez, Janice Siegford, Hao Gan, Yang Zhao, Daniel Berckmans, Robert T. Burns and Lingjuan Wang-Li

Received: 16 August 2023
Revised: 4 September 2023
Accepted: 6 September 2023
Published: 9 September 2023

1. Introduction

The supplementation of beef cattle can be used to reach nutrient requirements not met by forage alone or to increase animal production to achieve desired production goals [1]. Supplementation strategies influence farm profitability, and due to the recent cost increase in many commonly used supplement sources, precision supplementation is of major importance [2]. Precision livestock feeding, specifically through the use of sensor-based technologies, can be utilized to match nutrient supply to the individual's requirement in real time [3]. By increasing the precision of beef cattle supplementation, nutrient requirements can be met more precisely and excess nutrient use can be avoided [4]. Farm profitability improves when overall supplementation costs can be minimized, and animal performance can be optimized by avoiding excess nutrient consumption. Many supplementation strategies are designed to provide nutrients to the mean of the group rather than to an individual animal [5]. Measuring individual intake in production settings is difficult at best, but precision supplementation technologies allow this information to be gathered primarily in research settings [6,7]. The information gathered in research settings may then be disseminated to producers for application in production settings.

Understanding the relationship between feeding behavior and nutrient utilization is crucial to improving precision supplementation and total feeding strategies. A variety of precision feeding technologies have been developed to help understand feeding behaviors and collect real-time intake measurements [8]. Many of the precision feeding technologies are better utilized for total intake data collection and are relatively stationary, requiring a permanent connection to electricity or consistent access to networks for data collection. Older technologies such as Calan Gates (American Calan Inc., Northwood, NH, USA) require daily manual refilling and data collection. More recent technologies such as GrowSafe (GrowSafe Systems Ltd., Airdrie, AB, Canada) or SmartFeed (C-Lock Inc., Rapid City, SD, USA) offer automated intake data collection, but still require frequent refilling. In contrast, the Super Smart Feeder (SSF, C-Lock Inc., Rapid City, SD, USA), is a more mobile precision feeding technology that focuses more on supplementation; allows controlled feed allotment, in-field data collection, and in-field supplement delivery; and requires less manual refilling.

Cattle often require adaptation or training periods to acclimate and utilize precision feeding technologies, much like training to utilize conventional feed bunks [9]. Despite being crucial for accurate data collection, little information is available defining specific training or adaptation methods and the success of these methods. Both supplement intake and foraging behavior have been shown to be influenced by the age of cattle, leading to further questions regarding feeding behavior when utilizing precision feeding technologies [10]. The objective of this research was to assess the training and adaptation of beef cattle in three different production stages (suckling calves, weaned steers, replacement heifers) to the SSF.

2. Material and Methods

2.1. Experimental Site

Three experiments were conducted at the Piedmont Research and Extension Center, Pendleton, SC, USA, to evaluate the adaptation of beef cattle in various production stages to precision feeding technologies. All animal experimental procedures were reviewed and approved by the Clemson University Institutional Animal Care and Use Committee (AUP2021-0138, AUP2021-0044, and AUP2020-0041).

2.2. Super SmartFeeder (SSF)

The SSF is a solar-powered, automated, mobile precision feeder consisting of a four-chambered feed bin capable of dispensing up to four supplement types into four separate, individual feeding stalls (Figure 1). The presence of radio frequency identification tags assigned to each individual animal triggers feed dispersal as described by [11]. The SSF utilizes a cloud-based interface to collect and store data to allow users to determine individual animal supplement intake, the number of animal visits to the feeder, and the timing of animal visits. The SSF technology also allows researchers to limit individual animal supplement intake by assigning individual supplement allotments as well as exclude individuals from accessing the SSF.

Figure 1. Steers training to C-Lock SuperSmart Feeder. Image Credit: J. Luke Jacobs, Clemson University.

Proper set-up of the SSF unit is imperative to ensure accuracy. The SSF was calibrated prior to use per manufacturer recommendations. The SSF feeder was set to dispense small amounts of supplement (<0.1 kg) for each drop when the animal accessed the feeder. The weight of the supplement is recorded for each drop while the animals continue to access the SSF. The animals can only consume a maximum intake for each day as defined in the system software by the experimenter. The SSF solar panel was oriented to face south to ensure adequate charging. Additionally, each time the SSF unit was refilled or moved following initial calibration, the unit was leveled and feed drops of each feeding stall were manually calibrated per manufacturer recommendations on an independent scale to ensure accuracy within the SSF system.

The success of training for all experiments was defined as the total individual intake greater than or equal to the average percentage intake of total supplement offered to the respective group. Meeting this parameter helps identify individuals within their respective groups that are more likely to consistently utilize the SSF. Failure to train was defined as the total individual intake less than the average percentage intake of the total supplement offered to the respective group.

2.2.1. Experiment 1

Suckling calves (n = 31, age = 86 ± 13 d, BW = 133.5 ± 15.6 kg) were allowed access to all four SSF stalls for a 12 d training period in a 4.1-ha tall fescue pasture. The SSF stall gates remained down for the duration of training to prevent dams from accessing supplement and ensure individual access to the SSF. Suckling calves received 0.45 kg of a commodity mixed feed (J & D-Lancaster Inc., Lancaster, SC, USA) from d 0 through d 3, and were then transitioned to 0.45 kg of steam-rolled corn (Godfrey's Feed, Madison, GA; 87.4% DM, 7.3% CP, 83.8% TDN) on d 4 where they remained throughout the remainder of the training period. The commodity mixed feed was offered as a palatable starting feed to encourage suckling calf consumption from the SSF, since the suckling calves used for this experiment had no previous supplement exposure.

2.2.2. Experiment 2

Angus-crossed steers (n = 79, age = 188 ± 23 d, BW = 227.5 ± 35.8 kg) were weaned and placed into a 1.0-ha drylot with access to all four SSF stalls for a 13 d training period. Steers were allotted 2.27 kg of a commodity mixed feed (J & D-Lancaster Inc., Lancaster, SC, USA; 89.3% DM, 18.0% CP, 75.1% TDN) from the SSF daily. Additionally, steers were allotted 2.27 kg of the commodity mixed feed in concrete feed bunks at 1400 h daily throughout the training period and had ad libitum access to bermudagrass hay (86% DM, 15.6% CP, 58.9% TDN) and water. From d 0 through d 3, SSF stall gates were raised, and on d 4 gates were lowered to ensure individual access to the SSF. The SSF stall gates remained down throughout the remainder of the training period. Steers used for this experiment had no exposure to supplementation prior to the start of training.

2.2.3. Experiment 3

Replacement heifers (n = 63, age = 255 ± 20 d, BW = 267 ± 31.7 kg) were allowed access to the SSF for a 22 d training period in a 4.1-ha tall fescue pasture. Heifers were allotted 3.64 kg of a commodity mixed feed daily (J & D-Lancaster Inc., Lancaster, SC, USA; 90.0% DM, 16.3% CP, 68.7% TDN). Replacement heifers had no other access to supplements. From d 0 through d 5, SSF stall gates were raised, and on d 6 gates were lowered to ensure individual access to the SSF. The SSF stall gates remained down throughout the remainder of the training period. Heifers used in this experiment had previous supplementation exposure throughout the weaning period, prior to the initiation of training to the SSF.

2.3. Statistical Analysis

A chi-squared test was performed using the FREQ procedure of SAS Version 9.4 (SAS Inst. Inc., Cary, NC, USA) to determine differences in the frequency of training success and failure. Success was defined as a total individual intake greater than or equal to the average percentage intake of total supplement offered to the respective group. Non-feeders were included in all statistical analyses. The percent of maximum allotted supplement cattle consumed was analyzed using the GLM procedure of SAS with training outcome in the model. Individual daily intake was analyzed using the GLM procedure of SAS with day of training in the model. Least square means were generated and separated using the PDIFF option of SAS. Significance was determined at ($p < 0.05$). Individual animals were the experimental unit for all studies.

3. Results and Discussion

No statistical difference ($p = 0.590$) was observed for the frequency of training success or failure in suckling calves. An adaptation rate of 45% was observed for suckling calves. Their individual average percentage intake of total supplement offered relative to the group average of 41.84% is depicted in Figure 2A. An average intake of 66.04% ± 3.41% and 21.92% ± 3.10% of total offered supplement was observed for adapted and non-adapted calves, respectively ($p < 0.001$). Figure 2B depicts individual calf intake by day of training relative to the average daily intake of the group. Suckling calves had an overall average daily intake of 0.30 kg. Average daily intake differed by day ($p < 0.001$) for the sucking calf training period (Figure 2B). Average daily intake was lowest on d 1 and increased daily until d 3. On d 4, average daily intake decreased sharply, likely due to the change in feed offered by the SSF. However, consumption increased on d 5, peaking on d 6 and did not differ from the remainder of the training period. The literature discussing suckling calf adaptation or utilization of in-field precision supplementation technologies, such as the SSF, was unavailable. Collecting data on cattle in this production stage is difficult due to calf reliance on the dam. Suckling calf movement throughout paddocks is likely dictated by the dam as well. To help facilitate suckling calf interaction with the SSF, the cattle were placed in smaller 4.1-ha paddocks to increase the proximity of calves and dams to the SSF.

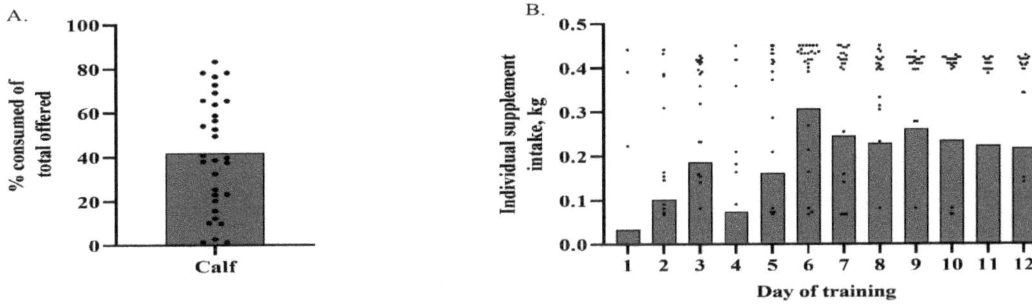

Figure 2. (**A**) Total percent of offered supplement consumed (kg) by individual suckling calves throughout training period relative to the group average of 41.84% (depicted by the bar). Individual dots above the line represent calves that successfully trained to the Super SmartFeeder. (**B**) Daily individual calf intake over the course of the training period. Individual calves are depicted as dots and the group average by day of training is depicted as a bar.

A greater ($p = 0.033$) number of successful training outcomes was observed compared to the failure-to-train outcomes for weaned steers in Experiment 2. Weaned steers exhibited an adaptation rate of 62%. Their individual average percentage intake of the total supplement offered relative to the group average of 49.78% is depicted in Figure 3A. An average intake of 70.18% ± 1.99% and 16.46% ± 2.54% of total offered supplement was observed for adapted and non-adapted calves, respectively ($p < 0.001$). Figure 3B depicts individual steer intake by day of training relative to the average daily intake of the group. Weaned steers had an overall average daily intake of 1.59 kg from the SSF. Average daily intake differed by day ($p < 0.001$) for steers and is illustrated in Figure 3B. Average daily intake was lowest on d 1. Steers reached their maximum average daily intake on d 10; however, there were no statistical differences from d 9 through d 11. A decrease in average intake occurred on d 12 of training but rebounded on d 13. The decrease was likely due to the SSF being refilled.

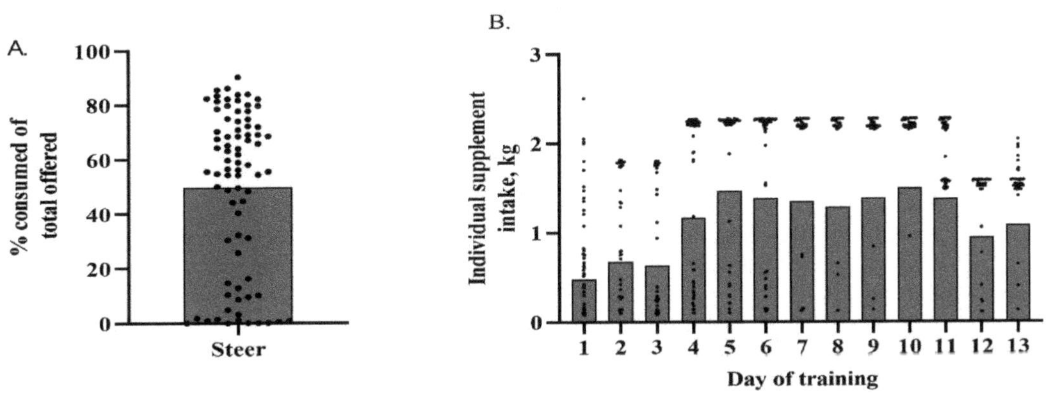

Figure 3. (**A**) Total percent of offered supplement consumed (kg) by individual weaned steer throughout training period relative to the group average of 49.78% (depicted by the bar). Individual dots above the line represent steers that successfully trained to the Super SmartFeeder. (**B**) Daily individual steer intake over the course of the training period. Individual steers are depicted as dots and the group average by day of training is depicted as a bar.

The frequency of successful training outcomes observed for replacement heifers was greater ($p < 0.001$) than the frequency of failure to training outcomes. Replacement heifers exhibited an adaptation rate of 73%. Their individual average percentage intake of total supplement offered relative to the group average of 70.64% is depicted in Figure 4A. An average intake of 90.02% ± 3.37% and 18.95% ± 2.05% of total offered supplement was observed for adapted and non-adapted heifers, respectively ($p < 0.001$). Figure 4B depicts individual heifer intake by day of training relative to the average daily intake of the group. The average daily intake of the training period was 3.23 kg. Replacement heifer average daily intake did not differ statistically by day ($p = 0.075$) throughout the training period. Numerical differences were observed and are depicted in Figure 4B. Replacement heifers had the lowest average daily intake on d 1 of training, and average daily intake increased numerically on d 2 and peaked on d 10. A sharp numerical decrease in average daily intake was observed on d 12 and decreased until d 13. This is likely due to the SSF being refilled. However, average daily intake appeared to increase to levels similar to what was observed prior to the numerical decrease on d 12.

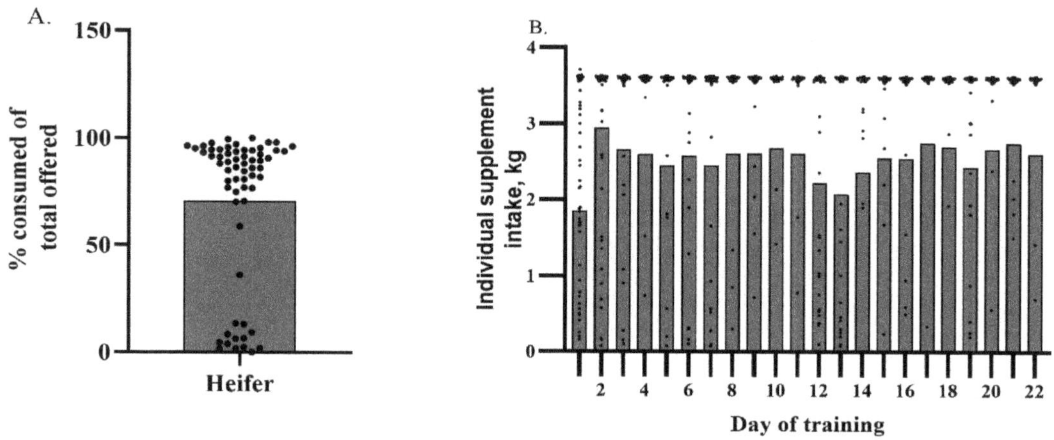

Figure 4. (**A**) Total percent of offered supplement consumed (kg) by individual replacement heifer throughout training period relative to the group average of 70.64% (depicted by the bar). Individual dots above the line represent heifers that successfully trained to the Super SmartFeeder. (**B**) Daily individual heifer intake over the course of the training period. Individual heifers are depicted as dots and the group average by day of training is depicted as a bar.

Although several studies regarding the utilization of precision feeding technologies have been published, little literature is available regarding training success and failure rates for beef cattle of any production stage with SSF units. Several reports describe training procedures. A study by Husz et al. [11] utilizing preconditioned beef steers (n = 418, 8–10 months of age) reported a 7 d training and adaptation period for steers in each year of a two-year study. Success was defined as a steer consuming 0.5 kg of supplement for a 3 d period. Unlike the current studies, steers were exposed to the SSF in groups of 40 to 45 animals to limit competition. Husz et al. [11] immediately removed trained steers from the group to further limit competition at the feeder. Steers not achieving the criteria within 7 d were considered 'non-feeders'. Throughout the study, Husz et al. [11] reported 31% of steers did not voluntarily use the SSF following the training period but did not report non-feeder numbers through the training period. Limiting initial competition of precision supplementation technologies like the SSF may improve the success of adaptation periods by allowing more opportunities to feed.

Similar to the weaned steer study, a study by Valliere et al. [12] reported a 24 d acclimation period for post-weaning lambs (n = 179) on pasture. Lambs were offered 0.23 kg of supplement daily from an SSF in a pasture setting. From d 0 through d 10, lambs had access to 0.45 kg of supplement in standard troughs, decreasing to 0.15 kg of supplement. From d 0 through d 5, Valliere et al. [12] reported 30 and 80 percent of lambs visited the SSF. However, lamb SSF visits varied between 64 and 78% after d 6. From d 6 to d 24, 72% of lambs visited the SSF daily. The success rate of post-weaning lambs reported by Valliere et al. [12] is greater than that observed in weaned steers in Experiment 2. Both the current studies and the study reported by Valliere et al. [12] suggest longer acclimation periods may prove helpful in improving the utilization of precision feeding technologies.

A study by Williams et al. [13] reported a 35 d training period to allow steers (n = 40, BW = 243 ± 23 kg) to acclimate to a similar automated feeder, SmartFeed (C-Lock Inc., Rapid City, SD), by offering 0.91 kg of supplement three times per week to encourage use of the feeding system. Williams et al. [13] reported a 12.5% non-feeder rate throughout the following study (n = 16) but did not report training results. A similar study by McCarthy et al. [14] reported a 14 d training period to a SmartFeed system utilizing yearling heifers (n = 126, BW = 400.4 ± 6.2 kg), and stated that non-training heifers were then utilized as control animals in the study following training. Another study by Stewart et al. [15] reported a 10 d acclimation period to allow mature ewes (n = 78) to acclimate the SSF. There was no mention of adaptation success or failure rates in these studies. These studies reported using older animals; however, the absence of training success and failure in the reports makes it difficult to compare the results of the current study using replacement heifers.

The age or production stage of cattle appears to influence adaptation time. In the current studies, replacement heifers that had been previously exposed to supplementation responded with a rapid increase in average daily supplement intake, reaching daily intakes greater than 90% of allotted supplement as early as d 2 of exposure to the SSF. Suckling calves and weaned steers naive to supplementation had slower increases in average daily supplement intake. Cattle that have never been exposed to supplementation of any kind not only have to adapt to the SSF but also the supplement itself, which may explain the increase in average daily supplement intake being less than that observed in the replacement heifer study.

4. Conclusions

Prior to data collection from precision feeding technologies such as the SSF, cattle must be allowed to train and adapt to utilizing such technologies to ensure the data collected are accurate. Adaptation rates of cattle appear to differ based on the production stage. Cattle previously exposed to supplementation (replacement heifers) appear to adapt quicker than cattle naive to supplementation (suckling calves and weaned steers). Suckling and weaned calves naive to supplementation may require training periods of 7 d or more, while older cattle with previous supplement exposure may be able to train in less than 7 d. However, allowing additional time during training may help identify animals that will utilize the SSF more consistently throughout the subsequent study. Introducing cattle to the SSF with the stall gates raised may hasten adaptation. However, in situations like the suckling calf study, this was not an option; therefore, longer training periods may be required. Animal proximity to SSF units may also help increase the success of adaptation. If possible, training or adaptation periods should be performed in smaller paddocks or drylots to increase animal interactions with the SSF. Placing SSF units on animal travel routes, i.e., between loafing, grazing, and watering areas, may help increase cattle interactions with the SSF. Further advancement and use of precision feeding technologies will enhance nutritional management and encourage more efficient nutrient utilization by ruminant animals. These advances will help improve supplementation recommendations, minimize wastage of expensive feedstuffs, enhance individual animal performance, and increase research opportunities utilizing precision supplementation technologies. More research regarding animal behavior and adaptation to precision feeding technologies is needed to better utilize these technologies and further understand the relationship between feeding behavior and nutrient utilization of ruminant animals.

Author Contributions: Conceptualization, J.L.J., M.J.H., J.G.A. and S.K.D.; Methodology, J.L.J., M.J.H., J.G.A. and S.K.D.; Validation, J.L.J., M.J.H., J.G.A. and S.K.D.; Formal Analysis, S.K.D.; Investigation, J.L.J.; Resources, S.K.D.; Data Curation, J.L.J. and S.K.D.; Writing—Original Draft Preparation, J.L.J.; Writing—Review and Editing, J.L.J., M.J.H., J.G.A. and S.K.D.; Visualization, J.L.J., M.J.H., J.G.A. and S.K.D.; Supervision, S.K.D.; Project Administration, J.L.J. and S.K.D.; Funding Acquisition, S.K.D. All authors have read and agreed to the published version of the manuscript.

Funding: Technical Contribution No. 7198 of the Clemson University Experiment Station. This material is based upon work supported by NIFA/USDA, under project number SC-1700580.

Institutional Review Board Statement: All animal experimental procedures were reviewed and approved by the Clemson University Institutional Animal Care and Use Committee (AUP2021-0138, AUP2021-0044, and AUP2020-0041).

Informed Consent Statement: Not applicable.

Data Availability Statement: Not applicable.

Acknowledgments: The authors would like to thank the farm staff at the Piedmont Research and Education Center, Pendleton, SC, USA, for their help in daily operations and support in research efforts.

Conflicts of Interest: The authors declare no conflict of interest.

References

1. Caton, J.S.; Dhuyvetter, D.V. Influence of energy supplementation on grazing ruminants: Requirements and responses. *J. Anim. Sci.* **1997**, *75*, 533–542. [CrossRef] [PubMed]
2. Meyer, A.M.; Gunn, P.J. Beef species symposium: Making more but using less: The future of the U.S. beef industry with a reduced cow herd and the challenge to feed the United States and world. *J. Anim. Sci.* **2015**, *93*, 4223–4226. [CrossRef] [PubMed]
3. Zuidhof, M.J. Precision livestock feeding: Matching nutrient supply with nutrient requirements of individual animals. *J. Appl. Poult. Res.* **2020**, *29*, 11–14. [CrossRef]
4. Schroeder, G.F.; Titgemeyer, E.C. Interaction between protein and energy supply on protein utilization in growing cattle: A review. *Livest. Sci.* **2008**, *114*, 1–10. [CrossRef]
5. Bowman, J.G.; Sowell, B.F. Delivery method and supplement consumption by grazing ruminants: A review. *J. Anim. Sci.* **1997**, *75*, 543–550. [CrossRef] [PubMed]
6. González, L.; Kyriazakis, I.; Tedeschi, L. Review: Precision nutrition of ruminants: Approaches, challenges and potential gains. *Animal* **2018**, *12*, 246–261. [CrossRef] [PubMed]
7. Islas, A.; Gilbery, T.C.; Goulart, R.S.; Dahlen, C.R.; Bauer, M.L.; Swanson, K.C. Influence of supplementation with corn dried distillers grains plus solubles to growing calves fed medium- quality hay on growth performance and feeding behavior. *J. Anim. Sci.* **2014**, *92*, 705–711. [CrossRef] [PubMed]
8. McCarthy, K.L.; Undi, M.; Becker, S.; Dahlen, C.R. Utilizing an electronic feeder to measure individual mineral intake, feeding behavior, and growth performance of cow-calf pairs grazing native range. *Transl. Anim. Sci.* **2021**, *5*, txab007. [CrossRef] [PubMed]
9. Chapple, R.S.; Lynch, J.J. Behavioral factors modifying acceptance of supplementary foods by sheep. *Res. Dev. Agric.* **1986**, *3*, 113.
10. Kincheloe, J.J.; Bowman, J.P.G.; Sowell, B.F.; Ansotegui, R.P.; Surber, L.M.M.; Robinson, B.L. Supplement intake variation in grazing beef cows. *Proc. Am. Soc. Anim. Sci. West. Sect.* **2004**, *55*, 331–334.
11. Husz, T.C.; Goad, C.; Ryan, R. Competition at an automated supplement feeder affects supplement intake and behavior of beef stocker steers. *Appl. Anim. Sci.* **2020**, *36*, 868–876. [CrossRef]
12. Valliere, N.K.; Wright, D.L.; Greiner, S.P.; Weaver, A.R. Evaluation of the Super Smartfeeder System (C-Lock, Inc) for lamb supplemental feeding in a pasture-based system. *J. Anim. Sci.* **2022**, *100*, 42. [CrossRef]
13. Williams, G.D.; Beck, M.R.; Thompson, L.R.; Horn, G.W.; Reuter, R.R. Variability in supplement intake affects performance of beef steers grazing dormant tallgrass prairie. *Prof. Anim. Sci.* **2018**, *34*, 364–371. [CrossRef]
14. Mccarthy, K.L.; Underdahl, S.R.; Undi, M.; Becker, S.; Dahlen, C.R. Utilizing an electronic feeder to measure mineral and energy supplement intake in beef heifers grazing native range. *Transl. Anim. Sci.* **2019**, *3*, 1719–1723. [CrossRef] [PubMed]
15. Stewart, W.C.; Murphy, T.W.; Page, C.M.; Rule, D.C.; Taylor, J.B.; Austin, K.; Pankey, C. Effects of increasing dietary zinc sulfate fed to primiparous ewes: I. Effects on serum metabolites, mineral transfer efficiency, and animal performance. *Appl. Anim. Sci.* **2020**, *36*, 839–850. [CrossRef]

animals

MDPI

Article

Broiler Mobility Assessment via a Semi-Supervised Deep Learning Model and Neo-Deep Sort Algorithm

Mustafa Jaihuni [1], Hao Gan [2], Tom Tabler [1], Maria Prado [1], Hairong Qi [3] and Yang Zhao [1,*]

1 Department of Animal Science, University of Tennessee, Knoxville, TN 37996, USA; mjaihuni@vols.utk.edu (M.J.); gtabler@utk.edu (T.T.); mprado@utk.edu (M.P.)
2 Department of Biosystems Engineering, University of Tennessee, Knoxville, TN 37996, USA; hgan1@utk.edu
3 Department of Electrical Engineering and Computer Science, University of Tennessee, Knoxville, TN 37996, USA; hqi@utk.edu
* Correspondence: yzhao@utk.edu; Tel.: +1-662-694-2267

Simple Summary: Tracking the movements of chickens is important for understanding their well-being. Traditional methods for measuring chicken mobility are time-consuming and cannot provide real-time information. In this study, we, the researchers, used a combination of artificial intelligence methods and computer algorithms to track individual chickens. Using these methods, it was possible to detect and track individual chickens in two pens. We analyzed the data to see how far and how fast each chicken moved every hour and every day. Compared to manual measurements, the combined model provided more accurate measurements of the mobility of each chicken and the entire group, even when some chickens were hidden, or the detection was not perfect. This study may serve to effectively track indicators critical for broiler production performance and welfare.

Abstract: Mobility is a vital welfare indicator that may influence broilers' daily activities. Classical broiler mobility assessment methods are laborious and cannot provide timely insights into their conditions. Here, we proposed a semi-supervised Deep Learning (DL) model, YOLOv5 (You Only Look Once version 5), combined with a deep sort algorithm conjoined with our newly proposed algorithm, neo-deep sort, for individual broiler mobility tracking. Initially, 1650 labeled images from five days were employed to train the YOLOv5 model. Through semi-supervised learning (SSL), this narrowly trained model was then used for pseudo-labeling 2160 images, of which 2153 were successfully labeled. Thereafter, the YOLOv5 model was fine-tuned on the newly labeled images. Lastly, the trained YOLOv5 and the neo-deep sort algorithm were applied to detect and track 28 broilers in two pens and categorize them in terms of hourly and daily travel distances and speeds. SSL helped in increasing the YOLOv5 model's mean average precision (mAP) in detecting birds from 81% to 98%. Compared with the manually measured covered distances of broilers, the combined model provided individual broilers' hourly moved distances with a validation accuracy of about 80%. Eventually, individual and flock-level mobilities were quantified while overcoming the occlusion, false, and miss-detection issues.

Keywords: broiler; welfare; mobility; YOLOv5; semi-supervised learning; neo-deep sort

Citation: Jaihuni, M.; Gan, H.; Tabler, T.; Prado, M.; Qi, H.; Zhao, Y. Broiler Mobility Assessment via a Semi-Supervised Deep Learning Model and Neo-Deep Sort Algorithm. *Animals* **2023**, *13*, 2719. https://doi.org/10.3390/ani13172719

Academic Editor: Alessandro Dal Bosco

Received: 25 July 2023
Revised: 11 August 2023
Accepted: 25 August 2023
Published: 26 August 2023

1. Introduction

The poultry industry growth continues skyrocketing globally, concurrent with the world population surge and the resultant overwhelming demands for ample, nutritious, and affordable protein sources. Meanwhile, this industry is struggling to create an ecosystem that would ensure sustainability and growth by reducing poultry production losses due to health and welfare issues, creating feasible livestock environments through precision technologies, and alleviating labor shortages in giant poultry markets, such as the USA [1].

Welfare plays a commendable role in rendering qualitatively healthy and quantitatively superior poultry productions [2]. Fundamentally, it is a multifaceted phenomenon that demonstrates chickens' physical conditions, living habitat feasibility, and mental situations. Birds require spacious environments to move with ease, access free feed and water sources, and socialize, while inhibiting infections, lameness, and stress throughout the rearing period. In this context, mobility tremendously affects individual birds' welfare levels [3]. Impaired or zero locomotion in birds may indicate issues such as insufficient nutrient consumption needed for growth, possible existing pain or stress, lameness, housing constraints, and even mortality. Hence, impaired locomotion has been highly identified in about 15–28% of birds in poultry farms [4]. Hence, mobility or locomotion studies play an important role in shedding light on the conditions that are effective factors for determining individual and flock-level welfare in birds.

The traditional Gait Scoring (GS) and Kinematics (GK) methods have been widely applied to assess locomotion status in individual chickens [5]. While the former method needs to be carried out by an expert tasked to observe and determine gait level one chicken at a time, the latter method utilizes statistical and algorithmic approaches to extract locomotion features, like walking ability, sitting, and standing postures, and correlates them with the predefined GS levels. As in the works of Aydin, Periera et al., and Doornweerd et al. [6–8], such features of individual birds are studied and correlated with their predetermined GS levels. Their results are influential in determining major lameness and locomotion problems in individual broilers. Due to welfare management urgency, it is more effective to provide timely, conclusive, subjective, and economical insights into the overall activities of individuals and flocks day in and day out.

On the other hand, several alternative artificial intelligence (AI) and/or Deep Learning (DL) methods have been proposed to tackle the locomotion problem and provide further information on other activities of chickens. The authors Nasiri et al., Fang et al., and de Alencar Nääs et al. [9–11] have developed very effective pose estimation and speed-based lameness and behavior detection methodologies using DL models. They have achieved high accuracies in correlating lameness GS levels and behavior classifications with broilers' skeletal positions and visual analysis, but we still need highly generalizable and practical methods in labs and commercial poultry farms; for instance, cameras used here capture lateral images of flocks, which greatly reduces the analyzability of individual chickens. Lin et al. and Fang et al. [12,13] have applied a Convolutional Neural Network (CNN)-based DL model to detect individual chickens in a shallow setup; consequently, their movements were calculated manually matching consecutive frames. In the work of Neethirajan [14], a CNN-based YOLOv5 model along with a deep sort algorithm were applied to track individual birds' trajectories and provide periodical movements. Although the study provided flock-level movement orientations, a detailed approach is more helpful in solving the occlusion problem and the consequent individual bird movement and identification issues. These studies are hardly generalizable to broiler farms where constant occlusions happen. Continuously tracking individual bird trajectories and flock-level commotions are thoroughly complicated. The utilization of large hierarchical datasets that would encompass different broiler growth phases and render a less biased DL model would also be crucial in such applications. Additionally, DL has been mostly applied in image classification tasks, as depicted in the previous works. The efficacy of DL models in image segmentation and object detection tasks is yet to be discovered.

Generally, the inadequacies in the quantity of data for developing DL models and the persistence of the occlusion-related identification problems greatly inhibit clear calculations of the individual birds' trajectories. On the other hand, to achieve economically beneficial and effective applications, one needs to consider the processing time, human resource requirement, and the inherent objectivity of the results. Henceforth, we aimed to tackle the mobility estimation of broilers while handling the occlusion instances, developing a more generalizable and robust model, i.e., the YOLOv5 deep sort model, and a new algorithm for large poultry farms while utilizing large datasets for DL model development. YOLOv5 is a vastly applied object detection DL model for computer vision. It is one of the most advanced versions of the YOLO family, surpassing older YOLO versions in terms of object detection precision. The YOLO models are primarily designed to perform object localization and classification in a single-stage regression process, thus outperforming counterpart CNN-based models, such as faster region-based or mask CNN, in terms of inference speed and memory efficiency, as well as provide comparable precision levels. Therefore, it is deemed one of the most effective models for real-time applications [15]. Meanwhile, the deep sort algorithm has been widely applied for the continuous tracking of detected objects. It is a robust algorithm that is convenient for real-time applications thanks to its fast run time speed of 20-30 frames per second (FPS) [16]. Overall, the objectives of this study were to utilize SSL for bringing more data into DL model development, add a new algorithm on top of the deep sort algorithm for solving the trajectory estimation and activity levels of individual broilers, and estimate flock level mobility.

2. Materials and Methods

The experiment was carried out in the Animal Science Department labs located in the Johnson Animal Research and Teaching Unit (JARTU), the University of Tennessee Knoxville, USA. It consisted of 28 chickens (Cobb 700, with a 1:1 male–female ratio) that were reared for a period of 54 days between October 18 and December 10, 2021. Day-old chicks were divided into 2 pens with 12 and 16 birds, respectively. We used birds at two growth rates: slow growing (<50 g/day) and standard (>65 g/day). Both groups of birds were reared under typical stocking densities, i.e., 24 kg/m^2 for slow growing and 32 kg/m^2 for standard, which translated to 12 birds/pen for slow-growing and 16 birds/pen for standard birds.

Each pen had a 100 cm $\times$ 150 cm pen with a standard camera (Amcrest UltraHD 5MP Outdoor POE Camera 2592 $\times$ 1944p) mounted at a 3 m height overlooking the pen, as depicted in Figure 1. The cameras recorded broiler movements continuously for 15 min per hour for 24 h. The 15 min period, i.e., the first quarter of every hour, was deemed statistically significant to represent the total mobility level of birds in an hour [17]. Additionally, it helped in effectively using computer storage while acquiring ample data for DL model development and mobility analysis. This setup collected data for 54 days continuously, encompassing the total life span of the broilers before they were moved to slaughterhouses at around 3 kg in weight. The video recordings were stored in standard hard drives for future analysis.

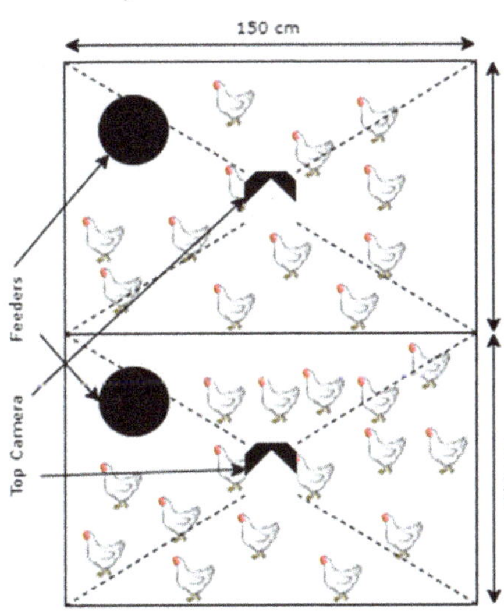

Figure 1. Pen dimensions, broilers, feeders, and camera positions.

2.1. Semi-Supervised Learning

With time and human resource constraints, semi-supervised learning (SSL) is an influential method in tackling big datasets for DL model development. It lies between the supervised and unsupervised learning counterparts, benefiting from their respective strengths. SSL utilizes the guided learning procedure, i.e., learning from labeled datasets, the former method, and the unguided methodology, i.e., predicting and learning simultaneously from unlabeled datasets in the latter one [18,19].

In this study, firstly, to develop the initial YOLOv5 model, we employed a dataset with 1650 labeled images extracted from the recordings of 5 random days. The video recordings when the broilers were 7, 17, 26, 36, and 41 days of age have been used to build this dataset. Consequently, with a ratio of 1 frame per minute, images were extracted from them; this ratio was chosen so that we obtained images that were temporally further away from one another to reflect different possible positions of the broilers in the pens.

This dataset helped in providing a CNN model with a descent accuracy level, between 50 and 80%; later, it was applied to the next batch of 2160 unlabeled images extracted from the recordings and corresponded to broilers being 4, 18, 30, 44, and 47 days of age. Collectively, in these two steps, the dates were selected highly dispersed throughout the experimental period to utilize recordings that provided unique and less correlated information about broilers in their different rearing phases. They rendered a final trained model with high generalizability and lesser bias in recognizing broiler mobility. Eventually, the new successfully predicted images were fetched to the partially developed YOLOv5 model and made it learn further and increase its accuracy. Henceforth, we expected that it would help in creating a more robust and highly accurate model with minimal time and labor mobilization.

2.2. YOLOv5 DL Model

The novelty of the YOLOv5 model lies in its architecture. It comprises backbone, neck, and head sections. The backbone section focuses on distinguishing the receptive or feature fields in an image and tries to reduce the number of model parameters to achieve agility and lower memory requirement [20]. Meanwhile, the neck section classifies important image features, which improve the localization success rate. The head section combines low- and high-level features to increase precision rates and provides the final losses of the model [21].

We used three YOLOv5 model performance metrics, namely mean average precision (mAP), precision, and the regression coefficient of determination (R^2), as depicted in Equations (1)–(3), respectively [22]. The mAP metric demonstrates the effectivity of the DL model by taking the mean of the average precision for each class under scrutiny. In this study, however, there was only one class, i.e., broiler, so the class number (Q) is equal to 1. Meanwhile, the precision metric is a measure of object detection and classification by the model, true positive is the successful detection of an object, and false positive shows misleading detections by the model. On the other hand, the R^2 value shows how efficiently the model predicts ($\hat{y}_i$) instances of actual data (y_i) [21].

$$\mathrm{mAP} = \sum_{q=1}^{Q} \mathrm{AP(q)} / \mathrm{Q} \quad (1)$$

$$\mathrm{Precision} = \frac{\mathrm{True\ Positive}}{\mathrm{True\ Positive} + \mathrm{False\ Positive}} \quad (2)$$

$$R^2 = 1 - \frac{\Sigma(y_i - \hat{y}_i)}{\Sigma(y_i - \hat{y})} \quad (3)$$

2.3. Deep Sort Algorithm

It is comprised of the Hungarian algorithm, the Kalman filter and the utilization of a bounding box, confidence, and deep features from frames detected by the CNN model.

The Hungarian algorithm and the Kalman filter are applied for position and velocity parameter tracking and predict and update their future status accordingly. The integration of deep features extracted from the detected frames increases the likelihood of tracking objects effectively, even when they are located very near or are occluded [23,24]. As a result of consecutive frame association, motion prediction, and deep features, the tracking algorithm can attach identification (ID) numbers to each detected object and track them if it can differentiate them effectively. One problem arises during dense occlusion instances where this algorithm would not be able to associate objects in newer frames. Hence, it counts them as new objects and assigns new IDs. Here, we developed and applied a new algorithm integrated with the deep sort algorithm (neo-deep sort) that resolved or mitigated this problem.

2.4. Neo-Deep Sort Algorithm

The problem of occlusion or lost detection instances poses a major hindrance to calculating overall individual broiler mobilities. The YOLOv5 deep sort algorithm would resolve this issue for the most part. However, as the broiler mobility and stocking density increase, it becomes a monumental task to overcome this issue. Hence, we tested the following algorithm to correlate new and old IDs assigned on detected objects by the deep sort algorithm. Essentially, the algorithm would detect when a previously detected ID is lost and, consequently, it tries to identify when one/more new IDs appear in the deep sort algorithm results. It then processes several steps to correlate the new ID with the lost ID or delete those instances that are falsely detected; the algorithm flowchart is shown in Figure 2, and it was implemented using Python. The algorithm utilized the position of these lost and new objects to correlate them. If they were located at a distance smaller than the threshold distance, then the two IDs would be considered the same. This limit would be selected based on a trial-and-error process [25]. We validated this approach with manual ID correlation and/or deletion and the neo-deep sort results.

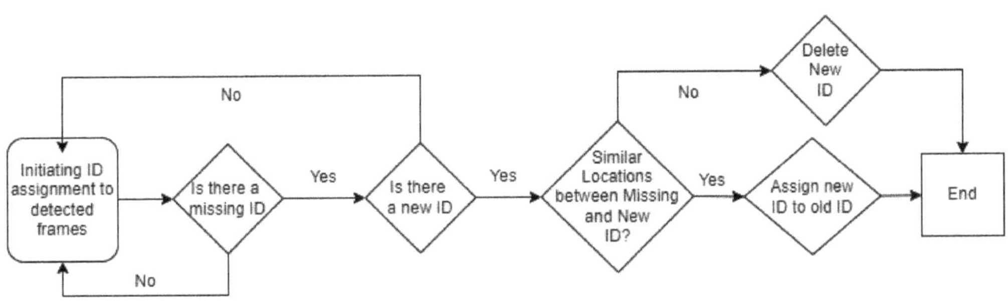

Figure 2. Neo-deep sort algorithm flowchart.

2.5. Final Model

The final model consists of the YOLOv5 model for object detection, the deep sort tracking algorithm, and the new algorithm to correlate and/or delete ID instances for solving the occlusion problem. We obtained the results after these 3 steps, and then the flock level and individual broiler's displacements and their respective speeds were extracted and categorized accordingly.

The results of the final model consisted of individual broiler's position in consecutive frames of a video. The model detected and tracked frames at an interval of about 1/100 s. Hence, even minuscule positional displacements of broilers were recorded. This can be an advantage of DL models and render challenging calculation burdens. Afterward, the coordinates of all the detected chickens in each frame were provided by the model in terms of maximum (x,y) values with the corresponding frame width and height. These values were used to assign a centroid coordinate (x_c,y_c) for each detected frame by Equation (4). The centroid values for consecutive frames were used to obtain a displacement value by the respective broiler. Eventually, the incremental displacement values would be added to obtain total displacements or moved distance of a broiler, as well as the flock mobility levels.

Broiler mobility determination was not a straightforward summation of displacements. We anticipated that small perturbations in body movements would result in the model reading the frames as a potential displacement. Therefore, we would try to validate the model's results with respect to baseline manual measurements as well. This might lead us to some degree of calibration of the model that would eradicate the misleading results.

$$(x_c, y_c) = \left(x + {}^{x_{width}}/_{2}, y + {}^{y_{height}}/_{2}\right) \quad (4)$$

3. Results and Discussion

3.1. Broiler Pens and Data Collection

The broilers' mobility was recorded for 15 min per hour daily for 54 days, rendering a total of 1268 recordings. These data were utilized for the training and development of the YOLOv5 model through the SSL approach. Eventually, broiler mobility analysis was carried out with the proposed YOLOv5 neo-deep sort model.

3.2. Semi-Supervised YOLOv5 Training

3.2.1. Primary YOLOv5 Model Training

The initial training of the model with the manually labeled images resulted in the validation mAP, precision, and R^2 levels, as shown in Figure 3. The graphs demonstrate a considerable increase in the success rates of the three metrics; though mAP, with an 81% success rate, may have more room for improvement. In other words, the precision and R^2 values might be high since the model predicts only one class of object, i.e., chicken, while the mAP may require more training data to achieve a success rate above 90% with higher confidence levels. Overall, the primary trained YOLOv5 model could be considered effective enough to be employed on the unlabeled dataset to achieve one of our objectives.

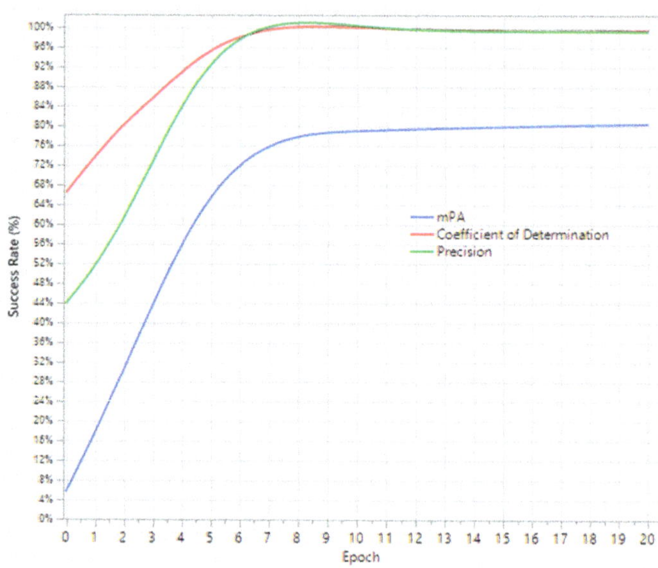

Figure 3. Training results of the primary YOLOv5 model.

3.2.2. Labeling Images with the Primary YOLOv5 Model

The primary trained YOLOv5 model was used to predict labels for these images, which successfully predicted broilers in 2153 of them. A sample of detected images and the corresponding labels are depicted in Figure 4. It can be deduced that while SSL provides highly reliable results by pseudo-labeling images, it introduces some level of error to the model as well. But it is safe to say that the advantage of this method far outweighs its misleading results. Additionally, manual labeling of the first dataset took days to complete, while the pseudo-labeling of the images took less than an hour to complete. The SSL method was truly effective in providing ample labeled training data to further enhance the object detection success rate of the DL model.

(**a**) (**b**)

Figure 4. (**a**) A manually labeled image. (**b**) An image labeled by the YOLOv5 model.

3.2.3. Final YOLOv5 Model Training

Finally, the model-labeled dataset was used to further train the existing trained YOLOv5 model. The newly labeled 2153 images were split into 80:20 sets for training and testing purposes, respectively. As a result of the second training, the YOLOv5 model's predictive capabilities enhanced, as shown in Figure 5. It can be clearly seen that the mAP level has increased to 98% from the previous value of 81%. Henceforth, the trained YOLOv5 model has higher detection accuracy compared with the trained models in similar studies by Fang et al. [13] and Neethirajan [14]. Hence, it was now ready to be applied along with the neo-deep sort algorithm to analyze broiler hourly mobility levels.

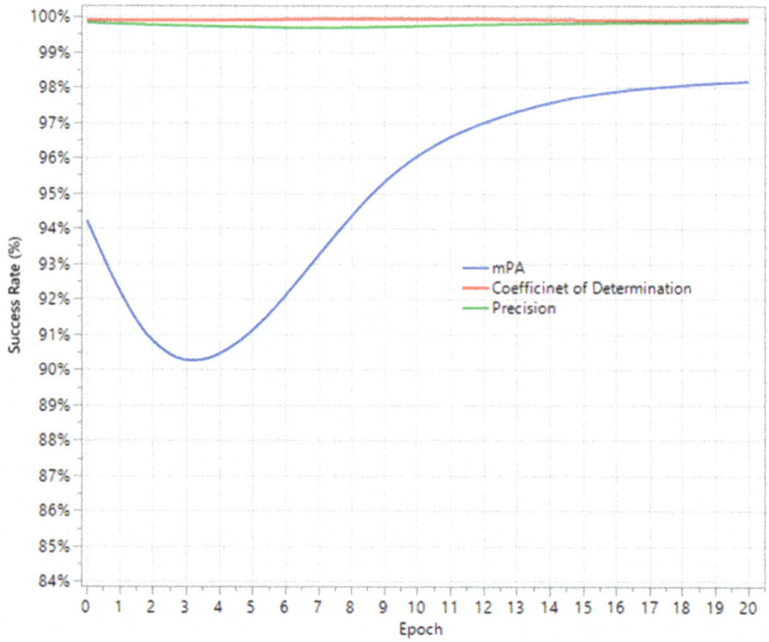

Figure 5. Second training results of the YOLOv5 model.

3.3. Final Model: YOLOv5 Neo-Deep Sort Application

The final model was applied to the video recordings of the broilers from two separate cameras overlooking respective Pen #1 (twelve broilers) and Pen #2 (sixteen broilers). As discussed earlier, the data under study consist of recordings of when broilers were 11, 18, 24, 30, 36, 41, and 47 days old. Therefore, we were able to see the broiler mobility levels at different ages throughout the rearing process. It is worth mentioning that the model was trained by the data from Pen #2, and the resultant trained model was applied to one familiar environment, Pen #2, and a completely new environment, Pen #1. This would show the ability of the trained model to generalize and perform a completely new dataset.

3.3.1. Broiler Detection Levels

The final model was able to detect broilers at different stages of their lifetime with reasonably high success rates. The general performance of our model is depicted in the distribution and boxplot graphs in Figure 6a–c. Overall, the number of broilers successfully detected in a frame shows a little skewed normal distribution with varying means of 9 and 14 for Pens 1 and 2, respectively. On average, the model was proportionate, 9/12 and 14/16, and was almost equally successful in detecting the number of broilers in both pens. But, in general, it performed better in Pen #2 over the course of 7 weeks. As seen in Figure 6c, the boxplots in the model were consistently performing better in detecting broilers in Pen #2 compared to Pen #1. Although they do not provide a higher success rate in the first batch of data from Pen #2, i.e., when the birds were 11 days old. They rendered better results in the consequent datapoints. On the other hand, in Pen #1, where the broiler stocking density is lower and the environment is newer to the model, the performance is comparatively less successful but still consistent and has reasonably high success rates.

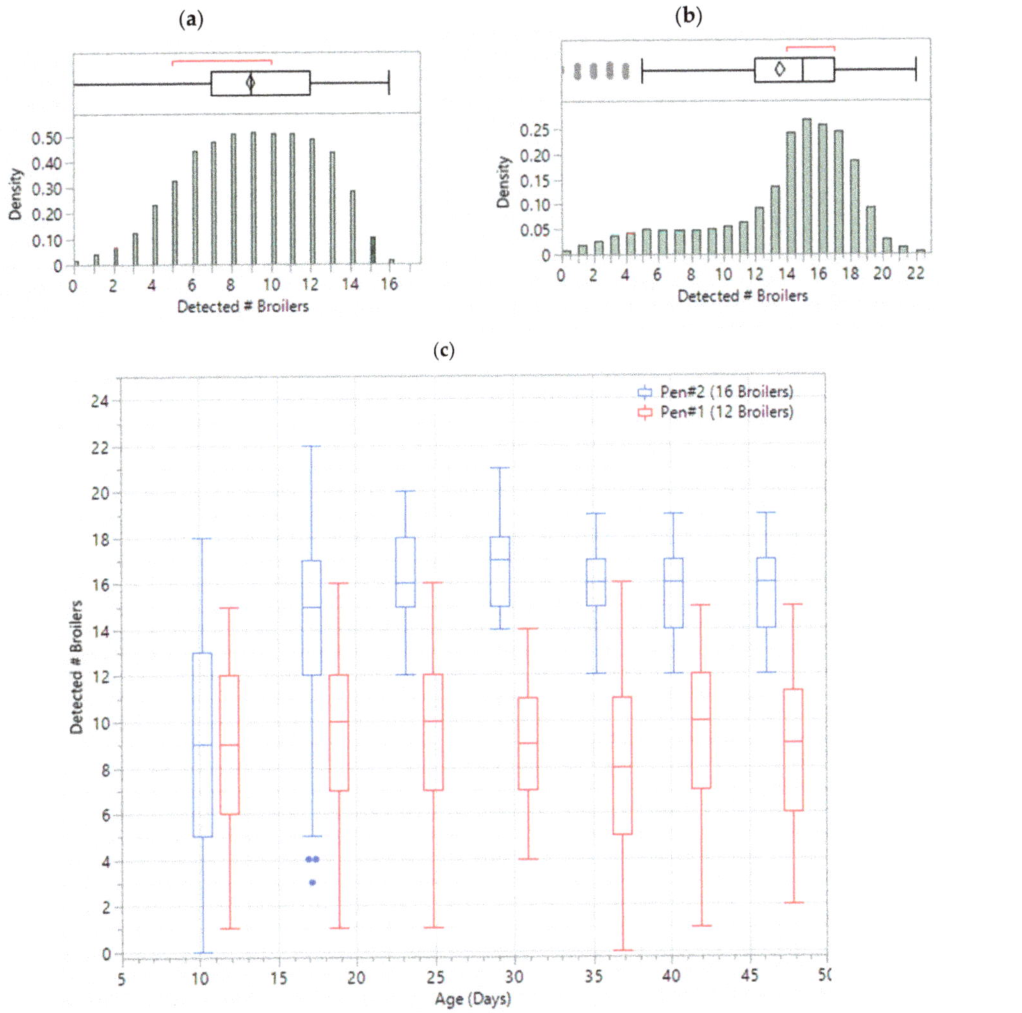

Figure 6. Broiler detection results by the final model. (**a**) Pen #1 frame success rate distribution. (**b**) Pen #2 frame success rate distribution. (**c**) Broiler success rate variation over different age periods.

3.3.2. Broiler Tracking Performance

After the broiler detection process, the tracking step, i.e., the deep sort algorithm, tried to extract the movement coordinates of the broilers from consequent frames. In some instances, the tracking sequence of a particular broiler might be lost for a short time due to occlusions, broilers gliding over, or mixing closely with each other. After reappearing, the model would start tracking those birds again but would deem them as new birds and hence assign a new ID. Here, the final model's continuous tracking ability of a broiler before losing it is shown in Figure 7a. On average, the model has tracked birds in different pens with varying success rates; it has tracked the birds better in Pen #2 compared to the results in Pen #1. In general, the model has tracked the broilers increasingly better as their ages grew. In other words, as broilers' weight and volume increased, it was relatively easier for the model to continuously track them. On average, the broilers were tracked at least 3 min (20% of the 15 min) into the monitoring period. This may provide enough information on the speed, displacement, and mobility levels of a particular broiler, even if we do not try to associate different IDs of a particular broiler.

(a)

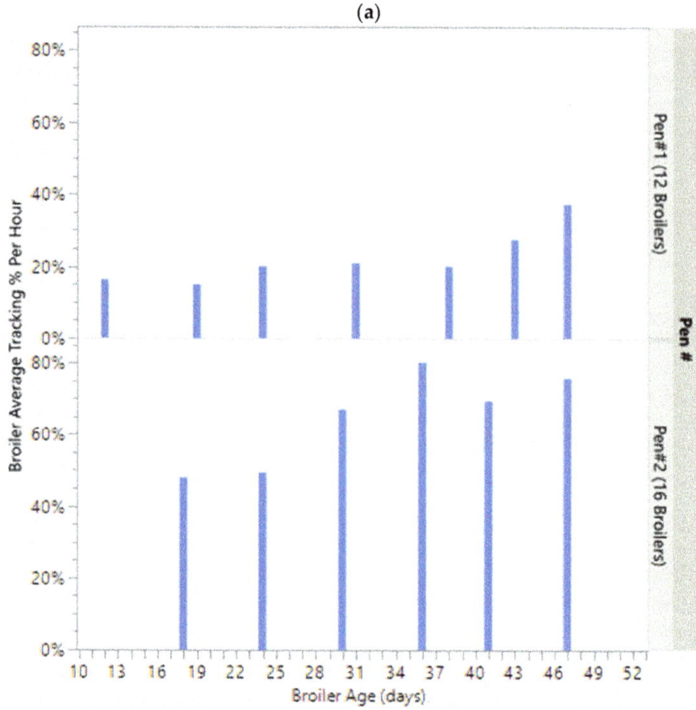

(b)

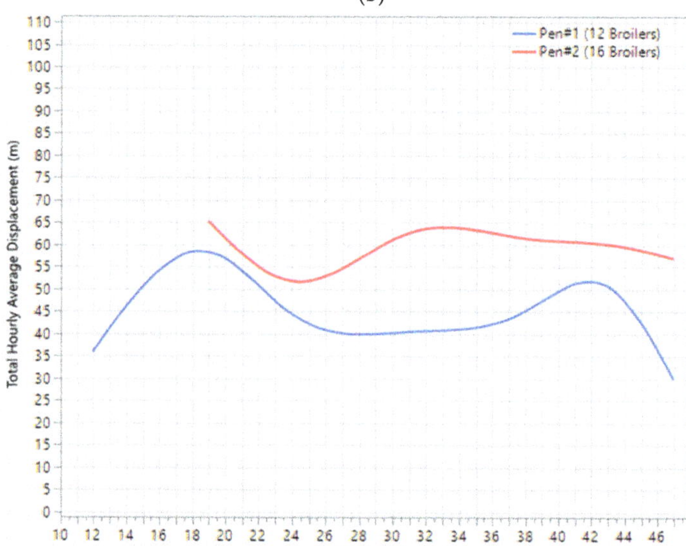

Figure 7. *Cont.*

(c)

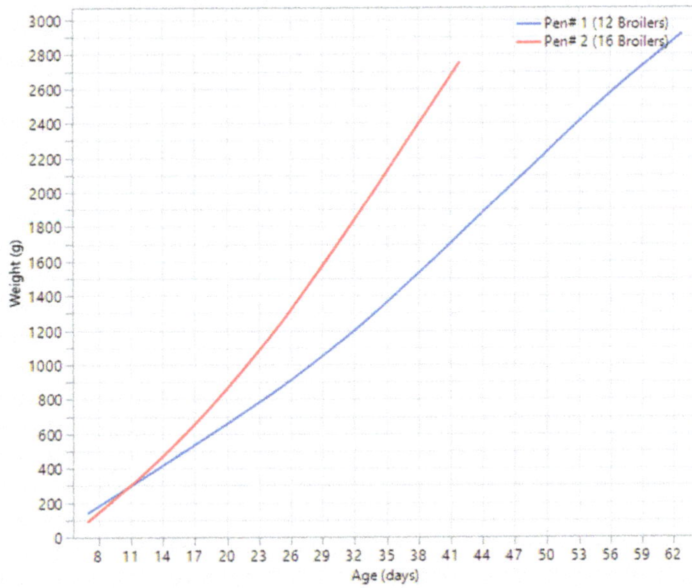

Figure 7. (**a**) Average broiler tracking level per hour. (**b**) Daily average displacement levels. (**c**) Average weight change in broilers as they grew.

3.3.3. Broiler Flock Mobility Level

The final model was utilized to track birds using the displacements between consecutive frames from video recordings separated by 1 s. This period was selected to decrease computation complexity by ignoring broiler perturbations that might have happened in less than a second. Consequently, the broilers' total mobility, including moved distances and speed, was calculated. On the other hand, our proposed algorithm was used to associate different broiler IDs, as discussed in the previous sections. Henceforth, we were able to categorize broilers' mobility comparatively efficiently at each hour of the day. This crucial result paved the way for continuous insights, enabling timely and effective interventions on birds with low or no mobility.

Total Displacement of Broilers at the Flock Level

The average daily covered distance by all the broilers and the corresponding average broiler weights in each pen are demonstrated in Figure 7b,c. As the broilers grew, their weights increased constantly until the saturation point at the end of the second month. But, the total displacement level had a different trend in both pens. The average daily displacements are highest when broilers are aged 18 d. Comparatively, Pen #2 has had a higher displacement level compared to Pen #1. It can be safely pointed out that higher stocking densities, such as in Pen #2, cause higher levels of mobility, as going from one point to another may require more walking and cause disturbance to the surrounding broilers. Additionally, the daily average covered distance trends give a better understanding of flock movements than in methods followed by similar works, such as Neethirajan's [14], which have avoided providing a quantitative result for this level of movements.

Flock Speed Levels

The average daily broiler speed levels show a similar distribution as the displacements explained in the previous subsection. As indicated in Figure 8a., broilers in both pens have shown very similar speed levels, although the mean speed in Pen #2 (with 16 broilers) is higher than Pen #1 (with 12 broilers). The lowest average speed happened when broilers were 24 to 25 days of age in both pens. Meanwhile, the highest broiler speeds were recorded when the broilers were very young, around the age of 18 d. In both pens, the speed levels have increased steadily from the age of 25 d to the age of 45 d. On the other hand, the average speed changes compared to the increase in the weights of the birds demonstrate a similar trend. The average speed in both pens decreases to a minimum, with birds having an average weight of 1200 g–1400 g. As weight continues to increase, so does the speed of the birds as they reach 3000 g in weight. Pen #1 birds seemed to move slower while the other pen's birds continued moving faster.

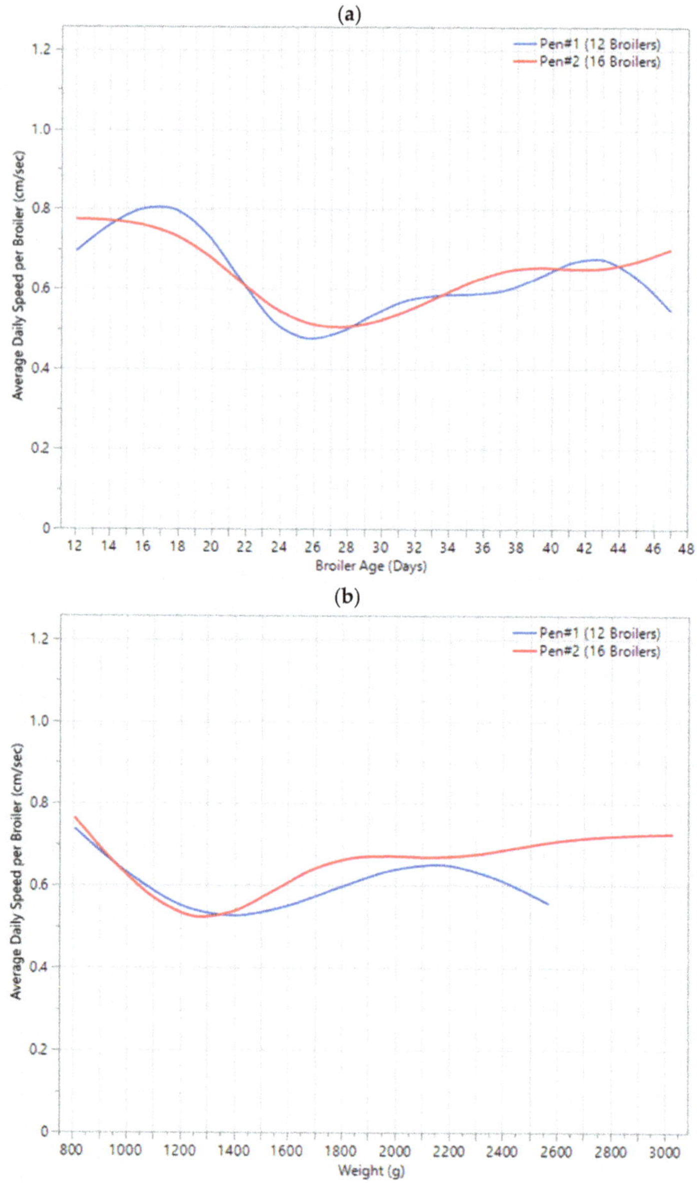

Figure 8. (**a**) Average speed change vs. broiler age. (**b**) Average speed change vs. weight.

New Algorithm Application

The proposed algorithm was applied to associate different IDs that were assigned to the same broiler during the tracking. Figure 9a shows a sample hourly displacement tracking of the birds on a specific day; for example, at 6:00 a.m., as per the preliminary results, most of the broilers have moved moderately but birds 12, 14, and 16 have shown no or very low mobility. But, after the algorithm was applied, it was able to associate different broiler instances to give a final picture of the mobility of the birds. In this hour, in Figure 9a, IDs 6, 9, 10, 12, 14, and 16 were associated with IDs 126, 118, 108, 87, 115, and 146, respectively. The deep sort tracking algorithm had tracked the broilers with the former IDs and had lost their track due to occlusion instances. After resuming the tracking when the respective birds had reappeared, the new IDs, in the former list, were attributed to these broilers. Hence, with the proposed algorithm, we associated these lost instances. As a result, the final mobility level of the broilers is calculated by adding the individual displacements of the associated IDs. As seen in Figure 9b, the above-mentioned birds' final displacement levels have changed after this process. For example, bird #6 had a displacement of about 10 cm, which increased to about 20 cm after ID association. As in the case of bird #14, even after ID association, the resultant mobility level and tracking appearance percentage still fell below 100%. Hence, we can conclude that even after the ID association process, we might not obtain a very high percentage of some broiler's appearances. But, even a 50% appearance level can give us a statistically sound indication of how much a broiler is mobile. Additionally, we can further investigate the sequential videos of that day and obtain a bigger picture of that broiler. On the other hand, the final hourly speed and covered distance by individual broilers provide a more detailed insight into their mobility levels with higher accuracies compared with similar works, such as Neethirajan [14], Nasiri et al. [9], and Fang et al. [13]. This may help farm managers have more control over their resources and day-to-day interventions.

(a)

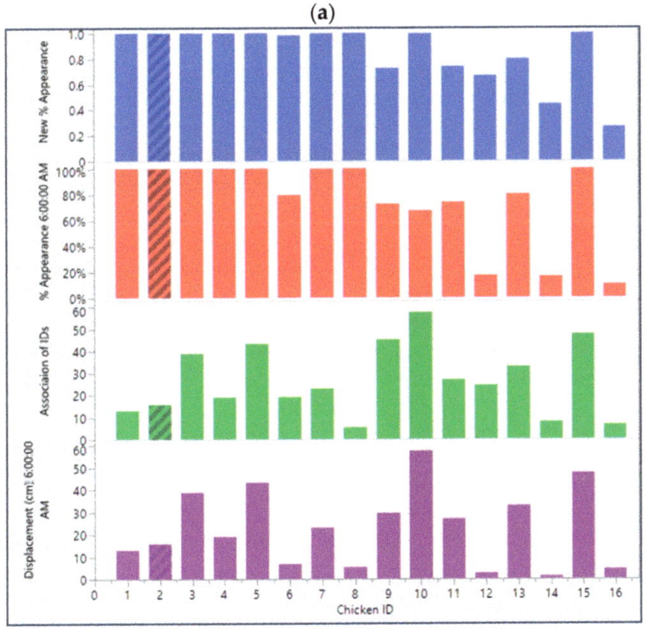

Figure 9. *Cont.*

(**b**)

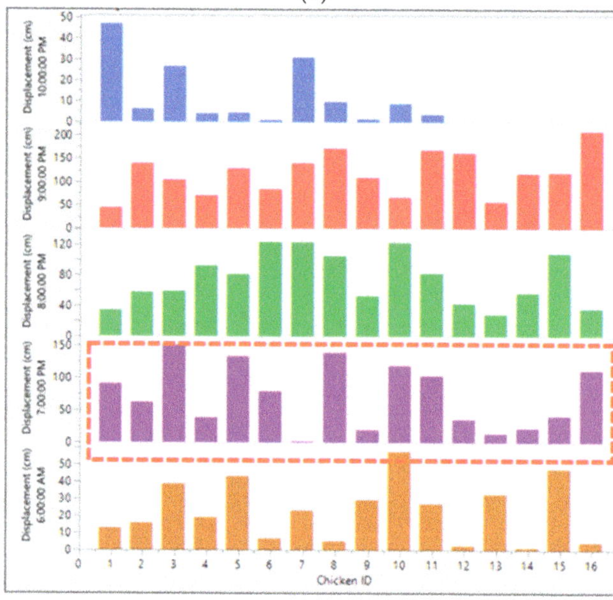

Figure 9. (**a**) Broilers' ID association process and the respective changes in displacement and % appearance. (**b**) Individual broilers with their hourly mobility levels.

4. Conclusions

Broilers' daily mobility levels significantly impact their daily behaviors, which could be detrimental to rendering high-quality meat to the market. In this study, we have utilized the semi-supervised YOLOv5 DL model alongside the deep sort algorithm for object detection and tracking, respectively. Additionally, we proposed a new algorithm that tackled deep sort tracking losses due to bird occlusion instances. A total of 7 days of hourly videos, corresponding to seven consecutive weeks, were studied. The SSL method paved the way to train the YOLOv5 model with considerably fewer human and time resources. It helped in increasing the YOLOv5 detection accuracy from 81% to 98%. The final YOLOv5 deep sort model was influential in tracking broilers continuously, at least around 20% of the time, but it had limitations in doing so during occlusion periods. Our proposed algorithm lessened this problem to some extent but was incapable of fully solving the tracking lost instances problem. Henceforth, in general, individual broiler and flock level displacements and speeds were computed throughout their growth period. This helped in understanding their mobility trends and categorizing them accordingly. This vision-based study of mobility indicators could be further developed by incorporating more effective algorithms to solve the lost information problem due to occlusions.

Author Contributions: Conceptualization, M.J. and Y.Z.; Methodology, M.J. and H.G.; Writing—original draft, M.J.; Writing—review & editing, H.G., T.T., M.P. and H.Q.; Supervision, Y.Z.; Project administration, Y.Z. All authors have read and agreed to the published version of the manuscript.

Funding: Financial support for this research was provided by the USDA-NIFA IDEAS program (Award No.: 2022-68014-36663) and AI TENNessee Initiative Seed Funds. The authors highly appreciate the assistance of the UT Animal Science Department, UT Joseph E. Johnson Research and Teaching Unit.

Institutional Review Board Statement: The animal study protocol was approved by the Institutional Animal Care and Use Committee of The University of Tennessee (Protocol number 2876).

Conflicts of Interest: The authors declare no conflict of interest.

References

1. Gallardo:, R.K.; Sauer, J. Adoption of Labor-Saving Technologies in Agriculture. *Annu. Rev. Resour. Econ.* **2018**, *10*, 185–206. [CrossRef]
2. Granquist, E.G.; Vasdal, G.; de Jong, I.C.; Moe, R.O. Lameness and its relationship with health and production measures in broiler chickens. *Animal* **2019**, *13*, 2365–2372. [CrossRef] [PubMed]
3. Gocsik, E.; Silvera, A.M.; Hansson, H.; Saatkamp, H.W.; Blokhuis, H.J. Exploring the economic potential of reducing broiler lameness. *Br. Poult. Sci.* **2017**, *58*, 337–347. [CrossRef]
4. Knowles, T.G.; Kestin, S.C.; Haslam, S.M.; Brown, S.N.; Green, L.E.; Butterworth, A.; Pope, S.J.; Pfeiffer, D.; Nicol, C.J. Leg disorders in broiler chickens: Prevalence, risk factors and prevention. *PLoS ONE* **2008**, *3*, e1545. [CrossRef]
5. Okinda, C.; Nyalala, I.; Korohou, T.; Okinda, C.; Wang, J.; Achieng, T.; Wamalwa, P.; Mang, T.; Shen, M. A review on computer vision systems in monitoring of poultry: A welfare perspective. *Artif. Intell. Agric.* **2020**, *4*, 184–208. [CrossRef]
6. Aydin, A. Using 3D vision camera system to automatically assess the level of inactivity in broiler chickens. *Comput. Electron. Agric.* **2017**, *135*, 4–10. [CrossRef]
7. Pereira, D.F.; Nääs, I.d.A.; Lima, N.D.d.S. Movement Analysis to Associate Broiler Walking Ability with Gait Scoring. *AgriEngineering* **2021**, *3*, 394–402. [CrossRef]
8. Doornweerd, J.E.; Kootstra, G.; Veerkamp, R.F.; Ellen, E.D.; van der Eijk, J.A.J.; van de Straat, T.; Bouwman, A.C. Across-Species Pose Estimation in Poultry Based on Images Using Deep Learning. *Front. Anim. Sci.* **2021**, *2*, 791290. [CrossRef]
9. Nasiri, A.; Yoder, J.; Zhao, Y.; Hawkins, S.; Prado, M.; Gan, H. Pose estimation-based lameness recognition in broiler using CNN-LSTM network. *Comput. Electron. Agric.* **2022**, *197*, 106931. [CrossRef]
10. Fang, C.; Zhang, T.; Zheng, H.; Huang, J.; Cuan, K. Pose estimation and behavior classification of broiler chickens based on deep neural networks. *Comput. Electron. Agric.* **2021**, *180*, 105863. [CrossRef]
11. de Alencar Nääs, I.; da Silva Lima, N.D.; Gonçalves, R.F.; Antonio de Lima, L.; Ungaro, H.; Minoro Abe, J. Lameness prediction in broiler chicken using a machine learning technique. *Inf. Process. Agric.* **2021**, *8*, 409–418. [CrossRef]
12. Lin, C.-Y.; Hsieh, K.-W.; Tsai, Y.-C.; Kuo, Y.-F. Monitoring chicken heat stress using deep convolutional neural networks. In Proceedings of the 2018 ASABE Annual International Meeting, Detroit, MI, USA, 29 July–1 August 2018.
13. Fang, C.; Huang, J.; Cuan, K.; Zhuang, X.; Zhang, T. Comparative study on poultry target tracking algorithms based on a deep regression network. *Biosyst. Eng.* **2020**, *190*, 176–183. [CrossRef]
14. Neethirajan, S. ChickTrack—A quantitative tracking tool for measuring chicken activity. *Measurement* **2022**, *191*, 110819. [CrossRef]
15. Redmon, J.; Divvala, S.; Girshick, R.; Farhadi, A. You Only Look Once: Unified, Real-Time Object Detection. In Proceedings of the 2016 IEEE Conference on Computer Vision and Pattern Recognition (CVPR), Las Vegas, NV, USA, 27–30 June 2016.
16. Parico, A.I.B.; Ahamed, T. Real Time Pear Fruit Detection and Counting Using YOLOv4 Models and Deep SORT. *Sensors* **2021**, *21*, 4803. [CrossRef] [PubMed]
17. Van Voorhis, C.W.; Morgan, B.L. Understanding power and rules of thumb for determining sample sizes. *Tutor. Quant. Methods Psychol.* **2007**, *3*, 43–50. [CrossRef]
18. van Engelen, J.E.; Hoos, H.H. A survey on semi-supervised learning. *Mach. Learn.* **2019**, *109*, 373–440. [CrossRef]
19. Ouali, Y.; Hudelot, C.; Tami, M. An Overview of Deep Semi-Supervised Learning. *arXiv* **2020**, arXiv:2006.05278.
20. Wang, H.; Zhang, S.; Zhao, S.; Wang, Q.; Li, D.; Zhao, R. Real-time detection and tracking of fish abnormal behavior based on improved YOLOV5 and SiamRPN++. *Comput. Electron. Agric.* **2022**, *192*, 106512. [CrossRef]
21. Liu, K.; Tang, H.; He, S.; Yu, Q.; Xiong, Y.; Wang, N. Performance Validation of Yolo Variants for Object Detection. In Proceedings of the 2021 International Conference on Bioinformatics and Intelligent Computing, Coimbatore, India, 6–7 March 2021; pp. 239–243.
22. Nepal, U.; Eslamiat, H. Comparing YOLOv3, YOLOv4 and YOLOv5 for Autonomous Landing Spot Detection in Faulty UAVs. *Sensors* **2022**, *22*, 464. [CrossRef]
23. Ge, Y.; Lin, S.; Zhang, Y.; Li, Z.; Cheng, H.; Dong, J.; Shao, S.; Zhang, J.; Qi, X.; Wu, Z. Tracking and Counting of Tomato at Different Growth Period Using an Improving YOLO-Deepsort Network for Inspection Robot. *Machines* **2022**, *10*, 489. [CrossRef]
24. Wojke, N.; Bewley, A.; Paulus, D. Simple Online and Realtime Tracking with a Deep Association Metric. In Proceedings of the 2017 IEEE International Conference on Image Processing (ICIP), Beijing, China, 17–20 September 2017.
25. Duggan, B.M.; Hocking, P.M.; Clements, D.N. Gait in ducks (anas platyrhynchos) and chickens (gallus gallus)-Similarities in adaptation to high growth rate. *Biol. Open* **2016**, *5*, 1077–1085. [CrossRef] [PubMed]

animals

MDPI

Article

Real-Time Monitoring of Grazing Cattle Using LORA-WAN Sensors to Improve Precision in Detecting Animal Welfare Implications via Daily Distance Walked Metrics

Shelemia Nyamuryekung'e [1,*], Glenn Duff [2], Santiago Utsumi [2,*], Richard Estell [3], Matthew M. McIntosh [3], Micah Funk [2], Andrew Cox [2], Huiping Cao [4], Sheri Spiegal [3], Andres Perea [2] and Andres F. Cibils [5]

1 Division of Food Production and Society, Norwegian Institute of Bioeconomy Research (NIBIO), PB 115, N-1431 Ås, Norway

2 Department of Animal and Range Sciences, New Mexico State University, Las Cruces, NM 88003, USA; glennd@nmsu.edu (G.D.); funkm@nmsu.edu (M.F.); arcox@nmsu.edu (A.C.); arperea@nmsu.edu (A.P.)

3 United States Department of Agriculture-Agriculture Research Service, Jornada Experimental Range, Las Cruces, NM 88003, USA; rick.estell@usda.gov (R.E.); mattmac@nmsu.edu (M.M.M.); sheri.spiegal@usda.gov (S.S.)

4 Department of Computer Science, New Mexico State University, Las Cruces, NM 88003, USA; hcao@nmsu.edu

5 United States Department of Agriculture Southern Plains Climate Hub, United States Department of Aagricultulre-Agriculture Rearch Services, Oklahoma and Central Plains Agricultural Research Center, El Reno, OK 73036, USA; andres.cibils@usda.gov

* Correspondence: shelemia@nibio.no (S.N.); sutsumi@nmsu.edu (S.U.); Tel.: +47-406-04-100 (S.N.); +1-575-646-2514 (S.U.)

Simple Summary: Global positioning system (GPS) coordinates are often used to calculate distance traveled, a useful metric for research and decision-making processes in livestock management. The study aimed to determine the accuracy of using LoRa-WAN sensors to measure the walking distances of grazing cattle in real time. The study compared the accuracy of distance computation using real-time LoRa-WAN sensed GPS alone or in combination with motion data from triaxial accelerometers. The analysis showed that the fusion of GPS and accelerometer data was more suitable for calculating walking distance in detecting animal welfare implications associated with immobility.

Abstract: Animal welfare monitoring relies on sensor accuracy for detecting changes in animal well-being. We compared the distance calculations based on global positioning system (GPS) data alone or combined with motion data from triaxial accelerometers. The assessment involved static trackers placed outdoors or indoors vs. trackers mounted on cows grazing on pasture. Trackers communicated motion data at 1 min intervals and GPS positions at 15 min intervals for seven days. Daily distance walked was determined using the following: (1) raw GPS data (RawDist), (2) data with erroneous GPS locations removed (CorrectedDist), or (3) data with erroneous GPS locations removed, combined with the exclusion of GPS data associated with no motion reading (CorrectedDist_Act). Distances were analyzed via one-way ANOVA to compare the effects of tracker placement (Indoor, Outdoor, or Animal). No difference was detected between the tracker placement for RawDist. The computation of CorrectedDist differed between the tracker placements. However, due to the random error of GPS measurements, CorrectedDist for Indoor static trackers differed from zero. The walking distance calculated by CorrectedDist_Act differed between the tracker placements, with distances for static trackers not differing from zero. The fusion of GPS and accelerometer data better detected animal welfare implications related to immobility in grazing cattle.

Keywords: precision livestock farming; precision livestock ranching; internet of things; long range wide area network

Citation: Nyamuryekung'e, S.; Duff, G.; Utsumi, S.; Estell, R.; McIntosh, M.M.; Funk, M.; Cox, A.; Cao, H.; Spiegal, S.; Perea, A.; et al. Real-Time Monitoring of Grazing Cattle Using LORA-WAN Sensors to Improve Precision in Detecting Animal Welfare Implications via Daily Distance Walked Metrics. *Animals* **2023**, *13*, 2641. https://doi.org/10.3390/ani13162641

Academic Editors: Brett Ramirez, Janice Siegford, Hao Gan, Yang Zhao, Daniel Berckmans, Robert T. Burns and Lingjuan Wang-Li

Received: 4 July 2023
Revised: 30 July 2023
Accepted: 9 August 2023
Published: 16 August 2023

1. Introduction

Store-on-board telemetry devices, including Global Positioning System (GPS) loggers and accelerometer sensors, have gained traction in the last two decades for studying livestock and wildlife grazing behavior. Integrating these unobtrusive telemetry devices has enabled the scientific community to gain insights into the behaviors of different animals in their natural settings [1–4]. For instance, studies performed on cattle grazing on extensive rangelands using GPS loggers have documented animal movement in correlation with foraging activities influenced by resource distribution, in their search for thermal comfort, to avoid predator presence in their environment or other essential activities [3,5–7]. Hence, the mobility of animals within their environment is crucial for their survival, as they respond to various biotic and abiotic stimuli.

Developments in communication technology have facilitated real-time data transmission from telemetry devices, helping advance Precision Livestock Farming/Ranching (PLF/R) applications. One such technology is the Long Range Wide Area Network (LoRa WAN), a wireless low-power data transmission system with bidirectional capabilities, enabling data packet transmission and configuration commands to be sent remotely [8–11]. These networks exhibit far-reaching coverage (up to 10 km) with a stronger signal strength compared to Wi-Fi and Bluetooth, penetrating insulated objects, and requiring minimal maintenance with a longer lifespan [11,12]. As a result, LoRa WAN offers an affordable and effective solution for implementing PLF/R on extensive rangelands with limited connectivity [11–13].

The Internet of Things (IoT) is at the heart of these applications, with sensory devices that process unique measurements of physiological and behavioral parameters on animals, gateways that enable data transmissions from sensors to the internet, cloud services that provide data storage and analysis, with an output application layering tailoring the data into usable information for a particular end-user [10,14–16].

Integrating PLF/R could facilitate the transition of traditional livestock production systems that place emphasis on maximizing animal output to aspirational management systems that optimize production efficiency by increasing animal welfare and whole-farm sustainability [17–19]. In addition, consumer pressure exerted through purchasing power in most developed countries has demanded livestock producers increase animal welfare transparency via enhanced traceability [16,20,21]. Also, the continuous monitoring application of the PLF/R technology in animal production would enable real-time management of the smallest manageable unit (sensor-based animal-specific attention) on the production front, enhancing management flexibility and minimizing environmental impacts with alleviation of intensive labor requirements [20,22,23].

Animal welfare is a multidimensional concept related to the repertoire of behaviors performed by an animal in its natural state that promote 'normal' biological functions [20,24]. Therefore, the success of the PLF/R application to address animal welfare will depend on the accuracy of sensors to detect changes associated with a deteriorating animal state [5,18,25]. For instance, using frequently acquired GPS data to compute daily distance traveled can have direct management implications in addressing animal welfare on extensive cattle production [17,26,27]. However, control of data quality and improvements in analytical procedures (algorithms for calculating daily distance traveled metrics) are paramount in facilitating the implementation of PLF/R applications to address issues stated previously.

For instance, the precision of GPS measurements is affected by several factors, including the type of device with internal limitations, the sensor environment placement, and the timing of data acquisition in relation to the orbiting satellites. Precision ranges for GPS sensors used in animal tracking have been reported to not exceed 30 m, provided that the units have a relatively unobstructed view of the sky [28,29]. However, the precision error varies among devices and can be further affected by obstructions in the communication pathways with orbiting satellites [29–32]. Therefore, flagging erroneous GPS locations is vital when incorporating the data into a computation informing a PLF/R application.

This study aimed to assess the effectiveness of real-time sensors using Long Range Wide Area Network (LoRa-WAN) communication for monitoring animal welfare through daily distance travel metrics. To achieve this, we evaluated the accuracy of calculating the daily distance travel metric using global position system (GPS) data alone or in combination with motion data derived from triaxial accelerometers. Three algorithms were tested to detect any differences between the static sensors placed outdoors (Outdoor, n = 6) or indoors with an obstructed view of satellites (Indoor, n = 5), vs. trackers mounted on grazing cows (Animal, n = 6). We hypothesized that the daily distance traveled by the animal tracker would be greater than the static trackers, with no significant difference between outdoor and indoor means. We also predicted that the daily distance traveled by the static trackers would be negligible and not differ from zero, as an animal welfare alert system would require immobility detection.

2. Materials and Methods

2.1. Experimental Site and LoRa Trackers Configuration

The study was conducted at the New Mexico State University's Clayton Livestock Research Center (NMSU CLRC), which is located 7 miles east of Clayton, New Mexico, USA, and covers a total area of 1.39 km^2 (320 acres). The research site consists of flat terrain, with a section of 0.79 km^2 (195 acres) of fenced land configured with a center pivot winter wheat (*Triticum aestivum* L.) irrigated pasture and a feedlot facility housed with hydraulic chutes for animal handling south of the section (Figure 1).

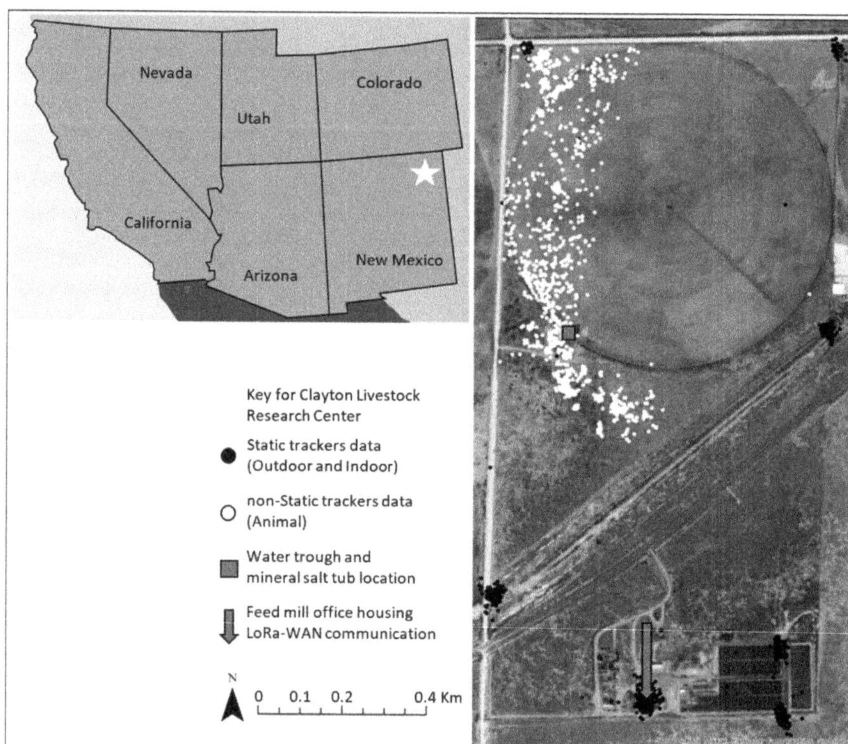

Figure 1. Top left: Reference map of the Southwest United States showing the research site marked with a star (Clayton, New Mexico). Right: Map of New Mexico State University's Clayton Livestock Research Center (CLRC) displaying GPS locations of static trackers and non-static trackers, along with other important landmarks on the research site.

Long Range Wide Area Network (LoRa WAN) was chosen as the mode of communication between an antenna and the trackers (widgets) [8,9,15]. The PLF/R system at the NMSU (CLRC) consisted of a single Kerlink® LoRa-WAN 915 MHz gateway (https://www.kerlink.com/, accessed on 15 October 2020) with an external high gain antenna mounted on the feed mill tower using a 30 m coaxial cable (Figure 1). The gateway was purchased with a licensed software platform (Thingpark) developed by Actility® (https://www.actility.com/, accessed on 15 October 2020), allowing data traffic monitoring and gateway functionality.

LoRa WAN-enabled Abeeway® (https://www.abeeway.com/, accessed on 10 October 2020) Industrial Trackers US915 were configured to communicate with a single gateway at the NMSU (CLRC) site. The trackers weigh 240 g, are housed in a waterproof casing, and operate using an internal Lithium-thionyl Chloride Type D battery (14 Ah/3.6 V). The trackers were equipped with position, motion, and temperature sensors, and a LoRa-WAN communication chip embedded in their motherboard. The trackers were then contained within a Pelican® R20 Ruck case strapped on an adjustable nylon collar belt to reinforce structural integrity and waterproofing capabilities [11].

The industrial trackers had a licensed software platform (Abeeway Device Manager 2.13.0) for the data surveillance (Map, Performance Monitor, and Uplinks data log) and tracker configuration under an annual subscription [11]. We opted for the "Activity tracking" configuration, with activity reporting as the "main operational" mode and periodic position message as a "side operation." The data collection interval was set at 1 min for motion detection using the tri-axial accelerometer sensors and 15 min intervals for position acquisition using the GPS-only option.

The process of GPS acquisition demands significant power, leading to a dilemma between prioritizing high-frequency, short-duration experiments, or low-frequency data collection for longer-duration studies [4]. According to the guidelines for LoRa WAN-enabled Abeeway® Industrial Tracker, the estimated device duration is 20 months when acquiring 24 positions per day (at an hourly frequency of GPS data collection). The selected motion and position acquisition scheduling for the study was hypothesized to provide a battery duration of approximately 5 months, making these trackers suitable for deployment on a working ranch with minimal livestock interaction.

2.2. *Study Deployment, Animals*

From 24 October to 17 November 2020, 6 randomly selected trackers (Outdoor) were positioned in fixed locations at incremental distances from the LoRa WAN gateway (Figure 1). Using adjustable belts, the trackers were secured on an existing fence line, approximately ~1 m above the ground and facing the antenna. In addition, five trackers were housed inside the feed mill office (Indoor), adjacent to the LoRa WAN gateway, to simulate obstructed GPS communication with orbiting satellites (Figure 1). The deployment lasted 24 days, but no data was collected for 15 days due to gateway software maintenance during the middle of deployment (27 October to 11 November 2020).

A follow-up deployment from 22 December to 31 December 2020, utilized the trackers mounted on mature cattle (Animal) with a subset of only six trackers out of the eleven randomly selected for the static phase utilized for analysis. Trackers were safely secured on the necks of the animals using adjustable nylon belts. The collared animals grazed on native grasslands with access to a portion of an irrigated winter wheat pasture, with ad libitum access to water and a mineral salt tub (Figure 1). Animal use was approved by the New Mexico State University Institutional Animal Care and Use Committee (protocol # 2019-008).

2.3. *Data Processing*

All the trackers' payloads were routed from the Thingpark server that populated the Abeeway Device Manager to a local New Mexico State University server for data retrieval and analysis. The data from the static phase were trimmed, with the initiation and termination dates excluded from the analysis. In addition, the dates when the gateway was under maintenance (27 October to 11 November 2020) were also excluded. A total of 7 days of deployment from the static trackers (Indoor, n = 5 and Outdoor, n = 6) were analyzed. In the follow-up phase with the trackers mounted on mature cattle (Animal, n = 6), animals were allowed three days to acclimate to collars and the new environment, followed by a 7-day data collection period (25 December to 31 December 2020).

GPS coordinates were projected to NAD 1983 UTM coordinate system (Zone 13 N) using ArcGIS software (ESRI 2018, ArcMap Desktop v. 10.6). Erroneous GPS locations were filtered using a z-score outlier detection analysis for the northing and easting coordinates separately, as described by Nyamuryekung'e et al. [33]. The z-score outlier detection analysis followed the conversion of daily projected coordinate values for an individual tracker into a normalized z-score, highlighting extreme score values with low probability under assumptions for a normal distribution of data points ($z > |4.5|$) [33]. Motion data were reported as counts (Motion Index, MI) of shock within the interval of data acquisition across the triaxial accelerometer using internal default threshold values and were represented as cumulative counts between successful GPS data (Figure 2).

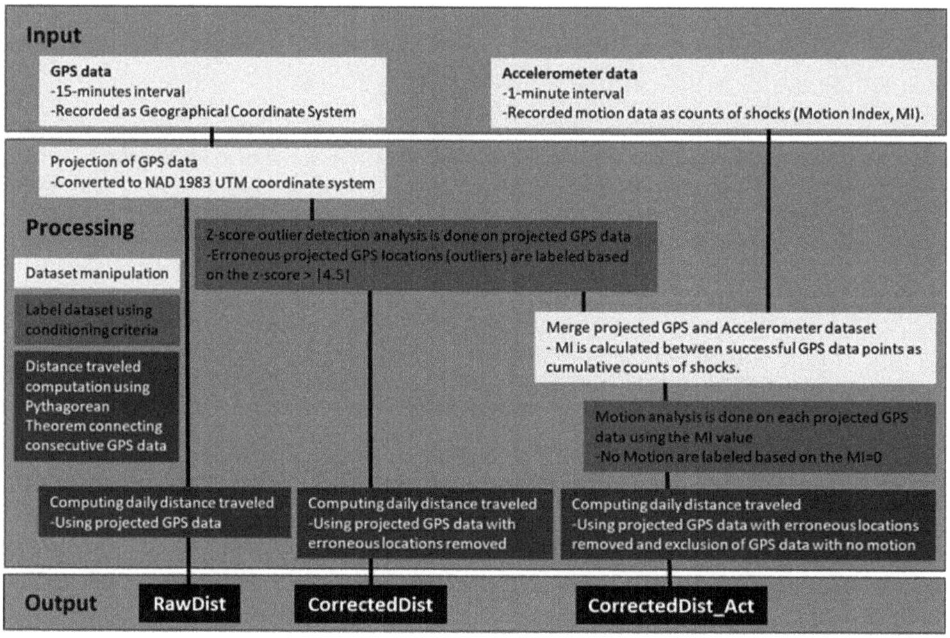

Figure 2. Data flow diagram illustrating input data from the LoRaWAN trackers and various dataset processing steps prior to calculating daily distance traveled using three algorithms: RawDist, CorrectedDist, and CorrectedDist_Act.

The three algorithms for daily distance traveled included using (1) raw GPS data with associated erroneous locations (RawDist), (2) GPS data with erroneous locations removed using z-score > |4.5| analysis (CorrectedDist), or (3) GPS data with erroneous locations removed combined with the exclusion of GPS data associated with no motion (MI = 0) from the triaxial accelerometer reading (CorrectedDist_Act) (Figure 2). Using projected GPS positions for the Static (Indoor and Outdoor) and non-Static (Animal) trackers, we calculated daily distances traveled (m) using three algorithms by summing the consecutive GPS distances calculated using the Pythagorean Theorem within a day, as described by Nyamuryekung'e et al. [34].

The daily distance traveled calculations followed the steps below with their mathematical equations:

1. Let consecutive GPS position be represented as $(x_1, y_1), (x_2, y_2), (x_3, y_3), \ldots, (x_n, y_n)$ where n is the total number of daily GPS positions.
2. Pythagorean Theorem for calculating the distance between two consecutive GPS positions (x_i, y_i) and $(x_{(i+1)}, y_{(i+1)})$:

$$\text{Distance} = \left(\sqrt{\left(x_i - x_{(i+1)}\right)^2 + \left(y_i - y_{(i+1)}\right)^2}\right)$$

3. The daily distance traveled within a day was calculated by summing the distance between all consecutive GPS positions within a day:

$$\text{Daily distance} = \left(\sqrt{(x_1 - x_2)^2 + (y_1 - y_2)^2}\right) + \left(\sqrt{(x_2 - x_3)^2 + (y_2 - y_3)^2}\right) + \ldots + \left(\sqrt{\left(x_{(n-1)} - x_n\right)^2 + \left(y_{(n-1)} - y_n\right)^2}\right)$$

Caution GPS coordinates must be projected to provide the daily distance traveled by the tracker in meters.

2.4. Data Analysis

An analysis based on descriptive statistics on each daily distance measurement (RawDist, CorrectedDist, and CorrectedDist_Act) was computed using the MEANS procedure in SAS 9.3 (SAS Institute, Cary, NC, USA). The data were grouped according to tracker placement (Indoor, Outdoor, or Animal) and the calendar date of the deployment. Mean, standard error, sample size, and minimum and maximum values for each daily distance measurement (RawDist, CorrectedDist, and CorrectedDist_Act) were computed for each unique combination of categorical grouping (Placement × Date).

The daily distance measurements (RawDist, CorrectedDist, and CorrectedDist_Act) were estimated using SAS 9.3 (SAS Institute, Cary, NC, USA). The MIXED procedure with a 'covtest' statement was used to analyze distances via one-way ANOVA comparing treatments of trackers' placement either inside a building (Indoor), on the field (Outdoor), or mounted on mature cows (Animal) grazing pastures at the research site (Ho: μIndoor = μOutdoor = μAnimal). The tracker's ID (n = 11), a categorical classification for each tracker ID and placement combination (n = 17), in addition to the dates (n = 14) of deployment, were modeled as random variables. Means were computed and compared via LSMEANS statement for each daily distance measurement (RawDist, CorrectedDist, and CorrectedDist_Act) between the treatments of tracker placement (Indoor, Outdoor, or Animal), with a 'pdiff' statement for pairwise comparison. In addition, t-tests were conducted within each model to determine if the daily distance metrics (RawDist, CorrectedDist, and CorrectedDist_Act) calculated for each tracker placement (Indoor, Outdoor, or Animal) differed from zero (Ho: μ = 0). Lastly, using the estimate statement, a comparison of the means between the static state (Indoor and Outdoor) and non-static trackers (Animal) was computed (Ho: μStatic = μnon-Static). For all procedures, differences were declared statistically detectable at $p \leq 0.05$.

3. Results

The descriptive analysis revealed variability in the accuracy of each daily distance measurement (RawDist, CorrectedDist, and CorrectedDist_Act) in combination with the treatment of the tracker's placement (Indoor, Outdoor, or Animal). Overall, the RawDist measurement computed using the indoor trackers had the highest means and standard error. In contrast, the daily distance calculated using CorrectedDist_Act for the outdoor trackers had the lowest means and standard error. However, animal trackers exhibited the least variability in the means when compared between the three daily distance measurements (RawDist, CorrectedDist, and CorrectedDist_Act) (Figure 3).

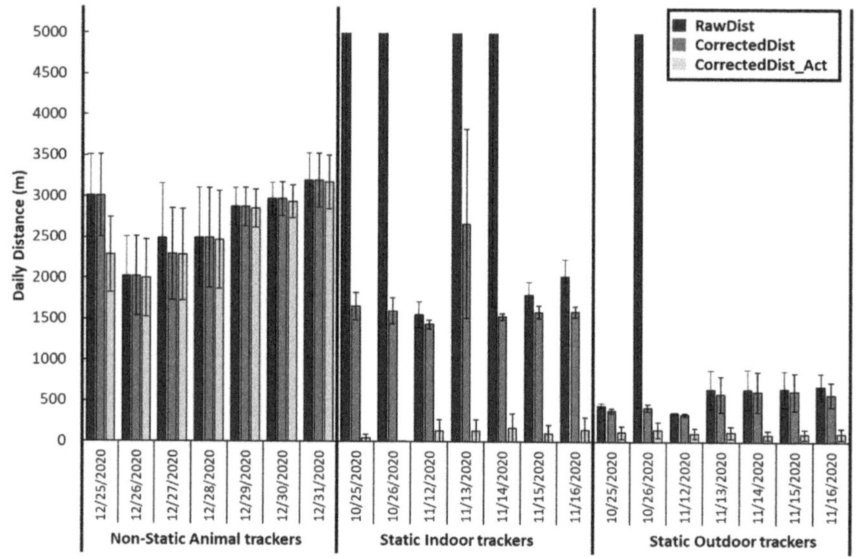

Figure 3. Mean daily distance with standard error bars calculated using three algorithms (RawDist, CorrectedDist, and CorrectedDist_Act), categorized by tracker placement (Animal, Indoor, or Outdoor) and the calendar date of deployment.

The results from our analysis indicated no significant relationship between RawDist measurement and the treatment of tracker placement ($p = 0.23$) when the assumption of normal distribution was violated (outliers detected) (Table 1). Due to the low precision, the means of the treatments (Indoor, Outdoor, or Animal) were not different from zero. In addition, there was no effect on the estimate comparison between static and non-static groupings of the means.

Table 1. Least square means ± standard errors for a daily distance calculated using three algorithms: (RawDist, CorrectedDist, and CorrectedDist_Act). Statistical analysis involved comparing distance computations across tracker placements (letters) for non-static vs. static trackers (μ non-static = μ static), and assessing the differences of means from zero (μ = 0).

Distance (m)	State	Placement	LSmeans ±Std Err		*p*-Value μ Non-Static = μ Static	μ = 0
RawDist *	non-Static	Animal	2724 ± 63,475	a	0.33	0.97
	Static	Indoor	150,644 ± 69,534	a		0.03
	Static	Outdoor	11,364 ± 66,543	a		0.86
CorrectedDist	non-Static	Animal	2695 ± 193	a	<0.01	<0.01
	Static	Indoor	1725 ± 211	b		<0.01
	Static	Outdoor	385 ± 202	c		0.06
CorrectedDist_Act	non-Static	Animal	2574 ± 186	a	<0.01	<0.01
	Static	Indoor	42 ± 203	b		0.84
	Static	Outdoor	170 ± 193	b		0.38

* Violation of statistical model assumptions associated with outlier presence.

The CorrectedDist measurement revealed a significance of treatment ($p < 0.01$), with animal trackers covering a greater daily distance than either indoor or outdoor placement (Table 1). However, indoor and outdoor means were different from each other ($p < 0.01$). In addition, indoor and animal placement trackers differed statistically from zero (Ho: μ = 0;

$p_{\text{Indoor}} < 0.01$, $p_{\text{Outdoor}} = 0.06$, $p_{\text{Animal}} < 0.01$). However, the estimate statement indicated a significant difference ($p < 0.01$) between the static and non-static groupings of the means.

With the last method of calculating the distance, the CorrectedDist_Act measurement was also affected by tracker placement ($p < 0.01$). Animal trackers covered a greater daily distance compared to either indoor or outdoor placement. In addition, indoor and outdoor means were not different from each other ($p < 0.62$). Only the animal trackers were significantly different from zero (Ho $\mu = 0$; $p_{\text{Indoor}} = 0.84$; $p_{\text{Outdoor}} = 0.38$; $p_{\text{Animal}} < 0.01$) (Table 1). The estimate statement also indicated a significant difference ($p < 0.01$) in the means grouping of Static vs. non-Static trackers.

4. Discussion

To achieve user-friendly PLF/R systems, data flow must be near real time for a farmer or rancher to monitor an individual animal's health, welfare, and yields [17,20]. However, it is worth noting that the development of the PLF/R platform depends on the advancement of precision and accuracy within sensors used in the IoT ecosystem. This is because the PLF/R system is data-driven, emphasizing the need for high-quality data [35]. While the future of PLF/R lies in data-driven techniques like machine learning (ML) and deep learning (DL) to identify data patterns, it is equally important not to overlook the value of mechanistic modeling approaches based on the conceptual understanding of system dynamics (hypothesis-centered) [18,36]. While robust data-driven approaches (ML and DL) often lack transparency in their predictions, mechanistic models use animal performance parameters for prediction. Therefore, a hybridized approach that integrates both data-driven and mechanistic modeling methodologies is warranted for enhanced PLF/R outcomes [18,36].

The distance traveled is a standard metric calculated from GPS coordinates with significant applications in research and management decision-making processes [3,4,26]. The objectives of this study were to test the reliability of the daily distance traveled metric calculation using three algorithms (GPS data alone or in combination with motion data) for detecting the differences between static trackers placed either outdoors or indoors with an obstructed view of orbiting satellites vs. trackers mounted on cows grazing on pasture. We hypothesized that the daily distance traveled metrics would be higher for the animal-mounted trackers than the static trackers and that there would be no difference between the two types of static trackers (Outdoor = Indoor), and that the daily distance traveled by the static trackers (Outdoor or Indoor) would not differ from zero.

4.1. GPS Accuracy Measurement on Static Trackers

An analysis of GPS data accuracy in this study showed that there were infrequent erroneous GPS locations that worsened the distance computational means output (RawDist) when they were integrated into the model. A related study that utilized GPS data from static indoor and outdoor trackers found that outdoor trackers had greater accuracy in remote sensing of GPS locations. This is because indoor trackers experience increased communication interference in acquiring satellite signals. The study's findings indicated that 95% of GPS data points fell within a radius of 15 and 40 m for outdoor and indoor trackers, respectively [33]. The study concluded that these trackers showed comparable accuracy to other devices available in the market, with a position bias calculated by excluding erroneous GPS positions using the Euclidean distance between the tracker's actual location (Stationary) and the projected GPS points averaging 5.20 and 17.76 m for outdoor and indoor trackers, respectively [33].

In pasturelands with a canopy cover, obstructed GPS data acquisition is common. For instance, previous studies collaring goats herded on rugged terrain, and those that calculated horizontal accuracy between trackers placed in the open field and under canopy cover both reported low accuracy for trackers with obstructed views of orbiting satellites [29,31]. Therefore, it is essential to include a pre-processing phase to detect outliers in GPS data before any further analysis [33,37].

4.2. Daily Distance Traveled Calculated Using RawDist

The computation of distance traveled using raw GPS data (RawDist) did not support our first hypothesis, as there was no detectable difference between the tracker placements. Additionally, statistical model assumptions associated with outliers were violated, so caution is necessary when interpreting these results. This underscores the need for internal algorithms to filter out extreme GPS outliers. RawDist had the highest numerical value compared to the other two computations (CorrectedDist and CorrectedDist_Act) used to calculate the distance traveled. McGavin et al. [27] also found that including GPS data with low accuracy increased the calculated distance. Furthermore, the RawDist measurement failed to support our second hypothesis, as the indoor trackers differed from zero, and there was no difference in the comparison between the static and non-static trackers. This failure of RawDist measurement suggests that this algorithm has low accuracy in detecting animal welfare implications associated with immobility.

4.3. Daily Distance Traveled Calculated Using CorrectedDist

The corresponding analysis for screening erroneous GPS positions using the z-score, as proposed by Nyamuryekung'e et al. [33], improved the normal distribution of the daily distance metric (CorrectedDist), leading to partial support for our first hypothesis, which predicted that the Animal tracker would cover a greater daily distance than the indoor and outdoor trackers. However, indoor and outdoor trackers differed from each other. CorrectedDist also failed our second hypothesis, as the distance means of the static trackers differed from zero. Due to the random errors in GPS positioning equipment, the static trackers registered a significant daily distance measurement that was different from zero, making the analysis sensitive to inflated error distance measurements in situations with a high temporal frequency of GPS acquisition and low GPS location accuracy. Ganskopp and Johnson [32] and McGavin et al. [27] also found a correlation between short GPS sampling intervals and overestimated distance calculation. However, CorrectedDist was able to detect differences between static and non-static trackers. Therefore, CorrectedDist has some utility for use in animal welfare metrics to detect differences between static and non-static states, but it might fail to statistically detect a zero daily distance measurement on static trackers.

4.4. Daily Distance Traveled Calculated Using CorrectedDist_Act

The final model, which excluded GPS data with erroneous locations and GPS data associated with no motion from the accelerometer reading (CorrectedDist_Act), supported our first hypothesis, showing that the Animal tracker covered a greater daily distance than both indoor and outdoor trackers, which did not differ from each other. Moreover, CorrectedDist_Act also supported our second hypothesis, with the daily distance covered by static trackers not differing from zero. As the static trackers were mostly stationary, activity data was almost non-existent, with only three trackers contributing to the daily distance (Indoor n = 1 and Outdoor n = 2). CorrectedDist_Act also detected the difference between static and non-static trackers. Therefore, CorrectedDist_Act is the recommended metric for daily distance measurement, with acceptable accuracy in detecting animal welfare implications associated with immobility.

4.5. Limitations of the Daily Distance Traveled Calculations

The selection of the GPS frequency acquisition has significant implications for calculating the distance traveled by animals. A high GPS frequency acquisition can lead to an overestimation of the distance traveled due to the inclusion of GPS positions with inherent position bias, as well as increased battery drainage of the trackers [4,32]. On the other hand, low GPS frequency acquisition may result in underestimating the distance traveled, as it may miss the sinuosity of the actual path taken by the animals [38]. Achieving a balance in selecting the GPS frequency acquisition is crucial to avoid both overestimation and underestimation of the true distance, and it is essential to consider the spatial extent to which the animals are operating. The chosen interval for GPS frequency acquisition in this study is believed to adequately represent the distance traveled by animals in extension operations [3].

Improvements in CorrectedDist_Act can be achieved by calibrating the triaxial accelerometer data to represent the animal's time budget, thus improving the accuracy of distance traveled estimation from GPS data by excluding a resting activity from the analysis [32]. Studies have shown that including resting and inactive GPS positions when calculating the distance traveled can artificially increase it by approximately 15.2% [32]. However, it is important to note that the purpose of this experiment did not involve time-budget calibration of the accelerometer data. The 1 min interval motion intensity measurements in this study revealed clear diurnal patterns consistent with grazing animals' behavior [33]. For example, intense grazing events are typically observed around dusk, which was evident from the accelerometer data with a peak of intensity around 1800 h [33]. Future analysis will explore decoding the activity messages into time budgets for the animals [39,40]. Additionally, accurate classification of the activity data can further enhance the filtering of erroneous GPS data, as suggested by Muminov et al. [37], who used maximum animal movement likelihood criteria based on activity classification to filter out erroneous GPS data.

5. Conclusions

The user interface or dashboard application is arguably the most valuable component in the Internet of Things (IoT) ecosystem, particularly in a Precision Livestock Farming and Ranching (PLF/R) platform. This is because the data collected by the sensory devices are transformed into information tailored to a specific end-user through the user interface platform. In a PLF/R system, the dashboard application assists ranchers in decision-making processes. Hence, the metrics presented in the dashboard application must possess both high accuracy and precision. These findings emphasize the necessary sensitivity required to develop bio-sensing algorithms that can alert managers of animal welfare concerns. Similarly, the CorrectedDist_Act model emphasizes the importance of combining GPS and accelerometer data when calculating the walking distance of grazing cattle. Furthermore, the results highlight the value of integrating multiple sources of independent sensor data for an improved interpretation of data derived from PLF/R tools.

Author Contributions: Conceptualization, S.N., S.U. and A.F.C.; methodology, S.N. and S.U.; software, S.N. and H.C.; validation, G.D., S.U., R.E., M.M.M., M.F., A.C., H.C., S.S., A.P. and A.F.C.; formal analysis, S.N.; investigation, S.N., M.F. and A.P.; resources, G.D., S.U., R.E., A.C., S.S. and A.F.C.; data curation, S.N. and H.C.; writing—original draft preparation, S.N.; writing—review and editing, G.D., S.U., R.E., M.M.M., M.F., A.C., H.C., S.S., A.P. and A.F.C.; visualization, S.N, and M.M.M.; supervision, S.U. and A.F.C. All authors have read and agreed to the published version of the manuscript.

Funding: This research was funded by the USDA-ARS Jornada Experimental Range Long-Term Agroecosystem Research Program (LTAR), USDA National Institute of Food and Agriculture, Hatch project 1000985 (A. Cibils), and USDA National Institute of Food and Agriculture Sustainable Agriculture Systems Coordinated Agricultural Project grant # 12726269. Partial support was provided by the Norwegian Institute of Bioeconomy Research (NIBIO).

Institutional Review Board Statement: The animal study protocol was approved by the Institutional Review Board (or Ethics Committee) of New Mexico State University Institutional Animal Care and Use Committee (protocol code 2019-008, 15 March 2019).

Informed Consent Statement: Not applicable.

Data Availability Statement: The raw data concerning the LoRa WAN sensors showcased in this study are available on request from the corresponding authors.

Acknowledgments: The authors extend their sincere gratitude to the management teams of the Clayton Livestock Research Center (CLRC) and the Chihuahuan Desert Rangeland Research Center (CDRRC) for their invaluable support throughout the experiment. Special thanks are also due to the U.S. Precision Livestock Farming 2023 committee members for their generous scholarship and for selecting the research to be featured in the special issue. Additionally, the authors express their appreciation to the editors and anonymous reviewers whose valuable feedback greatly contributed to enhancing the final version of the manuscript.

Conflicts of Interest: The authors declare no conflict of interest. The funders had no role in the design of the study; in the collection, analyses, or interpretation of data; in the writing of the manuscript; or in the decision to publish the results. Furthermore, references to specific trade names, proprietary products, or vendors do not imply any endorsement or warranty of the product by the USDA, nor does it indicate an exclusive approval over other products or vendors that could be equally suitable.

References

1. Gregorini, P. Diurnal Grazing Pattern: Its Physiological Basis and Strategic Management. *Anim. Prod. Sci.* **2012**, *52*, 416–430. [CrossRef]
2. Manning, J.K.; Cronin, G.M.; González, L.A.; Hall, E.J.S.; Merchant, A.; Ingram, L.J. The Effects of Global Navigation Satellite System (GNSS) Collars on Cattle (*Bos taurus*) Behaviour. *Appl. Anim. Behav. Sci.* **2017**, *187*, 54–59. [CrossRef]
3. Nyamuryekung'e, S.; Cibils, A.F.; Estell, R.E.; Vanleeuwen, D.; Spiegal, S.; Steele, C.; Gonz, A.L.; Mcintosh, M.M.; Gong, Q. Movement, Activity, and Landscape Use Patterns of Heritage and Commercial Beef Cows Grazing Chihuahuan Desert Rangeland. *J. Arid Environ.* **2022**, *199*, 104704. [CrossRef]
4. Bailey, D.W.; Trotter, M.G.; Knight, C.W.; Thomas, M.G. Use of GPS Tracking Collars and Accelerometers for Rangeland Livestock Production Research. *Transl. Anim. Sci.* **2018**, *2*, 81–88. [CrossRef]
5. Nyamuryekung'e, S.; Cibilis, A.F.; Estell, R.E.; McIntosh, M.; VanLeeuwen, D.; Steele, C.; Gonzalez, A.L.; Spiegal, S.; Avedaño-Reyes, L.; Rodríguez-Almeida, F.A.; et al. Foraging Behavior and Body Temperature of Heritage vs. Commercial Beef Cows in Relation Desert Ambient Heat. *J. Arid Environ.* **2021**, *193*, 104565. [CrossRef]
6. Clark, P.E.; Johnson, D.E.; Larson, L.L.; Louhaichi, M.; Roland, T.; Williams, J. Effects of Wolf Presence on Daily Travel Distance of Range Cattle. *Rangel. Ecol. Manag.* **2017**, *70*, 657–665. [CrossRef]
7. Brosh, A.; Henkin, Z.; Ungar, E.D.; Dolev, A.; Orlov, A.; Yehuda, Y.; Aharoni, Y. Energy Cost of Cows' Grazing Activity: Use of the Heart Rate Method and the Global Positioning System for Direct Field Estimation. *J. Anim. Sci.* **2006**, *84*, 1951–1967. [CrossRef]
8. Augustin, A.; Yi, J.; Clausen, T.; Townsley, W.M. A Study of Lora: Long Range & Low Power Networks for the Internet of Things. *Sensors* **2016**, *16*, 1466. [CrossRef]
9. Sanchez-Iborra, R.; Cano, M.D. State of the Art in LP-WAN Solutions for Industrial IoT Services. *Sensors* **2016**, *16*, 708. [CrossRef]
10. Germani, L.; Mecarelli, V.; Baruffa, G.; Rugini, L.; Frescura, F. An IoT Architecture for Continuous Livestock Monitoring Using Lora LPWAN. *Electronics* **2019**, *8*, 1435. [CrossRef]
11. McIntosh, M.; Cibils, A.; Nyamuryekung'e, S.; Estell, R.; Cox, A.; Duni, D.; Gong, Q.; Waterhouse, T.; Holland, J.; Cao, H.; et al. Deployment of a LoRa-WAN near Real-Time Precision Ranching System on Extensive Desert Rangelands: What We Have Learned. *Appl. Anim. Sci.* **2023**, *in press*.
12. Ayaz, M.; Ammad-Uddin, M.; Sharif, Z.; Mansour, A.; Aggoune, E.H.M. Internet-of-Things (IoT)-Based Smart Agriculture: Toward Making the Fields Talk. *IEEE Access* **2019**, *7*, 129551–129583. [CrossRef]
13. dos Reis, B.R.; Easton, Z.; Fuka, D. A LoRa Sensor Network for Monitoring Pastured Livestock Location and Activity. *Transl. Anim. Sci.* **2021**, *5*, txab010. [CrossRef] [PubMed]
14. Madakam, S.; Ramaswamy, R.; Tripathi, S. Internet of Things (IoT): A Literature Review. *J. Comput. Commun.* **2015**, *3*, 164–173. [CrossRef]
15. Navarro, E.; Costa, N.; Pereira, A. A Systematic Review of Iot Solutions for Smart Farming. *Sensors* **2020**, *20*, 4231. [CrossRef] [PubMed]
16. Nyamuryekung'e, S. Re-Imagining Rangeland Livestock Production in Response to the Evolving Demands of a Fast-Paced, Changing World. *Rangelands* 2023, *submitted*.
17. Bailey, D.W.; Trotter, M.G.; Tobin, C.; Thomas, M.G. Opportunities to Apply Precision Livestock Management on Rangelands. *Front. Sustain. Food Syst.* **2021**, *5*, 611915. [CrossRef]
18. Tedeschi, L.O.; Greenwood, P.L.; Halachmi, I.; Lezion, R. Advancements in Sensor Technology and Decision Support Intelligent Tools to Assist Smart Livestock Farming. *J. Anim. Sci.* **2021**, *99*, skab038. [CrossRef]
19. Greenwood, P.L.; Bishop-Hurley, G.J.; González, L.A.; Ingham, A.B. Development and Application of a Livestock Phenomics Platform to Enhance Productivity and Efficiency at Pasture. *Anim. Prod. Sci.* **2016**, *56*, 1299–1311. [CrossRef]
20. Berckmans, D. Precision Livestock Farming Technologies for Welfare Management in Intensive Livestock Systems. *OIE Rev. Sci. Tech.* **2014**, *33*, 189–196. [CrossRef]
21. Britt, J.H.; Cushman, R.A.; Dechow, C.D.; Dobson, H.; Humblot, P.; Hutjens, M.F.; Jones, G.A.; Ruegg, P.S.; Sheldon, I.M.; Stevenson, J.S. Invited Review: Learning from the Future—A Vision for Dairy Farms and Cows in 2067. *J. Dairy Sci.* **2018**, *101*, 3722–3741. [CrossRef]
22. Halachmi, I.; Guarino, M.; Bewley, J.; Pastell, M. Smart Animal Agriculture: Application of Real-Time Sensors to Improve Animal Well-Being and Production. *Annu. Rev. Anim. Biosci.* **2019**, *7*, 403–425. [CrossRef]

23. Rutten, C.J.; Velthuis, A.G.J.; Steeneveld, W.; Hogeveen, H. Invited Review: Sensors to Support Health Management on Dairy Farms Invited Review: Sensors to Support Health Management on Dairy Farms. *J. Dairy Sci.* **2013**, *96*, 1928–1952. [CrossRef]
24. Bracke, M.B.M.; Hopster, H. Assessing the Importance of Natural Behavior for Animal Welfare. *J. Agric. Environ. Ethics* **2006**, *19*, 77–89. [CrossRef]
25. Stachowicz, J.; Umstätter, C. Do We Automatically Detect Health- or General Welfare-Related Issues? A Framework. *Proc. R. Soc. B Biol. Sci.* **2021**, *288*, 20210190. [CrossRef]
26. McIntosh, M.M.; Cibils, A.F.; Estell, R.E.; Gong, Q.; Cao, H.; Gonzalez, A.L.; Nyamuryekung'e, S.; Spiegal, S.A. Can Cattle Geolocation Data Yield Behavior-Based Criteria to Inform Precision Grazing Systems on Rangeland? *Livest Sci.* **2022**, *255*, 104801. [CrossRef]
27. McGavin, S.L.; Bishop-Hurley, G.J.; Charmley, E.; Greenwood, P.L.; Callaghan, M.J. Effect of GPS Sample Interval and Paddock Size on Estimates of Distance Travelled by Grazing Cattle in Rangeland, Australia. *Rangel. J.* **2018**, *40*, 55–64. [CrossRef]
28. Ungar, E.D.; Henkin, Z.; Gutman, M.; Dolev, A.; Genizi, A.; Ganskopp, D. Inference of Animal Activity From GPS Collar Data on Free-Ranging Cattle. *Rangel. Ecol. Manag.* **2005**, *58*, 256–266. [CrossRef]
29. Agouridis, C.T.; Stombaugh, T.S.; Workman, S.R.; Koostra, B.K.; Edwards, D.R.; Vanzant, E.S. Suitability of a GPS Collar for Grazing Studies. *Trans. Am. Soc. Agric. Eng.* **2004**, *47*, 1321–1329. [CrossRef]
30. Turner, L.W.; Udal, M.C.; Larson, B.T.; Shearer, S.A. Monitoring Cattle Behavior and Pasture Use with GPS and GIS. *Can. J. Anim. Sci.* **2000**, *80*, 405–413. [CrossRef]
31. Buerkert, A.; Schlecht, E. Performance of Three GPS Collars to Monitor Goats' Grazing Itineraries on Mountain Pastures. *Comput. Electron. Agric.* **2009**, *65*, 85–92. [CrossRef]
32. Ganskopp, D.C.; Johnson, D.D. GPS Error in Studies Addressing Animal Movements and Activities. *Rangel. Ecol. Manag.* **2007**, *60*, 350–358. [CrossRef]
33. Nyamuryekung'e, S.; Cibils, A.F.; Estell, R.E.; Funk, M.; Mcintosh, M.M.; Cox, A.; Utsumi, S.A.; Cao, H.; Boucheron, L.; Gong, Q.; et al. Performance of Lora-WAN Sensors for Precision Livestock Tracking and Biosensing Applications. In Proceedings of the XXIV International Grassland Congress/XI International Rangeland Congress, Nairobi, Kenya, 25–29 October 2021; National Organizing Committee of 2021 IGC/IRC Congress, Ed.; Kenya Agricultural and Livestock Research Organization: Nairobi, Kenya, 2021.
34. Nyamuryekung'e, S.; Cibils, A.F.; Estell, R.E.R.E.; VanLeeuwen, D.; Steele, C.; Roacho-Estrada, J.O.; Rodriguez Almeida, F.; Gonzalez, A.L.; Spiegal, S.; Nyamuryekung'e, S.; et al. Do Young Calves Constrain Movement Patterns of Nursing Raramuri Criollo Cows on Rangeland? *Rangel. Ecol. Manag.* **2020**, *73*, 84–92. [CrossRef]
35. Akhigbe, I.; Munir, K.; Akinade, O.; Akanbi, L.; Oyedele, L.O. Iot Technologies for Livestock Management: A Review of Present Status, Opportunities, and Future Trends Bernard. *Big Data Cogn. Comput.* **2021**, *5*, 10. [CrossRef]
36. Ellis, J.L.; Jacobs, M.; Dijkstra, J.; van Laar, H.; Cant, J.P.; Tulpan, D.; Ferguson, N. Review: Synergy between Mechanistic Modelling and Data-Driven Models for Modern Animal Production Systems in the Era of Big Data. *Animal* **2020**, *14*, 223–237. [CrossRef]
37. Muminov, A.; Sattarov, O.; Lee, C.W.; Kang, H.K.; Ko, M.C.; Oh, R.; Ahn, J.; Oh, H.J.; Jeon, H.S. Reducing GPS Error for Smart Collars Based on Animal's Behavior. *Appl. Sci.* **2019**, *9*, 3408. [CrossRef]
38. Johnson, D.D.; Ganskopp, D.C. GPS Collar Sampling Frequency: Effects on Measures of Resource Use. *Rangel. Ecol. Manag.* **2008**, *61*, 226–231. [CrossRef]
39. Fogarty, E.S.; Swain, D.L.; Cronin, G.M.; Moraes, L.E.; Trotter, M. Behaviour Classification of Extensively Grazed Sheep Using Machine Learning. *Comput. Electron. Agric.* **2020**, *169*, 105175. [CrossRef]
40. Hosseininoorbin, S.; Layeghy, S.; Kusy, B.; Jurdak, R.; Bishop-Hurley, G.J.; Greenwood, P.L.; Portmann, M. Deep Learning-Based Cattle Behaviour Classification Using Joint Time-Frequency Data Representation. *Comput. Electron. Agric.* **2021**, *187*, 106241. [CrossRef]

animals

MDPI

Article

Estimating the Feeding Time of Individual Broilers via Convolutional Neural Network and Image Processing

Amin Nasiri [1], Ahmad Amirivojdan [1], Yang Zhao [2] and Hao Gan [1,*]

[1] Department of Biosystems Engineering and Soil Science, University of Tennessee, Knoxville, TN 37996, USA; anasiri@utk.edu (A.N.); aamirivo@vols.utk.edu (A.A.)
[2] Department of Animal Science, University of Tennessee, Knoxville, TN 37996, USA; yzhao@utk.edu
* Correspondence: hgan1@utk.edu

Simple Summary: Using automated approaches to investigate feeding behavior in broilers provides accurate, non-invasive, and large-scale data collection, real-time monitoring capabilities, and opportunities for advanced data analysis that would not be possible with manual observations. These benefits contribute to a better understanding of broilers' behavior for improving production efficiency and animal welfare, optimizing management practices, and promoting the profitability of poultry production. Hence, this study aimed to estimate the feeding time of individual broilers through an automated approach. First, the proposed algorithm detected the broilers' heads. Then, a Euclidean distance-based tracking algorithm tracked the detected heads. The developed algorithm can estimate the broiler's feeding time by identifying whether its head is inside the feeder area. The overall accuracy of each broiler's feeding time per visit to the feeding pan was 87.3%.

Abstract: Feeding behavior is one of the critical welfare indicators of broilers. Hence, understanding feeding behavior can provide important information regarding the usage of poultry resources and insights into farm management. Monitoring poultry behaviors is typically performed based on visual human observation. Despite the successful applications of this method, its implementation in large poultry farms takes time and effort. Thus, there is a need for automated approaches to overcome these challenges. Consequently, this study aimed to evaluate the feeding time of individual broilers by a convolutional neural network-based model. To achieve the goal of this research, 1500 images collected from a poultry farm were labeled for training the You Only Look Once (YOLO) model to detect the broilers' heads. A Euclidean distance-based tracking algorithm was developed to track the detected heads, as well. The developed algorithm estimated the broiler's feeding time by recognizing whether its head is inside the feeder. Three 1-min labeled videos were applied to evaluate the proposed algorithm's performance. The algorithm achieved an overall feeding time estimation accuracy of each broiler per visit to the feeding pan of 87.3%. In addition, the obtained results prove that the proposed algorithm can be used as a real-time tool in poultry farms.

Keywords: broiler; feeding time; YOLO; image processing

Citation: Nasiri, A.; Amirivojdan, A.; Zhao, Y.; Gan, H. Estimating the Feeding Time of Individual Broilers via Convolutional Neural Network and Image Processing. *Animals* **2023**, *13*, 2428. https://doi.org/10.3390/ani13152428

Academic Editor: Vincenzo Tufarelli

Received: 30 June 2023
Revised: 21 July 2023
Accepted: 25 July 2023
Published: 27 July 2023

1. Introduction

Monitoring broiler behaviors such as feeding, drinking, and perching is a crucial aspect of precision livestock farming to reflect their health status and provide early disease warning. In this regard, the feeding behavior of broilers plays a critical function in the breeding process. Deviation from the daily food consumption pattern is the early disease symptom. Thus, understanding poultry feeding behavior helps evaluate their use of feed resources, improve their health status, and provide vital economic and welfare implications for poultry production. As a result, new techniques are needed to extract the feeding behavior of broilers that are useful in warning about their health status and improving the breeding process.

Several studies have been conducted to investigate the feeding behavior of poultry under the influence of various environmental stimuli, management practices, and breeding systems. These studies often monitor poultry behaviors by manual observation or remotely [1]. Manual observation is an accurate and simple method to analyze the behavior of small samples and limited behavioral responses. However, manual observation is time-consuming and laborious, especially for large farms or monitoring multiple behaviors simultaneously. Therefore, it is necessary to develop automated methods to handle large sample sizes and multiple behaviors.

Broilers' behavior on the experimental or commercial farm can be examined by applying automated techniques (e.g., audio analysis, radio-frequency identification (RFID) devices, and image processing) to analyze the health and welfare of the poultry. For example, audio analysis can be applied as a poultry behavior-based early warning system, detecting growth rates and predicting health conditions. Collecting and analyzing the individual chickens' sounds can distinguish them from each other. In this respect, detecting which broiler makes the sound, collecting individual sounds on a farm, and ambient noise are significant challenges [2]. Wireless wearable sensors such as accelerometers and RFID microchips are primarily used to track people's location and activity remotely. The performance of the RFID system depends on the number of broilers and installed antennas. Due to the large number of broilers on commercial farms and the sensor cost in relation to the value of the individual bird, attaching an RFID device to every broiler is currently unrealistic. In addition, RFID systems can be employed for a limited number of broilers due to the time-consuming tasks of installing and recycling tags [2].

The image processing technique is an efficient, non-invasive, and cost-effective method for analyzing animal behavior. This technology includes image-capturing systems and various algorithms to recognize the behavior. Kashiha et al. [3] investigated real-time broiler distribution indicators using image-processing methods. These authors reported unusual feeding and drinking behaviors in broilers with 95% accuracy. Also, Nasiri et al. [4] proposed a computer vision-based system to recognize lameness in broilers. Despite using image processing techniques to analyze particular poultry behaviors, there needs to be more research on their application in investigating broiler feeding behaviors.

As a specialized version of image processing, video monitoring is a low-cost and straightforward method to detect feeding behavior. In this regard, effectively extracting information from surveillance videos is a fundamental problem. For each frame, the first step is to distinguish the broilers from the background. Segmentation thresholding based on histogram analysis, Otsu segmentation, the maximum entropy segmentation method, and multi-level threshold segmentation are among the methods applied to detect objects from the background [3,5,6]. Their performance depends on the differentiation between broilers from various background conditions and varying light conditions. Furthermore, it is difficult to distinguish them individually when broilers are huddled together. Each broiler can be detected in the video sequence by creating specific marks on every broiler and using pattern recognition techniques. Therefore, the automatic recognition of broilers through video footages is a fundamental issue, and selecting the appropriate feature is crucial.

The performance of image processing technology can be improved using machine learning methods. Valletta et al. [7] utilized PCA to extract pheasant egg characteristics and k-means clustering to identify individual pheasant eggs. Kasinathan et al. [8] used shape features extracted from various categories of insect images and machine learning algorithms, including artificial neural networks, support vector machine, k-nearest neighbors, naive Bayes, and convolutional neural networks, to classify the insect classes. Another study proposed an image processing and machine learning framework for leaf disease detection. In this framework, the k-means and principal component analysis approaches were applied to segment and extract the features from leaf images to evaluate the disease. Then, images were classified using RBF-SVM, SVM, random forest, and ID3 techniques [9]. Bai et al. [10] introduced a vision-based algorithm for picking point localization of tomatoes. The proposed algorithm extracted the shape, texture, and color features and applied the SVM classifier to recognize tomatoes. In this regard, deep learning algorithms, as a type of machine learning method, have been used to identify and classify objects in many applications, including animal identification and behavior recognition. For example, Chen et al. [11] proposed a method to define the sex of pigeons by enhancing images using image processing techniques combined with YOLO-v5. In another study, the YOLO-v5 structure-based method was developed to monitor cage-free hens' spatial distribution, including the number of birds in the perching, feeding, drinking, and nesting zones [12]. Li et al. [13] developed a Faster R-CNN-based algorithm to detect and track birds walking around the feeder as a feeding behavior indication. Guo et al. [14] evaluated the performance of different deep learning models such as ResNet, EfficientNet, and DenseNet to identify four broiler behaviors: (1) resting, (2) standing, (3) feeding, and (4) drinking. In their study, the accuracy of the best network for broiler behavior classification was 97%.

Monitoring broilers' feeding behaviors can help ensure an appropriate diet to support their growth. Additionally, observing feeding behaviors can provide insights into the health and welfare of broilers. For instance, any changes in appetite may reflect underlying health issues. As a result, the feeding assessment of individual broilers assists in implementing proper nutritional management strategies. Hence, monitoring the feeding behaviors of individual broilers is crucial for precision livestock farming, intending to ensure optimal growth and maintain health and welfare. Accordingly, the objective of the study was to develop an algorithm based on image processing techniques and a deep learning model with the aim of recognizing the individual feeding times of broilers. To the extent of the authors' knowledge, the present study was one of the first efforts to investigate the individual broilers' feeding time on a commercial farm. The performance of the developed algorithm was validated by comparing the achieved results with the manual observations in the labeled videos.

2. Materials and Methods

2.1. Acquisition of Broilers' Video

Data were collected at a commercial-scale broiler research farm (Tyson Foods, Inc., Huntsville, AR, USA) with 20,000 birds and a stocking density of 12.2 birds/m^2. A total of 12 surveillance cameras (Dahua Technology USA Inc., Irvine, CA, USA) were installed on the ceiling of the farm (approximately 3 m above ground) to collect RGB videos with a set speed of 15 frames per second (fps). The cameras were fixed and distributed uniformly along the two drinker lines on the farm. In addition, four feeder pans were also in the cameras' field of view. Videos were collected 24/7 for the entire cycle of several flocks in 2022.

2.2. Head Detection Model and Data Collection

Animals' feeding behavior can be chewing, biting, or putting the head in the feeder (named a feeding visit). Since it is difficult to tell whether broilers are chewing or biting, a feeding visit can be commonly decided by checking whether the broiler's head is in the feeder. Accordingly, this study defines broiler feeding behavior as broilers placing their heads in the feeder area. In this process, the feeder occupation time by the broiler's head is calculated as feeding time. Therefore, for each frame, it is obligatory to detect the broiler's head as a factor closely related to the detection of feeding behavior. In this study, a regression-based algorithm was used to address the broilers' head detection issue.

You Only Look Once (YOLO), proposed in 2016, formulates the object detection problem as a regression problem [15]. Compared to two-stage detectors, YOLO is speedy. YOLO calculates the region of interest and image classes in one algorithm implementation. First, a neural network is processed on the whole image. Then, the image is divided into different cells, and the objects in each cell are projected [16,17]. YOLO divides the input image into a grid and predicts bounding boxes and class probabilities directly from the grid cells. This approach makes YOLO faster and more efficient. Each cell predicts objects that have their centers within that cell. On the other hand, YOLO predicts multiple bounding boxes for each grid cell using a set of anchor boxes. In recent years, researchers have been utilizing YOLO to identify individual animals and their behavioral patterns, which can result in a better investigation of animals' behavioral mechanisms [18–22].

The database used in this study was created by selecting sample frames from surveillance video sequences to train and test the network. A more diverse database can be achieved by selecting frames with different postures of broilers (e.g., standing, lying, sitting, and with different lighting conditions). A total of 1500 images were selected for manual labeling. Figure 1 demonstrates an example of the labeled image. Then, the head of each broiler was labeled in the text format expected for training the YOLO-v3 model. In this study, the transfer learning method was adopted to solve the problem of the insufficient number of samples during the training process. For training the pre-trained YOLO on the COCO dataset, the dataset images were randomly divided into training and validation subsets with a ratio of 9:1.

(a) (b)

Figure 1. An instance of the labeling process for training the YOLO; (**a**) original image and (**b**) labeled image.

2.3. Broiler's Head Tracking Algorithm

YOLO object detection does not treat objects in every video frame the same. In other words, each broiler detected in the video frames is assigned a unique tag/identifier. Thus, it is necessary to use the tracking algorithm to assign a constant identifier for each detected broiler and for it to have high efficiency in the later stages. In each frame, the broiler tracking algorithm calculates the central point of the bounding box around the broiler's head marked by the trained YOLO model. Then, the algorithm delivers the central point along with the specific identifier to an array that stores the characteristics of the detected broilers in the last 10 frames. In the next step, the Euclidean distance between the coordinates of the central point of the broiler's head identified in the current frame is measured with the coordinates of the center points of the previous frame. The broiler receives its last stored identifier if the calculated distance is less than 20 pixels. If the distance exceeds 20 pixels, the algorithm identifies a new broiler that will receive a new identifier. This process is carried out for all the identified chicks in the current frame. Also, the maximum number of missed detections before the algorithm removes a tracking label is set to 10 frames.

2.4. Algorithm for Estimation of the Feeding Time

The feeding behavior of broiler chickens is related to a particular area where the broiler's head is placed inside the feeder. In this study, there is a need to define the feeding area in the monitoring scene to investigate broilers' feeding time. This area was manually determined in each video. In Figure 2, the blue circle represents the feeding area. The trained YOLO recognizes and marks the broiler head with a bounding box. Furthermore, the tracking algorithm tracks detected heads as long as they are inside the red area (Figure 2). When the center point of the head-labeled bounding box intersects and enters the feeding area (blue circle), feeding behavior may occur. The following index was used to determine whether a specific object covered a location (Equation (1)).

$$Index_{feed} = \frac{Area\ of\ head \cap Feeding\ area}{Feeding\ area} \quad (1)$$

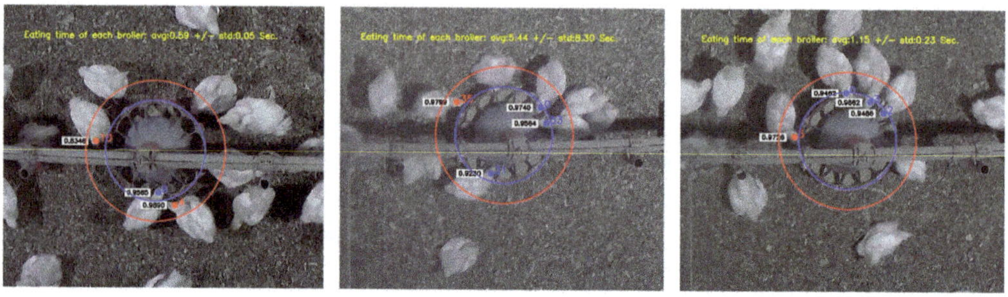

Figure 2. Samples of the developed algorithm's result. Broilers' heads with blue dots: feeding behavior; broilers' heads with red dots: non-feeding behavior.

An $Index_{feed}$ greater than 0 shows that the broiler has the head in the feeder. The feeding time of each broiler can be estimated by counting the number of frames that the broiler's head is inside the feeding area. Accordingly, the feeding behavior can be detected along with the feeding time. Figures 3 and 4 show the workflow and pseudocode of the developed algorithm for feeding time estimation.

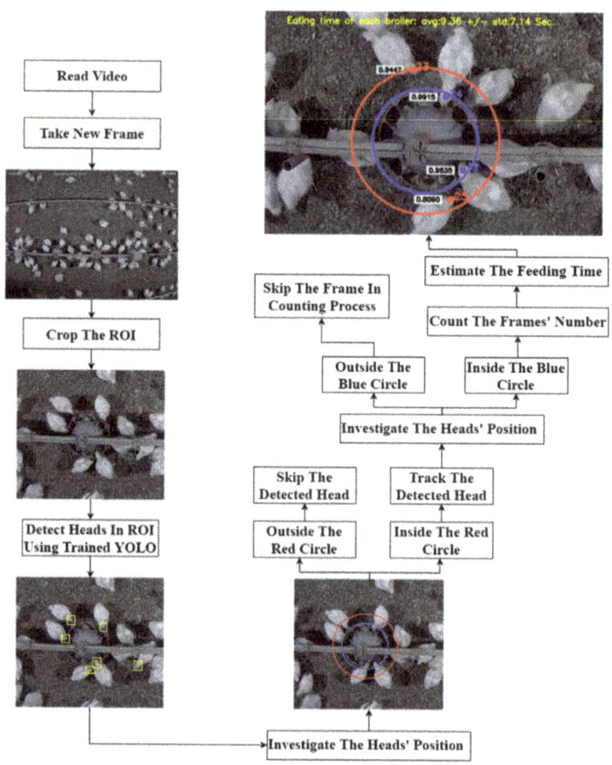

Figure 3. The workflow of the developed algorithm for feeding time estimation.

```
Input Data: Video
Result: Feeding time of each broiler (second)
Initialization;                                      # Load Head detection model (YOLO) & Video
Set HID as a dictionary;
    # HID: Head ID, save the broilers ID number(ID), Frame number that the head is detected(FN),
    head's center coordinates, and number of tracked frames(NT);
Set FT as an empty array;                            # FT: Feeding Time
Set BID as 0;                                        # BID: Broiler ID
Set FS;                                              # FS: The position of the feeder as a circle
Set FT;                                              # FT: a circle area for tracking
Set results as an empty array;
while True do
|   read new frame;
|   crop the ROI from the current frame;             # ROI: A rectangular area around the feeder
|   Detect heads in ROI as HDs;                      # HDs: Output of Head detection model
|   for HD in HDs do
|   |   calculate Cxy as the head's center coordinates;
|   |   # center coordinates of the drawn bounding box around the detected head
|   |   if Cxy is not inside FT then
|   |   |   continue
|   |   end if
|   |   Set SHD as False;                            # SHD: Same Head Detected
|   |   for items in HID do
|   |   |   calculate the Euclidean distance (Ed) between Cxy and the saved center points
|   |   |   coordinates;
|   |   |   if Ed < 20 then
|   |   |   |   Set SHD as True;
|   |   |   |   update head's center coordinates of tracked broiler and FN;
|   |   |   |   if Cxy is inside FS then;
|   |   |   |   |   Add 1 to NT of tracked broiler;
|   |   |   |   end if;
|   |   |   |   break for;
|   |   |   end if
|   |   end for
|   |   if SHD is False do
|   |   |   Generate a new ID;
|   |   |   Add the features of the new detected head to HID;
|   |   end if
|   ends for
|   do every 10 frames
|   |   for items in HID do
|   |   |   if the difference between the number of current frame and FN is 10 then
|   |   |   |   Append NF to results;
|   |   |   |   Del ID from HID;
|   |   |   end if
|   |   end for
|   end do
end while
```

Figure 4. Pseudocode of the developed algorithm for feeding time estimation. "#" shows the comments.

The Python language was applied to write the algorithm used in the present research under the TensorFlow deep learning framework. Also, practical training was conducted using a computer with 32-core processors, 64 GB of RAM, and Nvidia Quadro RTX5000 16 GB graphics card.

3. Results

3.1. Head Detection

In the training step, 192,000 iterations were performed, and after every 100 iterations, the model weights were saved. Figure 5 illustrates the values of mean average precision (mAP) and loss function for model training. The error was very high in the first 200 iterations, fluctuated significantly, then decreased. This model did not have any significant performance increase after about 100,000 iterations. The highest performance was attained at iteration 132,996, in which the mAP and loss function values were 0.9320 and 0.0303, respectively. The trained model can recognize the broiler's head when the image is captured. The detection result is determined by a bounding box that has a label and a number indicating the likelihood of belonging to this label. Figure 2 presents samples of the detection results.

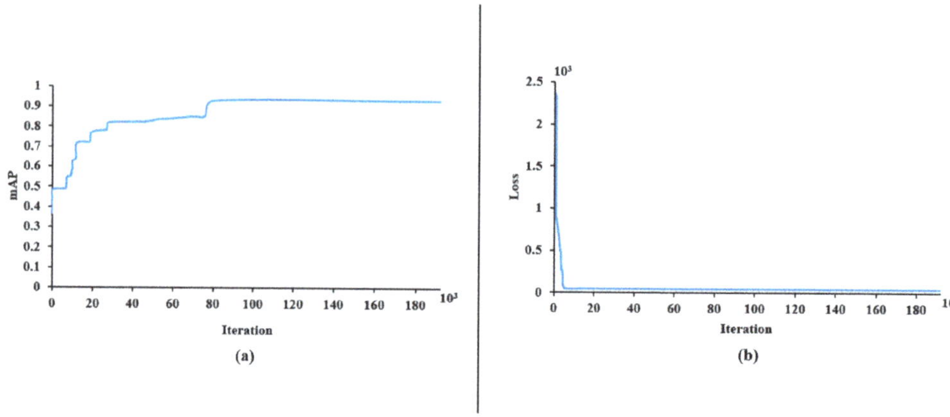

Figure 5. (**a**) Mean average precision (mAP) and (**b**) loss function for each iteration of the training process.

3.2. Evaluating the Developed Algorithm

The developed algorithm can continuously judge feeding behavior occurrence in every frame. The algorithm was evaluated by selecting 3 1-min videos from 3 different days (consisting of 2280 frames), which were annotated manually and accurately. The labeling process consisted of two steps: (1) counting the number of broilers' heads inside the feeding area (red area in Figure 2), to evaluate the algorithm's performance in detecting and tracking heads, and (2) counting the number of frames in which each broiler pecks the feeding pan, to determine the actual value of the feeding time. The number of frames each broiler spends feeding can be obtained by applying the proposed algorithm to each video frame. Finally, by dividing the obtained number by the fps value, the feeder occupation time by the broiler's head was calculated as feeding time.

Figure 6 presents the results of the diagnosis of feeding behavior for each video. For the first video, the number of detected heads was two more than the actual number. The same error can be seen in the other videos. This issue occurred due to an error in the detection model or tracking algorithm. It means that either the detection model failed to distinguish a specific head or the tracking algorithm assigned more than one identifier to one head. The developed algorithm achieved an overall head detection accuracy of 82.8%. The overall accuracy of the feeding time estimation and feeding time of each broiler per visit to the feeding pan was 97.9% and 87.3%, respectively. The developed algorithm demonstrated that each broiler spends a feeding time of 18.46 s per visit to the feeding pan. Some broilers occasionally spend a short feeding time. It is essential to mention that when the head of a specific broiler is inside the feeder, the feeding behavior may not occur, considered during the manual annotation of these videos. Distinguishing the exact feeding time when the broiler's head is inside the feeder is a challenging image processing task. The mentioned point can explain the reason for less than 90% accuracy in estimating feeding time per broiler. The obtained results prove that the developed algorithm, as an automatic and non-invasive tool for feeding time estimation, can be utilized in commercial poultry farms for effective management.

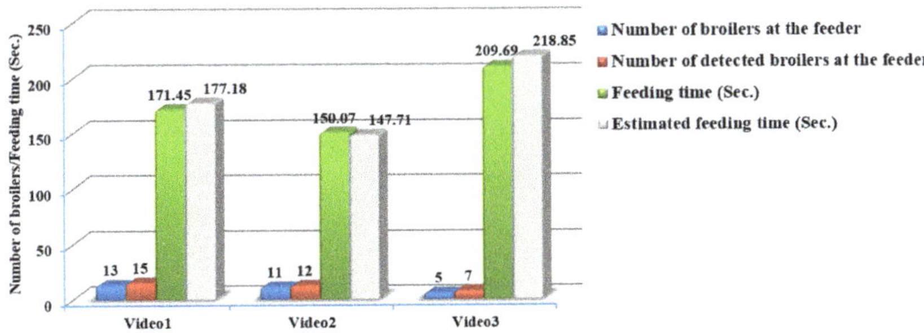

Figure 6. Results of evaluating the developed algorithm for three different 1-min videos.

4. Discussion

The study achieved an accuracy of 82.8% for broiler head detection and high accuracies of 97.9% and 87.3% for overall feeding time estimation and feeding time of each broiler per visit to the pan, respectively. Several studies have been conducted to investigate automated monitoring of animals' feeding behavior. Li, Zhao, Purswell, Du, Chesser Jr, and Lowe [1] investigated group-reared broilers' feeding and drinking behaviors with the image processing technique by determining the number of birds at the feeder and drinkers. The results demonstrated an 89–93% accuracy in determining the number of birds at the feeder. The limitation of this study was the inability of the proposed method to estimate the feeding time of individual broilers. Aydin and Berckmans [23] described a monitoring system to estimate the feeding behavior of broilers at the group level using pecking sound processing. They recorded the pecking sounds of 10 broilers with a microphone attached to the feeder. In their study, R^2 = 0.965 was reported for meal size estimation. In spite of significant results, analyzing individual broilers' sounds and feeding behavior was not performed in this study. In another effort, the Faster R-CNN model was employed to identify the feeding behavior of group-housed pigs [24]. In this study, the feeding behavior of four pigs on the experimental farm was investigated and the precision rate of 0.996 was obtained for feeding behavior recognition. Li et al. [25] monitored the feeding and drinking behaviors of broilers using an ultra-high-RFID-based system. The applied system included tags, antennas, a reader, and a data acquisition system. They investigated the feeding and drinking behaviors of 60 tagged broilers and achieved accuracies of 99.0% and 93.7% for determining time spent at the feeder and drinker, respectively. Despite satisfactory results, this system can monitor a limited number of broilers, and besides the sensor cost, installing the tags can result in stress and welfare issues. The Faster R-CNN model was also applied to recognize the number of pullets at each drinker and estimate the time spent at the drinker [26].

Table 1 summarizes some studies reporting automated monitoring of animals' behavior. The data collected from a research farm or specific tools such as an ultra-high-RFID system, a compartment, a tag installed on broilers, or each nipple drinker were commonly utilized in these studies. In addition, they have researched group-housed poultry behaviors. In comparison, the highlights of this study include that (1) the proposed algorithm in this study was performed on a commercial farm; (2) the algorithm estimated the feeding time of each broiler per visit to the feeding pan without disrupting the regular farm operations.

Table 1. Comparison of animals' behavior monitoring accuracy of different studies and the present study.

Study	Equipment	Dataset	Class	Main Processing Method	Accuracy (%)
Li, Zhao, Purswell, Du, Chesser Jr, and Lowe [1]	RGB camera	Experimental farm	Group broilers	Image processing/Linear model	89–93
Aydin and Berckmans [23]	Microphone	Experimental farm	Group broilers	Sound analysis/Linear model	0.965
Yang, Xiao, and Lin [24]	RGB camera	Experimental farm	Individual pigs	Image processing/Machine learning	0.996
Li, Zhao, Hailey, Zhang, Liang, and Purswell [25]	Ultra-high-RFID	Experimental farm	Group broilers	Statistical analysis	93.7 and 99.0
Current study	RGB camera	Commercial farm	Individual broilers	Image processing/Machine learning	87.3 and 97.9

However, it is worth mentioning that there are several limitations of this study and potential improvements that can be made in future studies. First, video data from this study were collected from only one commercial farm setting. Although the farm setting and the feeder pan are widely adopted among US broiler farms, more diverse data need to be used in order to conclude that our model is accurate for general commercial uses. In addition, training YOLO with a large image dataset collected from different farms can assist the model in generalizing well. Second, feeding area, defined as an area within a blue circle in Figure 2, was manually determined in this study. To make this algorithm fully automated, a simple feeder pan detection algorithm should be developed. Third, our algorithm considered a broiler as feeding when its head was in the feeding area. It is fine to define "feeding" behavior in this way; however, the method may not be accurate in estimating feed intake. In order to estimate true feed intake, the actual pecking behavior should be detected. Additional sensors, such as a microphone and pressure sensor, may be helpful for this purpose. Nevertheless, our study was one of the first efforts towards individual broiler feeding behavior analysis under commercial farm settings as mentioned in the highlights. We believe this study will be valuable as we continue moving towards individual broiler monitoring for precision management on commercial farms.

5. Conclusions

Understanding poultry feeding behavior is crucial for ensuring optimal growth, maintaining health and welfare, improving feed efficiency, managing economic implications, implementing appropriate nutritional management, and reducing the environmental impact of broiler production. Therefore, this study proposed an algorithm based on image processing techniques and a deep learning model to estimate broiler feeding time. Detection and tracking of the broiler's head are essential in judging daily behaviors such as feeding behavior. Hence, the developed algorithm applied the YOLO-v3 model and Euclidean distance-based tracking algorithm to determine whether feeding behavior occurs for each surveillance video frame. Achieving state-of-the-art accuracy in the feeding time estimation of individual broilers can assist in developing efficient equipment for monitoring the broilers' feeding behavior and providing valuable data for farm management. Our group continuously collects more data from various commercial farms to improve the accuracy of the developed algorithm and prepare the system for commercial farms globally.

Author Contributions: Conceptualization, A.N. and H.G.; methodology, A.N. and H.G.; software, A.N.; validation, A.N., Y.Z. and H.G.; formal analysis, A.N. and A.A.; investigation, A.A.; resources, H.G.; data curation, A.N. and A.A.; writing—original draft preparation, A.N. and A.A.; writing—review and editing, Y.Z. and H.G.; visualization, A.N. and A.A.; supervision, H.G.; project administration, Y.Z. and H.G.; funding acquisition, Y.Z. and H.G. All authors have read and agreed to the published version of the manuscript.

Funding: Research reported in this publication was supported by AgResearch at the University of Tennessee and the United States Department of Agriculture AFRI Program under the award number 2022-68014-36663.

Institutional Review Board Statement: Not applicable.

Informed Consent Statement: Informed consent was obtained from all subjects involved in the study.

Data Availability Statement: The data that support the findings of this study are available on request from the corresponding author.

Conflicts of Interest: The authors declare no conflict of interest.

References

1. Li, G.; Zhao, Y.; Purswell, J.L.; Du, Q.; Chesser, G.D., Jr.; Lowe, J.W. Analysis of feeding and drinking behaviors of group-reared broilers via image processing. *Comput. Electron. Agric.* **2020**, *175*, 105596. [CrossRef]
2. Li, N.; Ren, Z.; Li, D.; Zeng, L. Automated techniques for monitoring the behaviour and welfare of broilers and laying hens: Towards the goal of precision livestock farming. *Animal* **2020**, *14*, 617–625. [CrossRef] [PubMed]
3. Kashiha, M.; Pluk, A.; Bahr, C.; Vranken, E.; Berckmans, D. Development of an early warning system for a broiler house using computer vision. *Biosyst. Eng.* **2013**, *116*, 36–45. [CrossRef]
4. Nasiri, A.; Yoder, J.; Zhao, Y.; Hawkins, S.; Prado, M.; Gan, H. Pose estimation-based lameness recognition in broiler using CNN-LSTM network. *Comput. Electron. Agric.* **2022**, *197*, 106931. [CrossRef]
5. Guo, Y.; Zhu, W.; Jiao, P.; Chen, J. Foreground detection of group-housed pigs based on the combination of Mixture of Gaussians using prediction mechanism and threshold segmentation. *Biosyst. Eng.* **2014**, *125*, 98–104. [CrossRef]
6. Guo, Y.-Z.; Zhu, W.-X.; Jiao, P.-P.; Ma, C.-H.; Yang, J.-J. Multi-object extraction from topview group-housed pig images based on adaptive partitioning and multilevel thresholding segmentation. *Biosyst. Eng.* **2015**, *135*, 54–60. [CrossRef]
7. Valletta, J.J.; Torney, C.; Kings, M.; Thornton, A.; Madden, J. Applications of machine learning in animal behaviour studies. *Anim. Behav.* **2017**, *124*, 203–220. [CrossRef]
8. Kasinathan, T.; Singaraju, D.; Uyyala, S.R. Insect classification and detection in field crops using modern machine learning techniques. *Inf. Process. Agric.* **2021**, *8*, 446–457. [CrossRef]
9. Zamani, A.S.; Anand, L.; Rane, K.P.; Prabhu, P.; Buttar, A.M.; Pallathadka, H.; Raghuvanshi, A.; Dugbakie, B.N. Performance of machine learning and image processing in plant leaf disease detection. *J. Food Qual.* **2022**, *2022*, 1598796. [CrossRef]
10. Bai, Y.; Mao, S.; Zhou, J.; Zhang, B. Clustered tomato detection and picking point location using machine learning-aided image analysis for automatic robotic harvesting. *Precis. Agric.* **2023**, *24*, 727–743. [CrossRef]
11. Chen, S.; Tang, Y.; Zou, X.; Huo, H.; Hu, K.; Hu, B.; Pan, Y. Identification and detection of biological information on tiny biological targets based on subtle differences. *Machines* **2022**, *10*, 996. [CrossRef]
12. Yang, X.; Bist, R.; Subedi, S.; Chai, L. A deep learning method for monitoring spatial distribution of cage-free hens. *Artif. Intell. Agric.* **2023**, *8*, 20–29. [CrossRef]
13. Li, G.; Hui, X.; Chen, Z.; Chesser, G.D., Jr.; Zhao, Y. Development and evaluation of a method to detect broilers continuously walking around feeder as an indication of restricted feeding behaviors. *Comput. Electron. Agric.* **2021**, *181*, 105982. [CrossRef]
14. Guo, Y.; Aggrey, S.E.; Wang, P.; Oladeinde, A.; Chai, L. Monitoring Behaviors of Broiler Chickens at Different Ages with Deep Learning. *Animals* **2022**, *12*, 3390. [CrossRef] [PubMed]
15. Redmon, J.; Divvala, S.; Girshick, R.; Farhadi, A. You only look once: Unified, real-time object detection. In Proceedings of the IEEE Conference on Computer Vision and Pattern Recognition, Las Vegas, NV, USA, 27–30 June 2016; pp. 779–788.
16. Fang, W.; Wang, L.; Ren, P. Tinier-YOLO: A real-time object detection method for constrained environments. *IEEE Access* **2019**, *8*, 1935–1944. [CrossRef]
17. Wageeh, Y.; Mohamed, H.E.-D.; Fadl, A.; Anas, O.; ElMasry, N.; Nabil, A.; Atia, A. YOLO fish detection with Euclidean tracking in fish farms. *J. Ambient. Intell. Humaniz. Comput.* **2021**, *12*, 5–12. [CrossRef]
18. Ma, D.; Yang, J. Yolo-animal: An efficient wildlife detection network based on improved yolov5. In Proceedings of the 2022 International Conference on Image Processing, Computer Vision and Machine Learning (ICICML), Xi'an, China, 28–30 October 2022; pp. 464–468.
19. Prabhu, S.; Sreenath, M.; Malavika, V.; Om, H.; Swetha, S. Detection and Recognition of Animals Using Yolo Algorithm. In Proceedings of the 2023 International Conference on Distributed Computing and Electrical Circuits and Electronics (ICDCECE), Ballar, India, 29–30 April 2023; pp. 1–6.

20. Petso, T.; Jamisola, R.S., Jr.; Mpoeleng, D.; Bennitt, E.; Mmereki, W. Automatic animal identification from drone camera based on point pattern analysis of herd behaviour. *Ecol. Inform.* **2021**, *66*, 101485. [CrossRef]
21. Desai, B.; Patel, A.; Patel, V.; Shah, S.; Raval, M.S.; Ghosal, R. Identification of free-ranging mugger crocodiles by applying deep learning methods on UAV imagery. *Ecol. Inform.* **2022**, *72*, 101874. [CrossRef]
22. Yu, Z.; Liu, Y.; Yu, S.; Wang, R.; Song, Z.; Yan, Y.; Li, F.; Wang, Z.; Tian, F. Automatic detection method of dairy cow feeding behaviour based on YOLO improved model and edge computing. *Sensors* **2022**, *22*, 3271. [CrossRef]
23. Aydin, A.; Berckmans, D. Using sound technology to automatically detect the short-term feeding behaviours of broiler chickens. *Comput. Electron. Agric.* **2016**, *121*, 25–31. [CrossRef]
24. Yang, Q.; Xiao, D.; Lin, S. Feeding behavior recognition for group-housed pigs with the Faster R-CNN. *Comput. Electron. Agric.* **2018**, *155*, 453–460. [CrossRef]
25. Li, G.; Zhao, Y.; Hailey, R.; Zhang, N.; Liang, Y.; Purswell, J. An ultra-high frequency radio frequency identification system for studying individual feeding and drinking behaviors of group-housed broilers. *Animal* **2019**, *13*, 2060–2069. [CrossRef] [PubMed]
26. Li, G.; Ji, B.; Li, B.; Shi, Z.; Zhao, Y.; Dou, Y.; Brocato, J. Assessment of layer pullet drinking behaviors under selectable light colors using convolutional neural network. *Comput. Electron. Agric.* **2020**, *172*, 105333. [CrossRef]

animals MDPI

Article

Field Implementation of Forecasting Models for Predicting Nursery Mortality in a Midwestern US Swine Production System

Edison S. Magalhaes [1,*], Danyang Zhang [2], Chong Wang [1,2], Pete Thomas [3], Cesar A. A. Moura [3], Derald J. Holtkamp [1], Giovani Trevisan [1], Christopher Rademacher [1], Gustavo S. Silva [1] and Daniel C. L. Linhares [1]

1 Department of Veterinary Diagnostic and Production Animal Medicine, College of Veterinary Medicine, Iowa State University, Ames, IA 50011, USA
2 Department of Statistics, College of Liberal Arts and Sciences, Iowa State University, Ames, IA 50011, USA
3 Iowa Select Farms, Iowa Falls, IA 50126, USA
* Correspondence: edison@iastate.edu; Tel.: +1-515-708-5157

Simple Summary: Swine nursery mortality is highly impacted by the pre-weaning performance of the piglets. Even though the importance of the pre-weaning phase on the downstream post-weaning performance is acknowledged, predictive modeling has yet to be described in the swine industry to predict the downstream nursery performance of groups of pigs based on their previous pre-weaning phase. One obstacle to building such predictive models is that pieces of information concerning the factors impacting swine mortality are collected with separate record-keeping programs and stored in unconnected databases, creating multiple unutilized data stream clusters. Thus, in this study, we described the process of building a data-wrangling pipeline that automatically integrates diverse and dispersed data streams collected from one swine production company, creating then a master table that was utilized to predict the mortality of groups of pigs during the nursery phase.

Abstract: The performance of five forecasting models was investigated for predicting nursery mortality using the master table built for 3242 groups of pigs (~13 million animals) and 42 variables, which concerned the pre-weaning phase of production and conditions at placement in growing sites. After training and testing each model's performance through cross-validation, the model with the best overall prediction results was the Support Vector Machine model in terms of Root Mean Squared Error (RMSE = 0.406), Mean Absolute Error (MAE = 0.284), and Coefficient of Determination (R^2 = 0.731). Subsequently, the forecasting performance of the SVM model was tested on a new dataset containing 72 new groups, simulating ongoing and near real-time forecasting analysis. Despite a decrease in R^2 values on the new dataset (R^2 = 0.554), the model demonstrated high accuracy (77.78%) for predicting groups with high (>5%) or low (<5%) nursery mortality. This study demonstrated the capability of forecasting models to predict the nursery mortality of commercial groups of pigs using pre-weaning information and stocking condition variables collected post-placement in nursery sites.

Keywords: swine; mortality; data-wrangling; forecasting; machine-learning

Citation: Magalhaes, E.S.; Zhang, D.; Wang, C.; Thomas, P.; Moura, C.A.A.; Holtkamp, D.J.; Trevisan, G.; Rademacher, C.; Silva, G.S.; Linhares, D.C.L. Field Implementation of Forecasting Models for Predicting Nursery Mortality in a Midwestern US Swine Production System. *Animals* **2023**, *13*, 2412. https://doi.org/10.3390/ani13152412

Academic Editor: Jin-Ho Cho

Received: 3 July 2023
Revised: 19 July 2023
Accepted: 24 July 2023
Published: 26 July 2023

1. Introduction

The abundance of diverse and large-scale data streams often challenges the implementation of precision animal agriculture in livestock, which requires a multifaceted data-wrangling approach to investigate this complex livestock "big data" [1]. Using data management techniques and machine-learning models on these data can overcome its complexity for analytical purposes, such as forecasting. Although forecasting analysis in the livestock realm is acknowledged [2,3], this application has not yet been reported in the swine industry for mortality rate. Swine post-weaning mortality is a key performance indicator (KPI) utilized to measure the sustainability of swine production systems [4,5], divided into nursery and finisher mortality. Swine nursery mortality refers to the mortality of pigs in the first 5–8 weeks of the overall post-weaning phase (approximately 5.5 months), accounting for a large portion of the overall post-weaning mortality [6].

Information concerning the risk factors for swine mortality is routinely collected, such as health, environment, productivity, and infrastructure. However, integrating and merging these data streams is necessary for its collective utilization for targeting prediction or risk factor analyses. The development of means for data integration and analysis under field conditions allows the implementation of such data analysis approaches, as reported in previous studies [7–10]. Therefore, the objective of this study was to develop a data-wrangling pipeline within one swine production system to integrate and manage multiple data streams, enabling automated and near real-time data consolidation. Furthermore, the performance of multiple forecasting models was assessed on historical data, and the best model was tested on new data to predict the nursery mortality of prospective closeouts.

2. Materials and Methods

2.1. Overview and Study Design

This study utilized field data from a large U.S. swine production system in the Midwestern region. A total of six different and disconnected data streams related to 3242 groups of marketed pigs (over 13 million animals) slaughtered over three years, here referred to as closeouts, were collected for the analyses. The retrospective performance of both the pre- and post-weaning phases of production were imported and integrated into the respective closeouts' information, constructing a dataset (also known as the master table) containing breeding-to-market historical information for each closeout. The pre-weaning phase variables and stocking conditions data in this master table were utilized as predictors to forecast the downstream post-weaning mortality of each closeout on their initial 60 days in the post-weaning phase (nursery mortality), as demonstrated in Figure 1.

Closeouts were defined as the groups of pigs originating from the company's breeding herds. The pigs remained in the breeding herd until weaning at approximately 21 days of age. Following weaning, pigs were placed on feed at growing sites where the groups remained for around 5.5 months. The groups were managed all-in-all-out, meaning another group of pigs could only start once all the pigs from the previous groups had been marketed. The mortality of each closeout during the nursery phase was defined as the outcome variable of analysis in this study and was calculated as the following: (number of pigs at placement − number of pigs 60 days post placement) ÷ number of pigs at placement.

Closeouts originating from a single sow farm would have information concerning the performance of that breeding herd on the designated week. The productivity parameters assigned to the downstream weaned group represent the retrospective performance of that batch of pigs from farrow-to-wean, while the pre-farrow information (e.g., abortion rate) represents the reproductive performance on that farm on the week of the weaning event.

When a group of pigs originated from multiple sow farms (e.g., 2000 pigs placed in a growing site may have received 1000 pigs from two different sow farms), the variables concerning the pre-weaning phase for that specific group would be calculated by using a weighted average for the continuous independent variables, and the mixed classification was used for the whole group for disease classification statuses.

SAS® Version 9.4 (SAS Institute, Inc., Cary, NC, USA) was utilized to build data-wrangling pipeline algorithms, thus automating the processes of importing, managing, cleaning, and integrating the data streams. The integration of the six data streams resulted in a final master table for the 3242 closeouts that were utilized for comparing the performance of five different regression and machine-learning models for forecasting swine nursery mortality. After this step, the model with the best forecasting performance was utilized on a new dataset to validate the forecasting capability on prospective data, simulating ongoing near real-time forecasting.

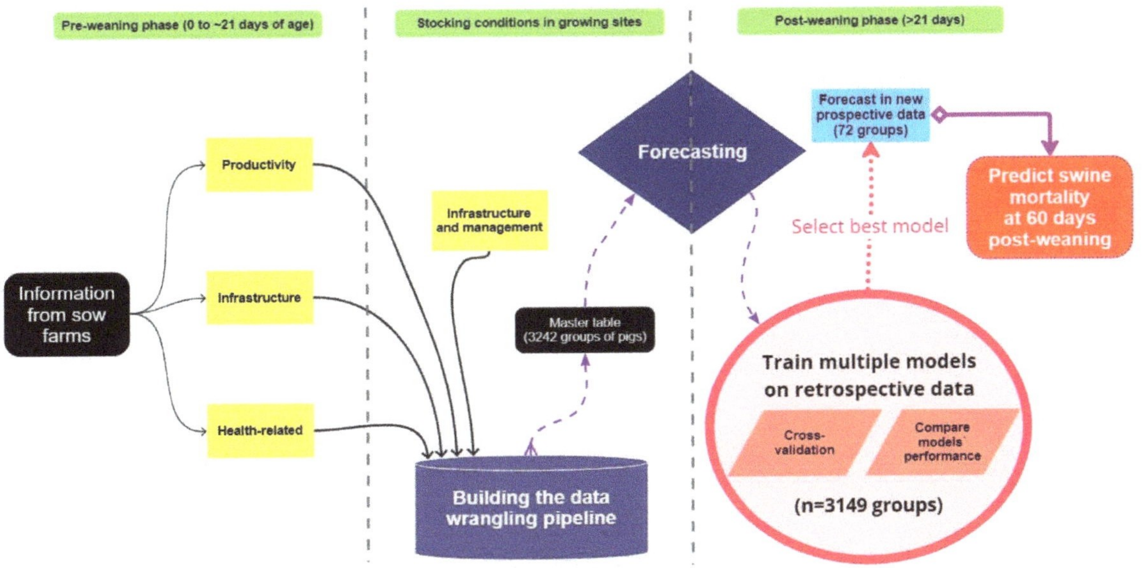

Figure 1. Flow chart explaining the process of integrating different data streams into a single master table to be used on the forecasting analyses. A data-wrangling pipeline was built to integrate information concerning the performance of groups of weaned pigs transferred to growing sites and their conditions at placement into a single master table. This table was utilized to train forecasting models on predicting the mortality of the weaned groups throughout the initial 60 days in the post-weaning phase. The model with the best performance was then applied to forecast the mortality of prospective new groups of weaned pigs.

2.2. Data-Wrangling Pipeline

The six different data streams available for the development of the master table were: (1) pre-weaning phase (i.e., breeding herd) productivity and health data; (2) post-weaning phase (i.e., growing phase) productivity data; (3) closeouts' health status reports; (4) pig transportation records; (5) stocking conditions reports; and (6) management procedure records. The SAS algorithms developed in this study used a similar methodology to that described by Magalhaes et al. (2022) [10], where the processes of matching and merging different data streams were conducted based on an identifier (time and location of events) and through the development of PROC Statements algorithms (PROC MERGE, PROC SET, PROC SQL, PROC SORT, PROC UNIVARIATE, and PROC FREQ). The swine production system provided access to the aforementioned data, where a data workflow was developed using Microsoft Power Automate (Microsoft Corporation, Redmond, WA, USA) and SAS to automate the data-wrangling processes in this study. Once the master table was built, the dataset contained information for 3242 closeouts of pigs, originating from 42 breeding herd sources and weaned into 529 different growing sites. The information from each of the six data streams was matched and merged to each respective closeout of pigs marketed in this study period (i.e., each closeout's historical data from breeding-to-market).

2.3. Comparing Forecasting Models Based on Training Data

The initial step after completing the master table was to select the breeding herd variables from the pre-weaning phase of production and parameters that represent the stocking conditions of the weaned groups into growing sites (i.e., characteristics at placement). Among all variables in the master table, 42 parameters were utilized as predictors in the forecasting analyses (Table 1). The nursery mortality was log-transformed after verifying that its distribution was not normal, thus, utilizing the log-mortality as the response variable. The classes of each categorical variable included in the model are described in Table 2.

Table 1. Variables selected from the master table for the forecasting analyses.

Data Streams	Variable Type ‡	Variables
(1) Breeding Herd Productivity and Health *	Rate	Service repeat rate; Abortion rate; Services per inventory; Proportion of gilts bred; Last week weaned sows bred rate; Proportion of sows pregnant at 105 days; Farrowing rate; Stillborn rate; Mummies rate; Pre-weaning mortality; Pre-natal losses; Sow death rate; Sow culls rate
	Count	Number of services; Number of farrows; Sows inventory
	Average	Wean-to-service interval; Total born; Born alive; Parity at the farrow; Gestation length; Interval between farrows; Pigs weaned/sow; Piglet wean age; Non-productive days; Productive sow days; Litter/female/year; Mated inventory; Pigs/weaned/female/year
(2) Growing Phase Productivity †	Rate	Nursery mortality (mortality on the initial 60 days post placement in a growing site)
(3) Closeouts Health Status *	Category	Group status for porcine reproductive and respiratory syndrome (PRRS) at placement; Group status for *Mycoplasma hyopneumoniae* (MhP) at placement
(4) Pig Transportation *	Time	Weaning movement year; Weaning movement week
	Count	Number of animals transported
(5) Stocking Conditions *	Category	Type of flow; Type of ventilation;
	Count	Number of origins; Time to fill the site; Breeding herd origins
(6) Management Procedure *	Category	Type of PRRS vaccine; Type of piglet medication at weaning; Breeding herd type of mass medication protocol

(1)–(6) Data streams utilized; † Outcome variable; * Predictor Variables. ‡ Type of variables. Variable type classified as "Average" refers to the average number of count events occurring for the batch of groups weaned. For example, "Total Born" variable represents the average number of total piglets born per farrow over a total number of farrowing events in a week. More details about the classes of the categorical variables are described in Table 2.

The 42 variables included in the forecasting analysis were selected based on their potential as factors related to the quality of weaned pigs (i.e., health status and productivity performance) and the overall conditions at placement in growing sites (i.e., infrastructural and management factors). Also, only variables that were provided in the master table at the moment when weaned pigs were placed into growing sites were included in the model.

To forecast the log-mortality, five models were investigated: multiple linear regression model (MLR), LASSO regression, support vector machine (SVM), neural network (NNet), and random forest (RF). The evaluation criteria for each forecasting model included Root Mean Squared Error (RMSE), Mean Absolute Error (MAE), and Coefficient of Determination (R^2). Using the R package 'caret' [11], and specifically the 'train' function, the optimal parameters of LASSO regression, SVM, and NNet were selected based on the smallest RMSE by doing three repetitions of 5-fold cross-validation, and the optimal parameters of RF were selected based on the smallest out-of-bag (OOB) error.

In order to evaluate the prediction performance of each forecasting model, a leave-one-out cross-validation was performed, where, for each record, the training set was the dataset excluding that record. The trained model was then used to predict the log-mortality of the excluded record. The best model was selected based on higher R^2 and lowest RMSE and MAE values.

Table 2. Description of the categorical variables included in the forecasting model.

Data Type	Variable †	A *	B *	C *	D *	E *	F *
Breeding herd health	PRRS status	Epidemic	Endemic	Mixed	Negative	-	-
	MhP status	Epidemic	Endemic	Mixed	Negative	-	-
Stocking conditions *	Type of flow [3]	DS-M	DS-S	Y	S	-	-
	Type of ventilation	Tunnel Barn	Curtain Barn	-	-	-	-
Management procedure *	PRRS vaccine	Vaccine A	Vaccine B	-	-	-	-
	Piglet medication [1]	Enrofloxacin	Tulathromycin	Ceftiofur	Florfenicol	Mixed	None
	Sow medication [2]	CTC [4]	Lincomycin	Tilmicosin	Mixed	None	-

* Categories of each variable; [1] Type of medication treatment in piglets; [2] Type of medication treatment in sows; [3] DS-M: Double stock moved; DS-R: Double stock remained; Y: Nursery-to-finisher flows; S: Single stock flows; [4] CTC: Chlortetracycline; † Categorical variables from Table 1.

2.4. Performance of the Selected Model on Independent Validation Data

After comparing the performance of the different forecasting models on the retrospective dataset of 3242 groups, which refers to groups stocked into nursery sites between week 29 of 2019 through week 5 of 2022 (i.e., marketed between January 2020 to August 2022), a new dataset containing 72 new closeouts weaned into nursery sites between weeks 6 and 12 of 2022 (i.e., marketed between August and September of 2022) was obtained through the data-wrangling pipeline. The forecasting model was then utilized on this naïve data to predict the nursery mortality of the groups, and the forecasting performance of the selected model was measured using the same metric of the same step (R^2, RMSE, and MAE). Also, the predicted vs. actual nursery mortality values were classified into relatively "high nursery mortality" (>5%) or "low nursery mortality" (<5%) groups, as the company providing the data used the same classification as their target mortality values. The performance of the SVM model on accurately predicting closeouts with high or low nursery mortality was assessed in terms of accuracy (Ac), sensitivity (Se), Specificity (Sp), positive predicted value (PPV), and negative predicted value (NPV), calculated based on the difference between the predicted vs. actual mortality of the 72 groups.

3. Results

3.1. Data-Wrangling Pipeline

When assessing data completeness for the 3242 groups, a total of 93 closeouts (2.87%) were excluded due to a lack of information for all the characteristics included in the master table, resulting in a final dataset composed of 3149 closeouts and 42 explanatory variables to be used in the forecasting analyses.

3.2. Comparing Forecasting Models

The overall performance for all forecasting models is reported in Table 2. Notably, the machine learning models performed better than the regression models, where the RF and SVM models demonstrated the best overall prediction performance, similar to other livestock-related studies comparing the performance of multiple forecasting models [3,12,13]. Furthermore, the SVM outperformed the other models (Table 3) measured in terms of R^2 (0.731) and lower errors measured by RMSE (0.406) and MAE (0.284).

Table 3. Performance of the forecasting models on predicting swine nursery mortality.

Model [1]	Parameters [2]		
	R^2	RMSE	MAE
MLR	0.385	0.614	0.475
LASSO	0.392	0.611	0.471
RF	0.725	0.421	0.313
SVM	0.731	0.406	0.284
NNet	0.533	0.566	0.393

[1] MLR: Multiple Linear Regression; LASSO: LASSO regression; RF: Random Forest; SVM: Support Vector Machine; NNet: Neural Network. [2] RMSE: Root Mean Square Error; MAE: Mean Absolute Error; R^2: r-square.

Thereafter, the predicted values for each closeout using the SVM model were averaged by week for the data collected in this study (Figure 2), where it was observed that the SVM predicted values were underestimated compared to the actual nursery mortality values of the closeouts. Despite this, both the average weekly predicted and actual mortalities followed similar seasonal trends over time, which can be explained by the seasonal activity of major diseases impacting the swine industry [14,15].

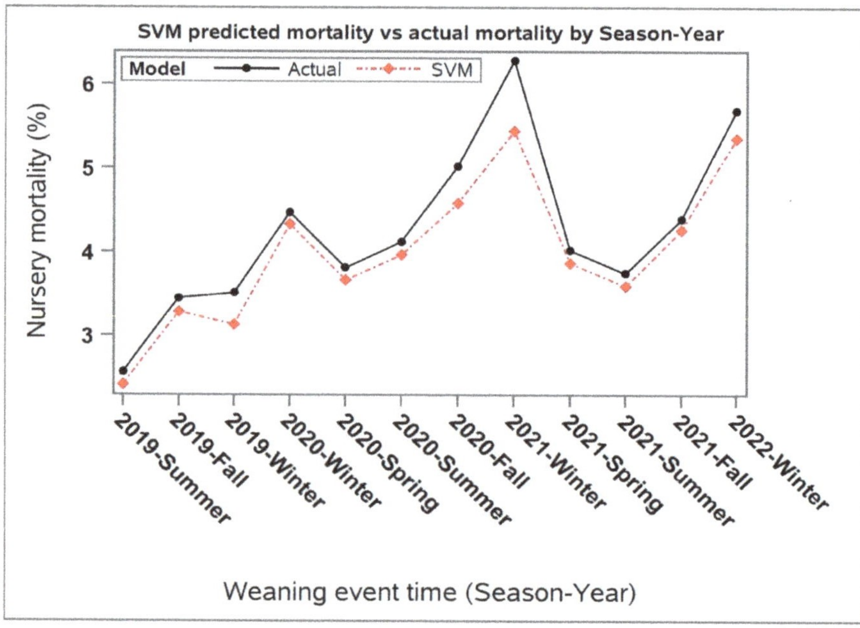

Figure 2. Average predicted nursery mortality versus actual mortality over season-year for Support Vector Machine (SVM) forecasting model using the results of the cross-validation step (3149 closeouts). Season-Year corresponds to the time during the study period when the pigs were weaned.

3.3. Performance of the Selected Model

Identified as the superior model, SVM was prospectively applied to new data consisting of 72 closeouts (Figure 3), representing one month of closeouts, to predict the nursery mortality of the new groups. The overall forecasting performance of the SVM model was lower than the training database's performance on the cross-validation procedure (R^2 = 0.554 and 0.731, respectively). However, even though the prediction of mortality in new groups was already simulated in the training database during the cross-validation procedure, the prediction performance was inferior when applying the same model to a smaller sample of a prospective dataset. It is important to note that the training step was conducted in a much larger dataset, while the testing of the SVM model was conducted in a smaller dataset.

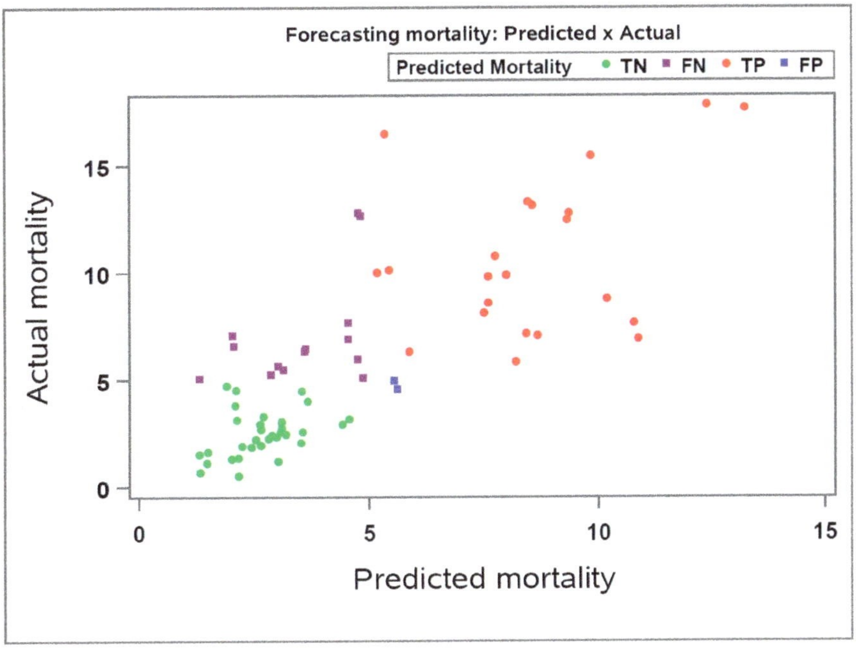

Figure 3. Correlation plot between the observed and predicted nursery mortality using the SVM model on 72 new closeouts. True Negative (TN): Predicted as Low mortality (<5%) and actual was Low (<5%); False Negative (FN): Predicted as Low mortality (<5%) and actual was High (>5%); True Positive (TP): Predicted as High mortality (>5%) and actual was High (>5%); False Positive (FP): Predicted as High mortality (>5%) and actual was Low (<5%).

Despite the SVM's decreased performance on naïve data when categorizing both predicted and actual nursery mortality of the 72 closeouts into high (>5%) or low (<5%) nursery mortality, a high accuracy value (77.78%) was observed for the SVM on correctly predicting the closeouts as high or low nursery mortality. Also, we observed that most of the groups are located in the positive diagonal axis of the chart, which is the desired area in terms of prediction (Figure 3).

The values for sensitivity (62.16%), specificity (94.29%), positive predicted value (92.00%), and negative predicted value (70.21%) also demonstrated an acceptable prediction performance, especially for precisely predicting groups with "high nursery mortality" rates (i.e., at high risk). Overall, the SVM model accurately predicted 62.16% of the closeouts with relatively "high nursery mortality" and 94.29% with relatively low mortality. In other words, even though the SVM model did not predict all groups that had "high nursery mortality" as high (false negatives), the model had a high positive predicted value, indicating that 92.00% of the closeouts predicted as "high nursery mortality" were observed as actually high.

For the categorical variables ($n = 7$) included in the forecasting model, when comparing the frequency distribution between the number of closeouts with high and low mortality groups compared to their respective predicted values (Figure 4), the forecasting model overestimated the number of groups with low predicted mortality (i.e., right-side transparent bars are longer than the right-side solid bars). On the other hand, the forecasting model underestimated the actual number of closeouts with "high nursery mortality" for all classes of the categorical variables illustrated (i.e., left-side transparent bars are shorter than the left-side solid bars).

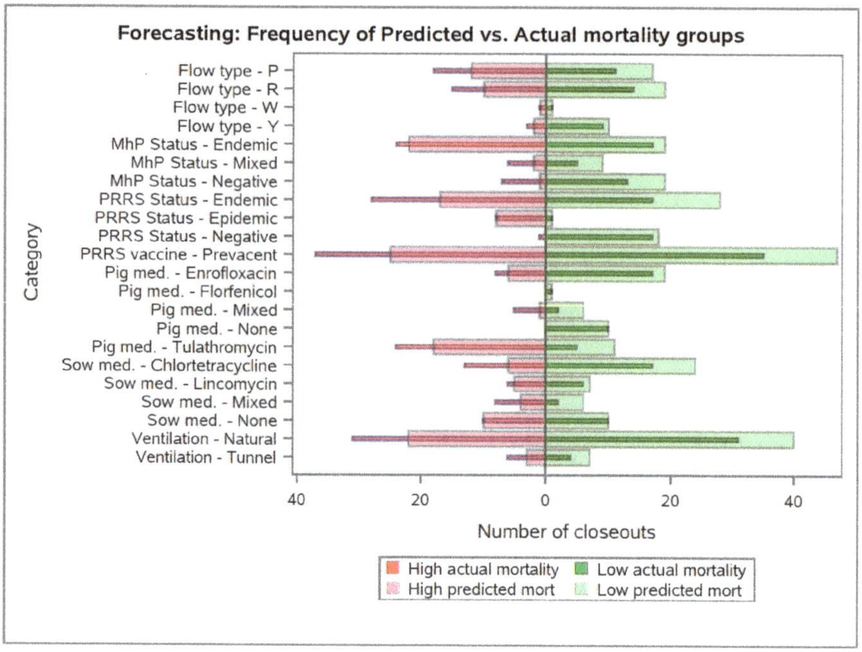

Figure 4. Frequency distribution of the categorical variables included in the forecasting model. Left and right solid bars refers to the number of closeouts with actual high or low nursery mortality (>5%), respectively. Left and right transparent bars refer to the number of closeouts predicted as high or low nursery mortality (<5%), respectively.

Notably, for specific classes within the categorical variables (e.g., "Pig med.—Tulathromycin" or "PRRS Status—Epidemic"), the proportion of groups predicted as "high nursery mortality" were higher than the number of groups predicted as "low nursery mortality". This hypothesis is supported by the common knowledge that PRRS infections in breeding herds generate downstream PRRS-epidemic weaned pigs [10,16–20], which are expected to be more challenged throughout the post-weaning phase. Also, the use of tulathromycin to treat piglets in breeding herds indicates that health-challenged pigs were weaned, as this is a frequently prescribed antibiotic in swine due to its ability to modulate the immune system, as well as an effective treatment against key respiratory diseases [21].

On the other hand, for some factors such as "MhP—Negative" or "Pig med.—None", the largest proportion of the groups of pigs were predicted as low nursery mortality groups, which can be explained by the fact that *M. hyopneumoniae* infection in weaned pigs can increase growing pig mortality [22], thus, negative pigs are expected to have higher survivability. Also, the presence of groups of weaned pigs that were not treated with medication during the lactation can indicate groups with higher quality that did not need this procedure.

Altogether, the results demonstrated in Figure 4 indicate the influence of specific factors on the overall prediction. However, this study was not designed to investigate the influence of these specific factors on nursery mortality, as this type of approach requires a causal inference analysis [23,24], which was not the scope of this study.

4. Discussion

The algorithms developed in this study for the data-wrangling pipeline allowed the integration of information previously stored independently and underutilized for analysis purposes, combining the dispersed predictors in multiple data streams into a single master table. This approach of combining multiple data streams to investigate post-weaning performance was previously described in other studies [7,8,25–27].

Multiple machine learning and regression algorithms were applied to the master table to compare their forecasting performance in predicting swine nursery mortality. Also, other studies in the livestock realm demonstrated the application of similar models for predicting important KPIs of productivity [2,3,28–33].

Assuming that the swine production system maintains the format of the data streams utilized to build the master table over time, the algorithms can be utilized to integrate and prepare new incoming information for prospective analyses, including forecasting and causal inference, as it is seen that incompatibility of data streams is one of the major challenges in data integration [34].

The results of both the data-wrangling pipeline procedure and the forecasting models' comparison allowed the training of the best model on retrospective data and further testing on new data, simulating the ongoing application of forecasting models on future data, in other words, utilizing the pre-weaning phase and stocking condition variables to predict the future mortality of closeouts.

The algorithms developed in this study can support swine practitioners in their decision-making process to strategically allocate resources (or not) for groups with predicted high nursery mortality. Notably, the predictive performance of the models refers specifically to the dataset collected in this study and to the time analyzed. In other words, the performance may change over time within this company as swine nursery mortality is impacted by multifactorial components that are dynamically interacting over time and period [4,5,10], limiting the external validity of this study to other field conditions.

Although there is an opportunity for improving the prediction of the exact values of nursery mortality (i.e., continuous outcome), there is a trade-off between prediction error and the utility of the predicted value when using binary vs. continuous outcome. For example, more relevance was given by the production system in this study to identify relatively high nursery mortality groups instead of predicting their exact mortality values.

The lower sensitivity results of this study can be explained by limiting the inclusion of predictor variables related only to the pre-weaning phase and conditions of weaned pigs at placement in growing sites (stocking conditions variables), as post-weaning infectious and non-infectious factors are likely to increase swine mortality as well [5]. However, as the goal of this study was to forecast nursery mortality at the beginning of the post-weaning phase (at placement), a trade-off of losing accuracy in terms of prediction but allowing early intervention is expected.

On the other hand, the model demonstrated a high performance when predicting groups that would have high nursery mortality (high positive predicted value), thus indicating that sow farm variables related to the quality of the piglets at weaning can drive their mortality throughout the post-weaning phase as demonstrated by other authors [35–38].

5. Conclusions

Forecasting swine nursery mortality can support decision-makers in allocating resources or interventions toward precision swine health and productivity management. This study demonstrated the capability of building system-specific algorithms that allows the development of an automated data-wrangling pipeline, which enables ongoing and near real-time forecasting. Also, this study demonstrated the ability to utilize breeding herd characteristics and data concerning the stocking conditions of weaned pigs placed in nursery sites as predictors for forecasting nursery mortality. Despite the overall acceptable performance for predicting groups at high nursery mortality risk, there is an opportunity for improving the model's performance by including more predictors and other machine-learning models.

Author Contributions: Conceptualization, D.C.L.L., E.S.M., G.S.S., C.R., G.T. and D.J.H.; methodology, D.C.L.L., E.S.M., G.S.S., C.W., G.T. and D.J.H.; software, E.S.M. and D.Z.; validation, C.W. and G.S.S.; formal analysis, E.S.M., D.Z., C.W. and G.S.S.; investigation, C.A.A.M., P.T., C.R. and E.S.M.; resources, C.A.A.M. and P.T.; data curation, C.A.A.M., P.T. and E.S.M.; writing—original draft preparation, E.S.M., D.Z. and C.W.; writing—review and editing, E.S.M., D.J.H., C.R., D.C.L.L., G.S.S., P.T., C.A.A.M. and G.T.; visualization, E.S.M. and D.Z.; supervision, E.S.M. and D.C.L.L.; project administration, E.S.M. and D.C.L.L.; funding acquisition, D.C.L.L., E.S.M., G.S.S., G.T., D.J.H. and C.W. All authors have read and agreed to the published version of the manuscript.

Funding: This study was funded by the U.S. Department of Agriculture—National Institute of Food and Agriculture (USDA-NIFA) grant #022-68014-36668, and the C. R. Henderson Fund for Excellence in Predictive Inference and Its Applications.

Data Availability Statement: The data presented in this study are available on reasonable request from the corresponding author [E.S.M.]. The data are not publicly available due to privacy.

Acknowledgments: The support from Iowa Select Farms was fundamental for conducting this study. Also, the support of the research and extension team at the Veterinary Diagnostic and Production Animal Medicine at Iowa State University was essential in all steps of this study. In addition, special thanks to the faculty and graduate team at the Department of Statistics at Iowa State University for supporting this research.

Conflicts of Interest: The authors declare no conflict of interest.

References

1. Morota, G.; Ventura, R.V.; Silva, F.F.; Koyama, M.; Fernando, S.C. Big Data Analytics and Precision Animal Agriculture Symposium: Machine Learning and Data Mining Advance Predictive Big Data Analysis in Precision Animal Agriculture. *J. Anim. Sci.* **2018**, *96*, 1540–1550. [CrossRef]
2. Murphy, M.D.; O'Mahony, M.J.; Shalloo, L.; French, P.; Upton, J. Comparison of Modelling Techniques for Milk-Production Forecasting. *J. Dairy Sci.* **2014**, *97*, 3352–3363. [CrossRef] [PubMed]
3. Nguyen, Q.T.; Fouchereau, R.; Frénod, E.; Gerard, C.; Sincholle, V. Comparison of Forecast Models of Production of Dairy Cows Combining Animal and Diet Parameters. *Comput. Electron. Agric.* **2020**, *170*, 105258. [CrossRef]
4. Gebhardt, J.T.; Tokach, M.D.; Dritz, S.S.; DeRouchey, J.M.; Woodworth, J.C.; Goodband, R.D.; Henry, S.C. Postweaning Mortality in Commercial Swine Production II: Review of Infectious Contributing Factors. *Transl. Anim. Sci.* **2020**, *4*, 485–506. [CrossRef] [PubMed]
5. Gebhardt, J.T.; Tokach, M.D.; Dritz, S.S.; DeRouchey, J.M.; Woodworth, J.C.; Goodband, R.D.; Henry, S.C. Postweaning Mortality in Commercial Swine Production. I: Review of Non-Infectious Contributing Factors. *Transl. Anim. Sci.* **2020**, *4*, 462–484. [CrossRef] [PubMed]
6. Bush, E. *Swine 2012 Part I: Baseline Reference of Swine Health and Management in the United States, 2012*; United States Department of Agriculture: Washington, DC, USA, 2015.
7. Agostini, P.D.S.; Manzanilla, E.G.; De Blas, C.; Fahey, A.G.; Da Silva, C.A.; Gasa, J. Managing Variability in Decision Making in Swine Growing-Finishing Units. *Ir. Vet. J.* **2015**, *68*, 20. [CrossRef]
8. Goumon, S.; Faucitano, L. Influence of Loading Handling and Facilities on the Subsequent Response to Pre-Slaughter Stress in Pigs. *Livest. Sci.* **2017**, *200*, 6–13. [CrossRef]
9. Passafaro, T.L.; Van De Stroet, D.; Bello, N.M.; Williams, N.H.; Rosa, G.J.M. Generalized Additive Mixed Model on the Analysis of Total Transport Losses of Market-Weight Pigs. *J. Anim. Sci.* **2019**, *97*, 2025–2034. [CrossRef]
10. Magalhães, E.S.; Zimmerman, J.J.; Thomas, P.; Moura, C.A.A.; Trevisan, G.; Holtkamp, D.J.; Wang, C.; Rademacher, C.; Silva, G.S.; Linhares, D.C.L. Whole-Herd Risk Factors Associated with Wean-to-Finish Mortality under the Conditions of a Midwestern USA Swine Production System. *Prev. Vet. Med.* **2022**, *198*, 105545. [CrossRef]
11. Kuhn, M. Building Predictive Models in R Using the Caret Package. *J. Stat. Softw.* **2008**, *28*, 1–26. [CrossRef]
12. Arulmozhi, E.; Moon, B.E.; Basak, J.K.; Sihalath, T.; Park, J.; Kim, H.T. Machine Learning-Based Microclimate Model for Indoor Air Temperature and Relative Humidity Prediction in a Swine Building. *Animals* **2021**, *11*, 222. [CrossRef] [PubMed]
13. Semakula, J.; Corner-thomas, R.A.; Morris, S.T.; Blair, H.T.; Kenyon, P.R. Application of Machine Learning Algorithms to Predict Body Condition Score from Liveweight Records of Mature Romney Ewes. *Agriculture* **2021**, *11*, 162. [CrossRef]
14. Trevisan, G.; Linhares, L.C.M.; Crim, B.; Dubey, P.; Schwartz, K.J.; Burrough, E.R.; Main, R.G.; Sundberg, P.; Thurn, M.; Lages, P.T.F.; et al. Macroepidemiological Aspects of Porcine Reproductive and Respiratory Syndrome Virus Detection by Major United States Veterinary Diagnostic Laboratories over Time, Age Group, and Specimen. *PLoS ONE* **2019**, *14*, e0223544. [CrossRef]
15. Trevisan, G.; Linhares, L.C.M.; Crim, B.; Dubey, P.; Schwartz, K.J.; Burrough, E.R.; Wang, C.; Main, R.G.; Sundberg, P.; Thurn, M.; et al. Prediction of Seasonal Patterns of Porcine Reproductive and Respiratory Syndrome Virus RNA Detection in the U.S. Swine Industry. *J. Veter.-Diagn. Investig.* **2020**, *32*, 394–400. [CrossRef] [PubMed]

16. Almeida, M.N.; Rotto, H.; Schneider, P.; Robb, C.; Zimmerman, J.J.; Holtkamp, D.J.; Rademacher, C.J.; Linhares, D.C.L. Collecting Oral Fluid Samples from Due-to-Wean Litters. *Prev. Vet. Med.* **2020**, *174*, 104810. [CrossRef]
17. Alvarez, J.; Sarradell, J.; Kerkaert, B.; Bandyopadhyay, D.; Torremorell, M.; Morrison, R.; Perez, A. Association of the Presence of Influenza A Virus and Porcine Reproductive and Respiratory Syndrome Virus in Sow Farms with Post-Weaning Mortality. *Prev. Vet. Med.* **2015**, *121*, 240–245. [CrossRef]
18. Dong, J.G.; Yu, L.Y.; Wang, P.P.; Zhang, L.Y.; Liu, Y.L.; Liang, P.S.; Song, C.X. A New Recombined Porcine Reproductive and Respiratory Syndrome Virus Virulent Strain in China. *J. Vet. Sci.* **2018**, *19*, 89–98. [CrossRef]
19. Fablet, C.; Rose, N.; Grasland, B.; Robert, N.; Lewandowski, E.; Gosselin, M. Factors Associated with the Growing-Finishing Performances of Swine Herds: An Exploratory Study on Serological and Herd Level Indicators. *Porc. Health Manag.* **2018**, *4*, 6. [CrossRef]
20. Holtkamp, D.J.; Kliebenstein, J.B.; Neumann, E.J.; Zimmerman, J.J.; Rotto, H.F.; Yoder, T.K.; Wang, C.; Yeske, P.E.; Mowrer, C.L.; Haley, C.A. Assessment of the Economic Impact of Porcine Reproductive and Respiratory Syndrome Virus on United States Pork Producers. *J. Swine Health Prod.* **2013**, *21*, 72–84.
21. Pomorska-Mól, M.; Kwit, K.; Czyżewska-Dors, E.; Pejsak, Z. Tulathromycin Enhances Humoral but Not Cellular Immune Response in Pigs Vaccinated against Swine Influenza. *J. Vet. Pharmacol. Ther.* **2019**, *42*, 318–323. [CrossRef]
22. Silva, G.S.; Yeske, P.; Morrison, R.B.; Linhares, D.C.L. Benefit-Cost Analysis to Estimate the Payback Time and the Economic Value of Two Mycoplasma Hyopneumoniae Elimination Methods in Breeding Herds. *Prev. Vet. Med.* **2019**, *168*, 95–102. [CrossRef] [PubMed]
23. Rosa, G.J.M.; Valente, B.D. Breeding and Genetics Symposium: Inferring Causal Effects from Observational Data in Livestock. *J. Anim. Sci.* **2013**, *91*, 553–564. [CrossRef] [PubMed]
24. Bello, N.M.; Ferreira, V.C.; Gianola, D.; Rosa, G.J.M. Conceptual Framework for Investigating Causal Effects from Observational Data in Livestock. *J. Anim. Sci.* **2018**, *96*, 4045–4062. [CrossRef] [PubMed]
25. Agostini, P.S.; Fahey, A.G.; Manzanilla, E.G.; O'Doherty, J.V.; De Blas, C.; Gasa, J. Management Factors Affecting Mortality, Feed Intake and Feed Conversion Ratio of Grow-Finishing Pigs. *Animal* **2014**, *8*, 1312–1318. [CrossRef]
26. Oliveira, J.; Yus, E.; Guitián, F.J. Effects of Management, Environmental and Temporal Factors on Mortality and Feed Consumption in Integrated Swine Fattening Farms. *Livest. Sci.* **2009**, *123*, 221–229. [CrossRef]
27. Larriestra, A.J.; Maes, D.G.; Deen, J.; Morrison, R.B. Mixed Models Applied to the Study of Variation of Grower-Finisher Mortality and Culling Rates of a Large Swine Production System. *Can. J. Vet. Res.* **2005**, *69*, 26–31.
28. Aiken, V.C.F.; Fernandes, A.F.A.; Passafaro, T.L.; Acedo, J.S.; Dias, F.G.; Dórea, J.R.R.; de Magalhães Rosa, G.J. Forecasting Beef Production and Quality Using Large-Scale Integrated Data from Brazil. *J. Anim. Sci.* **2020**, *98*, 1–12. [CrossRef]
29. Alonso, J.; Castañón, Á.R.; Bahamonde, A. Support Vector Regression to Predict Carcass Weight in Beef Cattle in Advance of the Slaughter. *Comput. Electron. Agric.* **2013**, *91*, 116–120. [CrossRef]
30. Golden, C.E.; Rothrock, M.J.; Mishra, A. Comparison between Random Forest and Gradient Boosting Machine Methods for Predicting Listeria Spp. Prevalence in the Environment of Pastured Poultry Farms. *Food Res. Int.* **2019**, *122*, 47–55. [CrossRef]
31. Kamphuis, C.; Mollenhorst, H.; Feelders, A.; Pietersma, D.; Hogeveen, H. Decision-Tree Induction to Detect Clinical Mastitis with Automatic Milking. *Comput. Electron. Agric.* **2010**, *70*, 60–68. [CrossRef]
32. Shine, P.; Murphy, M.D.; Upton, J.; Scully, T. Machine-Learning Algorithms for Predicting on-Farm Direct Water and Electricity Consumption on Pasture Based Dairy Farms. *Comput. Electron. Agric.* **2018**, *150*, 74–87. [CrossRef]
33. Zhang, F.; Upton, J.; Shalloo, L.; Murphy, M.D. Effect of Parity Weighting on Milk Production Forecast Models. *Comput. Electron. Agric.* **2019**, *157*, 589–603. [CrossRef]
34. Lenzerini, M. Data Integration: A Theoretical Perspective. In Proceedings of the Twenty-First ACM SIGMOD-SIGACT-SIGART Symposium on Principles of Database Systems, Madison, WI, USA, 3–5 June 2002; pp. 233–246.
35. Davis, M.E.; Sears, S.C.; Apple, J.K.; Maxwell, C.V.; Johnson, Z.B. Effect of Weaning Age and Commingling after the Nursery Phase of Pigs in a Wean-to-Finish Facility on Growth, and Humoral and Behavioral Indicators of Well-Being. *J. Anim. Sci.* **2006**, *84*, 743–756. [CrossRef] [PubMed]
36. Leliveld, L.M.C.; Riemensperger, A.V.; Gardiner, G.E.; O'Doherty, J.V.; Lynch, P.B.; Lawlor, P.G. Effect of Weaning Age and Postweaning Feeding Programme on the Growth Performance of Pigs to 10 Weeks of Age. *Livest. Sci.* **2013**, *157*, 225–233. [CrossRef]
37. Main, R.G.; Dritz, S.S.; Tokach, M.D.; Goodband, R.D.; Nelssen, J.L. Increasing Weaning Age Improves Pig Performance in a Multisite Production System. *J. Anim. Sci.* **2004**, *82*, 1499–1507. [CrossRef]
38. Collins, C.L.; Pluske, J.R.; Morrison, R.S.; McDonald, T.N.; Smits, R.J.; Henman, D.J.; Stensland, I.; Dunshea, F.R. Post-Weaning and Whole-of-Life Performance of Pigs Is Determined by Live Weight at Weaning and the Complexity of the Diet Fed after Weaning. *Anim. Nutr.* **2017**, *3*, 372–379. [CrossRef]

MDPI
St. Alban-Anlage 66
4052 Basel
Switzerland
www.mdpi.com

Animals Editorial Office
E-mail: animals@mdpi.com
www.mdpi.com/journal/animals

www.ingramcontent.com/pod-product-compliance
Lightning Source LLC
Chambersburg PA
CBHW061452210926
51850CB00008B/1695
* 9 7 8 3 7 2 5 8 1 0 4 1 3 *